Edward Kellogg Dunham

**Histology**

Normal and Morbid

Edward Kellogg Dunham

**Histology**
*Normal and Morbid*

ISBN/EAN: 9783337405403

Printed in Europe, USA, Canada, Australia, Japan

Cover: Foto ©berggeist007 / pixelio.de

More available books at **www.hansebooks.com**

# HISTOLOGY:

# NORMAL AND MORBID.

BY

EDWARD K. DUNHAM, Ph.B., M.D.,

PROFESSOR OF GENERAL PATHOLOGY, BACTERIOLOGY, AND HYGIENE IN THE UNIVERSITY
AND BELLEVUE HOSPITAL MEDICAL COLLEGE, NEW YORK.

---

ILLUSTRATED WITH 363 ENGRAVINGS.

LEA BROTHERS & CO.,
NEW YORK AND PHILADELPHIA.
1898.

# PREFACE.

In presenting to the student of medicine so condensed a volume upon normal and morbid histology an explanation of the author's purpose may, perhaps, not be amiss.

It appears to the writer that the most important lesson to be derived from a study of the tissues in health and in disease is a knowledge of the constant and potent *activities* of the cells to which those tissues owe both their origin and usefulness. When the body develops under normal conditions those cells build up the tissues, gradually modifying their *formative* activities so as to occasion a diversity of structure in the various parts of the body. During this developmental epoch, and after maturity is attained, the activities which are grouped as *functional*, and which it is the lot of the tissues to maintain, are also carried on by the cells.

But in order that these manifold cellular activities shall be of the usual or "normal" character, the conditions under which they are carried on must not depart greatly or for any considerable length of time from a certain usual, but rather indefinite standard. If those conditions are materially altered, the cellular activities become modified, and the functions they perform suffer aberration, as a result of which structural changes in the cells and tissues may ensue.

It is this close relation between cellular activity and structure which unifies the subjects usually kept distinct under the titles of normal and pathological histology, for it is evident that there is no natural separation between those subjects.

In the preparation of this manual the author has steadfastly kept in view such a conception of the relations between cellular activity and structure. To carry out this purpose it did not appear necessary to describe the various changes wrought in the individual organs or tissues by unusual conditions. It seemed to him that a general statement of the alterations in structure attributable to modified cellular activity would enable the student to interpret such departures from the normal as he might observe in particular specimens, provided he was familiar with the normal structures of the body. In this belief the writer has devoted most of his space to a description of the normal structures, and has contented himself with only a brief account of the histology of the more prevalent morbid processes. He was encouraged in this course by the consciousness that in individual cases the application of the principles involved might be more successfully made by the instructors under whose guidance these studies were pursued. For the sake of clearness, however, examples of morbid structure have been selected from various parts of the body to illustrate the different phases of the processes that were being outlined.

Those histological methods and data which are utilized for the purpose of clinical diagnosis have been almost entirely omitted, because they are fully described in special works on that subject and are not strictly within the limits assigned to this more elementary book.

Occasional reference has been made to technical journals on histology. Those which contain abstracts of the current literature on that subject, and which will, therefore, be of greatest use to the student, are: *The Journal of the Royal Microscopical Society*, *Zeitschrift für wissenschaftliche Mikroskopie*, and *Centralblatt für allgemeine Pathologie und pathologische Anatomie*. The student is also referred to Mallory and Wright's *Pathological Technique*, Lee's *Microtomist's Vade Mecum*, and to the more recent German revised edition, *Grundzüge der mikroskopischen Technik*, by Lee and Mayer.

It may be that well-founded exceptions will be taken to some of the explanations of morbid processes which are here offered; but it is the author's hope that he has not advanced theoretical views with sufficient emphasis to mislead the student. Should the general plan of the work meet with a kindly reception, it will be his endeavor to correct, in a future edition, such errors and omissions as may be revealed by friendly criticism.

E. K. D.

NEW YORK, October, 1898.

# CONTENTS.

|  | PAGE |
|---|---|
| INTRODUCTION | 17 |

## PART I.

### NORMAL HISTOLOGY.

#### CHAPTER I.
THE CELL . . . . . . . . . . . . . . . . 27

#### CHAPTER II.
THE ELEMENTARY TISSUES . 41

#### CHAPTER III.
THE EPITHELIAL TISSUES 45

#### CHAPTER IV.
THE CONNECTIVE TISSUES 63

#### CHAPTER V.
TISSUES OF SPECIAL FUNCTION 82

#### CHAPTER VI.
TISSUES OF SPECIAL FUNCTION (Continued) 94

#### CHAPTER VII.
THE ORGANS . . . 106

## CHAPTER VIII.
THE CIRCULATORY SYSTEM . . . . . . . . . . . . . . . . . . 108

## CHAPTER IX.
THE BLOOD AND LYMPH . . . . . . . . . . . 122

## CHAPTER X.
THE DIGESTIVE ORGANS . . . . . . . . 128

## CHAPTER XI.
THE LIVER . . . . . . . . . . . . . 146

## CHAPTER XII.
THE URINARY ORGANS . . 153

## CHAPTER XIII.
THE RESPIRATORY ORGANS . . . . 168

## CHAPTER XIV.
THE SPLEEN . . . . . 176

## CHAPTER XV.
THE DUCTLESS GLANDS . . . . . . . . . . 180

## CHAPTER XVI.
THE SKIN . . . . . . . . . . . . . . . . . 195

## CHAPTER XVII.
THE REPRODUCTIVE ORGANS 207

## CHAPTER XVIII.
THE CENTRAL NERVOUS SYSTEM . . . . . . . . . . 234

## CHAPTER XIX.
THE ORGANS OF THE SPECIAL SENSES . . . . . . . . . . . 252

# PART II.

## HISTOLOGY OF THE MORBID PROCESSES.

### CHAPTER XX.
DEGENERATIONS AND INFILTRATIONS … 265

### CHAPTER XXI.
ATROPHY … 284

### CHAPTER XXII.
HYPERTROPHY AND HYPERPLASIA … 288

### CHAPTER XXIII.
METAPLASIA … 291

### CHAPTER XXIV.
STRUCTURAL CHANGES DUE TO AND FOLLOWING DAMAGE … 293

### CHAPTER XXV.
TUMORS … 341

# PART III.

## HISTOLOGICAL TECHNIQUE.

### CHAPTER XXVI.
PRACTICAL SUGGESTIONS FOR THE CARE AND USE OF THE MICROSCOPE—MICROSCOPICAL TECHNIQUE … 397

# HISTOLOGY:

## NORMAL AND MORBID.

### INTRODUCTION.

During life all parts of the human body are the seat of constant activity. This is a fact too readily overlooked by the student who gains his knowledge of the structures of the body by a study of the tissues after death. To make that study of use to him in his medical thinking he should constantly bear in mind that he is viewing the mechanism of the body while it is at rest, and, furthermore, that the methods employed in the study of the minute structure of the parts not only arrest the normal activities of those parts, but expose them to mutilation. He must, therefore, constantly supplement the knowledge of structure he gains by his histological studies by recalling to mind and applying that which he has acquired by a study of physiology, habitually associating his ideas of structure and functional activity, until he can hardly think of what a structure *is* without at once recalling what it *does*. This he cannot do till he has mastered at least the general outlines of systematic anatomy and of physiology. Those two fundamental subjects are brought together by an intelligent study of the minute structure of the body, histology, which, for this reason, has also and appropriately been called physiological anatomy.

But the student of medicine must go beyond this. To the conception of the body during health, which he has formed by this thoughtful method, he must then add a conception of the influence exerted, both on the structure and activities of the body, by abnormal conditions which disturb or thwart the usual working of that complex mechanism. The more closely he can make those conceptions agree with observed facts, the more perfect will become his ability to interpret the physical signs and symptoms of disease, and the clearer will grow his insight into the causes and tendencies of the processes of which they are an expression.

In all his studies he must seek not merely to train his powers of observation; he must endeavor to cultivate his ability to interpret what he sees; to deduce the processes and causes that have wrought the results he perceives, and to compare those deductions with the conceptions of living things he has already formed, so that his ideas may remain in perfect accord with one another as his grasp of the subject enlarges. By so doing he may hope to create a lifelike mental picture of the body both in health and during disease.

The activities of the body involve changes in the substances of which it is composed. Some of these changes are always destructive in character—that is, they result in chemical rearrangements which convert more complex combinations of less stable nature into simpler combinations of greater stability. Such chemical changes, whether they take place within the body or in external nature, among organic or inorganic substances, are always accompanied by a liberation of energy hitherto locked up or stored in latent or potential form in the compounds of higher complexity. It is this liberated or kinetic energy which is utilized by the bodily mechanism for the performance of internal or external work. When directed in various ways and operating through different structures, this energy occasions visible movement, appears as heat, etc., or passes again into the latent form in the elaboration of more complex chemical substances from those of simpler constitution.

These associated transformations of matter and energy involve a continual loss to the bodily economy. The stock of energy is diminished during the execution of external work and by the dissipation of heat. The store of useful chemical substances is reduced by their progressive conversion into compounds that are insusceptible of further utilization, and which, in many cases, may act injuriously upon the structures of the body. Under normal conditions such substances are eliminated from the body.

It is evident, then, that the body is constantly suffering a loss of both energy and matter. This loss must be made good if the activities of the body are to be maintained, and this is accomplished, during health, through the absorption of fresh material, containing latent energy, from the food taken into the body.

The activities of the body are not the same in all its parts. They are all alike in one particular—namely, that each part must maintain its own nutrition, incorporating the food-materials that are accessible to it and using them in such a way as to keep its structure in a normal condition. But, aside from this duty which is common to all, each part has a duty to perform for the good of the whole organism; and, as we shall see, this duty often appears to be paramount, the activities which it necessitates being carried on even if they involve a sacrifice in the nutrition or structure of the individual part.

Each part of the body has some particular kinds of work assigned to it, which constitute its functions, and which it performs for the benefit of the whole body. The development and life-history of each part has direct reference to those functions, through which it co-operates with all the other parts in maintaining the integrity and normal activities of the whole body, all the parts being interdependent upon each other and subservient to the general needs.

The foregoing considerations prepare us for the fact that the structure of the various parts of the body differs in its details. The study of those finer details can only be pursued with the aid of the microscope, for the microscopical constituents of the tissues are the elements which confer upon them their particular properties and powers. This study is called histology.

Investigation has shown that there is one form of tissue-element which is always present in all parts of the body. This is the cell. It does not always possess the same form or internal structure, but in all its variations the same general plan of construction is adhered to. These cells are the essentially active constituents of the tissues. It is within them that the transformations of matter and energy are chiefly carried on, and it is due to their activities that the tissues forming the body are elaborated and enabled to perform their several functions. These marvellous powers possessed by the cell have created our conception of life, and, in spite of eager study, remain inscrutable. We do not know why a living cell differs from a dead cell, but we do know that the mysterious vital powers are only derived from pre-existent living cells and are not antagonistic to the chemical and physical laws governing unorganized matter.

All the cells of the body are descendants of a single cell, the egg, from which they arise by successive divisions, and throughout the existence of the body they retain some of the characters of the original cell. But as the body develops the cells of the different parts display divergent tendencies, which finally result in the formation of a considerable variety of tissues, grouped in various ways to form organs or systems of very different kinds of utility to the whole organism. This divergent development is known as differentiation and results in a specialization of the different parts of the body. Its study constitutes embryology, but it will make the comprehension of histology easier if some of the simpler and broader facts derived from a study of development are first briefly summarized.

A new individual arises through the detachment of a single cell, the ovum (Fig. 1), from the parent organism. This cell divides

FIG. 1.

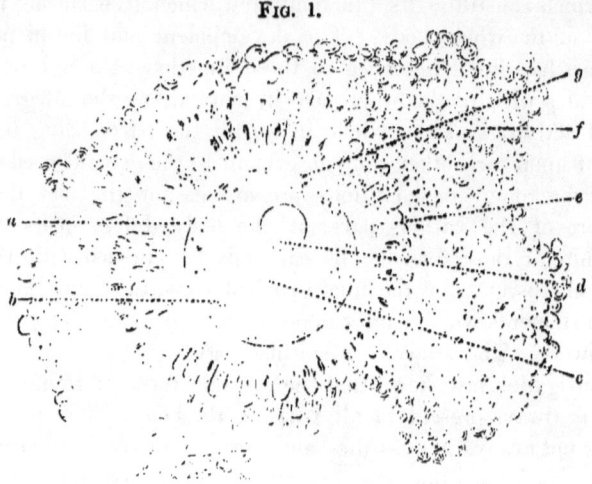

Section of human ovum and its immediate surroundings within the ovary. (Nagel.) a, zona pellucida; b, cytoplasm of the ovum; c, granules and globules of stored food materials within the cytoplasm, collectively known as the metaplasm or deutoplasm; d, germinal vesicle or nucleus of the ovum containing, in this case, two germinal spots or nucleoli; e, zone of epithelial cells immediately surrounding the ovum; f, cells of the discus proligerus; g, perivitelline spaces separating the zona pellucida from the cytoplasm of the ovum.

into two cells, which, even at this stage of development, differ slightly from each other. These daughter-cells in turn divide in two, and this process of division is continued, each cell giving rise to two new cells, until a considerable aggregate of cells has resulted (Fig. 2). Then the cells assume a definite arrangement into layers. Some become disposed in a superficial layer enclosing the rest of the cells and a body of fluid. This layer is called the primitive ectoderm. The remaining cells accumulate in an irregular laminar mass beneath the primitive ectoderm at the site of the future embryo. This mass of cells is the primitive entoderm. Thus, at this stage of development, there is a cellular sac, containing fluid, with a reinforcement of its wall at the region occupied by the primitive entoderm (Fig. 3).

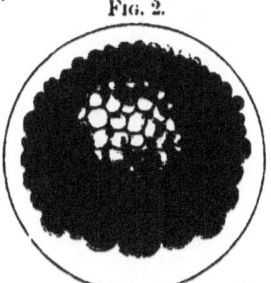

Fig. 2.

Segmented egg of Petromyzon Planeri ; Surface view of the collection of cells. The nuclei are invisible. (Kupffer.)

Subsequent to these events a third layer of cells becomes interposed between the primitive ectoderm and entoderm. Most of its

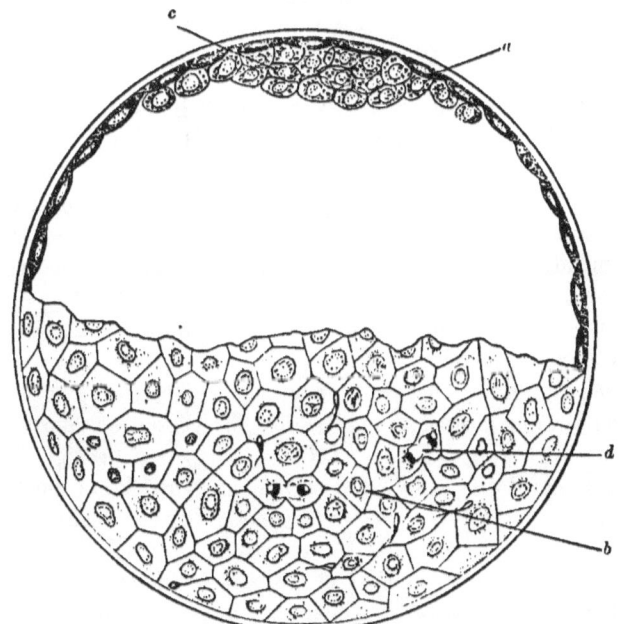

Fig. 3.

Ovum of rabbit: *a*, primitive ectoderm in section; *b*, primitive ectoderm, surface view; *c*, primitive entoderm ; *d*, dividing cell of the ectoderm. (van Beneden.)

cells are derived from those of the primitive ectoderm, but the primitive entoderm may also participate in its formation. This third layer is called the mesoderm. Soon after its formation, the mesoderm divides at the sides of the embryo into two layers—a parietal, which joins the under surface of the ectoderm, and a visceral, attached to the upper surface of the entoderm. The space between these two layers is occupied by fluid, and is destined to form the future body-cavities. In the axis of the embryo the three earlier layers remain in continuity, forming a cellular mass around the site of the future spinal column (Fig. 4).

FIG. 4.

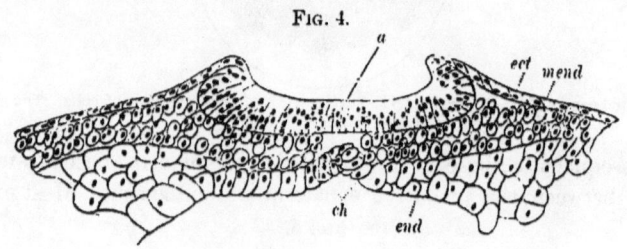

Embryo of Necterus in cross-section. (Platt.) *ect.*, ectoderm; *mend.*, mesoderm; *end.*, entoderm; *a*, neural groove; *ch*, site of future spinal column.

From these three embryonic layers of cells the body of the fœtus is developed. The entoderm, with the visceral or lower layer of the mesoderm, turns downward and inward to meet its fellow of the opposite side and form the alimentary tract. The ectoderm and parietal or upper layer of the mesoderm also turn downward and inward, outside of the alimentary tube, and join those of the other side to form the walls of the body.

Meanwhile, the upper surface of the ectoderm over the axis of the embryo becomes furrowed. The edges of this furrow grow upward, deepening the groove between them, and finally arch over it and coalesce, forming a canal around which the central nervous system is developed (Fig. 5). Traces of this canal persist through life as the central canal of the spinal cord and the ventricles of the brain.

The embryonic layers have a deeper significance than the mere furnishing of the architectural materials from which the body is built up. They are evidences of a distinct differentiation in the development of the cells of which they are composed. The ectoderm gives rise to the functional part of the nervous system and to the epithelial structures of the skin and its appendages. The cells of the mesoderm elaborate the muscular tissues and that great group known as the connective tissues, and the entoderm contains the cells that build up the linings of the digestive tract, including its glands, and of the respiratory organs. It appears, then, that this division of the cells of the embryo into three layers marks a distinct difference in the destinies of the cells composing those layers. This distinction persists through life, the tissues arising from a given layer showing, in general, a closer relationship to each other than the tissues arising from different layers. But this relationship is not always revealed by a similarity in structure, for the latter is determined by the functions the tissues are destined to perform, and tissues of like function acquire a similarity in structure. Thus, for example, the neuroglia in the central nervous system resembles

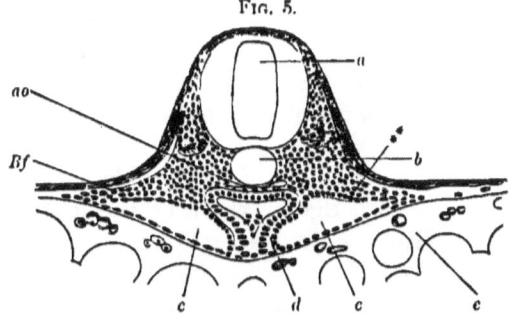

Fig. 5.

Cross-section of fish embryo. (Ziegler.) *a*, neural canal, cells enclosing it not represented; *b*, chorda dorsalis, site of future spinal column; *ao*, aorta; *Bf*, external layer of mesoderm; *c, c*, body-cavity; *d*, alimentary canal, not yet completely closed * *, passes through the external layer of the mesoderm to its inner surface; *e*, deutoplasm, or yolk of egg.

some of the connective tissues, although one develops from the ectoderm and the other from the mesoderm; and the ganglion cells of the central nervous system differ greatly in structure from the epithelium of the skin, nails, etc., and the cells of the neuroglia, notwithstanding the fact that they all spring from the cells of the ectoderm. The explanation is to be sought in the similarity of the usefulness of neuroglia and connective tissue and the difference in the functions of ganglion cells and those of the other tissues emanating from the ectoderm.

During the early stages of development the cells of the germinal layers are very similar in character, although, as we have seen, their potential qualities are quite diverse. As growth proceeds, they begin to vary in size, shape, and internal structure in the different parts of the fœtus. Their relative positions become modified. The primitive organs are defined and the tissues of which they are composed become elaborated.

The elaboration of the tissues is wrought by the cells, which display what is called their formative powers in the production of materials of various sorts which lie between them, and are called the intercellular substances. The amount and kind of intercellular substance vary, each form of tissue having its own peculiarities in this respect, dependent upon the *rôle* it is to play in the general economy. Some of the tissues perform functions which require the active processes that can be carried on only in cells, and in these the intercellular substances are either small in amount and apparently structureless, as in epithelium, or their place is taken by a tissue of separate origin, while the cells, relieved of the necessity for exercising their formative powers in this direction, become highly specialized to meet the functional demands imposed upon them. This development is met with in the muscular and nervous tissues.

Other tissues of the body are of use mainly because of their physical properties, such as rigidity, elasticity, tensile strength, pliability, etc. These tissues, collectively called the connective tissues, are essentially passive. They require little or no cellular activity for the performance of their functions, and it is in the elaboration of these tissues that the cells exercise their most marked formative powers during the development of the body, causing the deposition of intercellular substances which possess the requisite physical characters—rigidity and elasticity in the case of bone, pliability and tensile strength in the case of ligamentous structures, etc. As these substances are perfected, the cells decrease in activity, until they merely preside over the integrity of the intercellular substances they have already produced.

It may be well to point out here a distinction that divides the tissues of active cellular function into two groups. The first group, including the various modifications of epithelium, displays its activity in the elaboration of material products, taking the form of either new cells which are continually being produced, or of certain chemical substances which appear as a secretion. The second group, comprising the muscular and nervous tissues, exercises its functional activities in the storage of latent energy in such substances of unstable chemical nature and in such a manner that it can be liberated when required and directed toward the accomplishment of some definite purpose. The functions of both groups require an active intracellular metabolism, resulting in the formation of particular chemical substances. In this they are alike. But in the first group the production of those substances is, in itself, the functional purpose of the process, while in the second group those substances are merely a means for holding energy in the latent condition. If we may so express ourselves, the first group utilizes energy for the elaboration of material, the second group elaborates material for the utilization of energy.

In the adult, under normal conditions, each kind of cell, if it reproduce at all, gives rise to cells only of its own kind. But when the conditions are morbid, a sort of reversion may take place, the progeny of a given cell then showing less evidence of specialization than the parent cell. Such reverted cells, or their descendants, may never develop into more specialized cells, or they may regain the original degree of specialization possessed by the first cell, or, finally, they may become specialized along some divergent line of development, giving rise to a tissue that is nearly or remotely akin to that from which they started, according to the degree of reversion which has taken place. The reversion appears never to extend further back than the degree of specialization that is marked by the formation of the three embryonic layers in the history of development; for example, epithelium which springs from either the entoderm or ectoderm does not revert to a primitive condition from which it can develop into bone or some other form of connective tissue normally derived from the mesoderm. Examples of reversion will be met with in the chapters on Inflammation, Tumors, and Metaplasia.

# PART I.

## NORMAL HISTOLOGY.

### CHAPTER I.

#### THE CELL.

As has been stated in the introductory chapter, the cells of the body are not all alike. Most of them have undergone modifications fitting them for the performance of some definite function, and the majority of them are in consequence not appropriate objects for a study of the general characters of a cell. The extent to which this modification has affected the visible structure of the cell is, however, very different in the different tissues, and in some of them the cells retain so much of their original embryonic appearance as to closely resemble the unspecialized cell.

This is true of the cells of some varieties of *epithelium*. But, though in appearance they give little evidence of specialization, in their functional activities they display very marked modifications of the powers of the *primitive cell*. Some of those powers, perhaps the *nutritive*, perhaps the *secretory*, have become exaggerated, while others, *e. g.*, the *locomotory*, or *reproductive*, have fallen into abeyance, or suffered almost total extinction.

On the other hand, it is obvious that such cells as constitute the whole body of unicellular animals must retain all the powers essential to a living cell in relatively equal states of development. No one of them can be extinguished or thrown out of its proper balance with respect to the others if the cell is to remain normal. And yet, even among the unicellular organisms, certain parts of the cell may be very evidently specialized for the performance of particular functions. For example, the cilia of infusoria have the power of executing much more rapid movements than the other parts of the same cell. And it is probable that all protozoa, i. e. unicellular animals, possess similar, though less obvious and internal, heterogeneity of constitution.

The less the degree of specialization or differentiation in the structure of an organism, the less highly developed is the functional activity of which it is capable, and the less perfect its ability to cope with possible unfavorable environment. The value to the whole organism of a diversity in its parts is, therefore, unquestionable, and the higher we go in the animal kingdom, the greater we find the development of this diversity, coupled with a more and more perfectly adjusted co-operative interdependence of the different parts of the body.

In the protozoa the single cell does all the work of the whole organism. In the multicellular animals, the metazoa, this work is distributed among the component cells of the body, each of which has developed an efficiency for performing its special work that would be incompatible with a wider range of duties.

It is quite impossible to find in nature any example of a cell devoid of all individual peculiarities attributable to differentiation or specialization. We must, therefore, study several varieties of

FIG. 6.

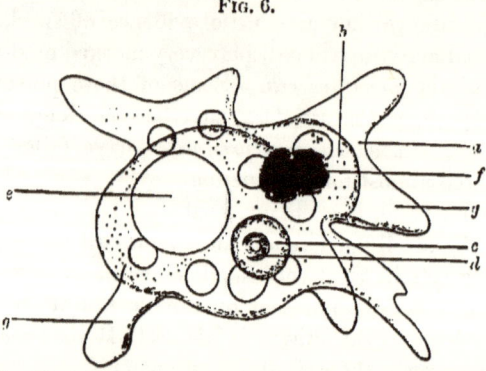

Amœba pellucida. (Frenzel.) *a*, ectoplasm; *b*, endoplasm; *c*, nucleus; *d*, nucleolus; *e*, large contractile vacuole; *f*, incorporated foreign body; *g, g*, pseudopodia.

cell in order to gain an ideal conception of such a cell. This accomplished, we may consider those cells which occur in nature as special modifications of that type.

Perhaps the simplest cell leading an independent existence is the protozoon, amœba (Fig. 6). This animal is widely distributed in moist earth, upon the surfaces of aquatic plants, and in the soil at the margins of ponds and sluggish streams.

The body of the amœba consists of a gelatinoid substance which has received the name protoplasm, or, more definitely, cytoplasm. Within this cytoplasm and sharply defined from it is a round or oval, vesicular body, called the nucleus, which in turn contains one or more particularly conspicuous granules, the nucleoli.

The most superficial layer of the cytoplasm appears perfectly clear, colorless, and homogeneous. It envelops the rest of the cytoplasm, which has a granular appearance. The clear peripheral portion is distinguished as the "hyaloplasm," or "ectoplasm;" the granular internal portion as "spongioplasm," or "endoplasm." The terms hyaloplasm and spongioplasm are also used in a different and more restricted sense, as will presently appear.

When viewed under the microscope, the granules of the cytoplasm are seen to possess a constant, slight, vibratile motion, the Brownian movement, to which is added now and then a flowing movement from one part of the cell to another. At intervals there is a protrusion of the ectoplasm at some point, extending for some distance from the body of the cell, a pseudopodium. This may soon be retracted again, merging with the rest of the ectoplasm, or some of the endoplasm may flow into the central portion of the pseudopodium, converting it into a broad extension of the cell-body. This may subsequently be withdrawn, or the whole mass of cytoplasm, with the nucleus, may flow into the pseudopodium, gradually increasing its size, until the whole cell occupies the original site of the pseudopodium. In this way the animal executes a slow, creeping locomotion.

These pseudopodial movements and the locomotion occasionally incident to them appear to be wholly spontaneous, *i. e.* dependent upon internal conditions of which we have no knowledge. They may, however, be influenced by external circumstances. Certain substances evidently attract the amœba, others are either matters of indifference to it or repel it. If a pseudopodium comes in contact with some particle in the surrounding medium, it may retreat from it, appear indifferent to it, or be attracted and proceed to incorporate it. This is accomplished by the cytoplasm flowing around the foreign body and coalescing on its further side so as to enclose it. It is then conveyed to the body of the cell, either by cytoplasmic currents, by the withdrawal of the pseudopodium containing it, or by the streaming of the cell-body into that protrusion. The fate of the particle thus incorporated depends upon its nature. If it be serviceable as food, it is gradually digested and absorbed, or such parts of it as are digestible are so utilized, and the remainder, no longer of use to the amœba, is extruded from its body.

These phenomena reveal powers of perception and selection on the part of this cell which are very closely akin to the intelligence of more complex organisms. They also demonstrate its power of assimilating material from without, to serve as nourishment and the source of the energy which it expends in executing its movements and in carrying on the chemical processes pertaining to its internal economy.

At intervals, there appears within the endoplasm a small, clear, spherical spot. This gradually increases in size and constitutes a little drop of fluid, sharply defined from the surrounding cytoplasm. After it has attained a certain size, it suddenly disappears, the cytoplasm around it coalescing and leaving no trace of its existence. Such a clear space, filled with fluid, within the body of a cell is called a vacuole, and those which are suddenly obliterated, contractile vacuoles. Their purpose is not clearly understood, but probably has to do with a primitive circulatory or respiratory function, since contractile vacuoles are not observed in the cells of higher organisms where those functions are carried on by more elaborate mechanisms.

Eventually the amœba reproduces its kind by dividing into two similar cells, each of which grows into a likeness to the parent individual.

Let us now compare the amœba with some other varieties of cell, in order to learn what they all have in common.

The amœba has an outer, soft, transparent layer of cytoplasm, the ectoplasm. This is not present in all cells. In many the granular cytoplasm has no envelope, but appears to be quite naked. In other varieties it is enclosed in a distinct membrane.

In the great majority of cells the active streaming of the cytoplasm and the pseudopodial protrusions described in the amœba are wanting, but the Brownian movement of the granules is more constantly present. The cells have fixed positions and their food is brought to them, usually in solution, so that the more active movements so essential to the welfare of the amœba would be superfluous.

For a similar reason, as already intimated, they can dispense with the contractile vacuole.

We learn, then, that when we reduce the cell to its simplest terms, it consists of a mass of cytoplasm enclosing a nucleus. To these we must probably add a third essential constituent, the centrosome, which is a minute granule situated in the cytoplasm. It is so small that its presence has not been established in all cells, its detection in many cells being extremely difficult because of the general granular appearance of the cytoplasm in which it lies. It plays such an important part, however, in the division of those cells in which it has been studied, that the inference that it is an essential part of all cells appears justified.

These three constituents, the cytoplasm, nucleus, and centrosome, appear to be the essential organs of a cell among which its activities are distributed (Fig. 7). We do not know how they do their work,

Fig. 7.

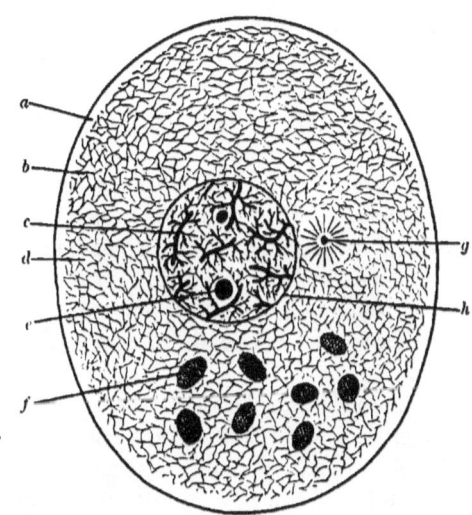

Schematic diagram of a cell: *a*, cytoplasm composed of hyaloplasm; *b*, spongioplasm; *c*, chromosome, composed of "chromatin," and forming a part of the intranuclear reticulum; between these chromatic fibres is the achromatin; *d*, hyaloplasm in the meshes of the spongioplasm; *e*, one of the two nucleoli represented in the diagram; *f*, one of eight bodies constituting the metaplasm represented; *g*, centrosome, with radiate arrangement of the surrounding spongioplasm; *h*, nuclear membrane.

but we have a general conception of the distribution of the work performed by the whole cell among these three organs.

1. The cytoplasm, which usually makes up the chief bulk of the cell, especially in those varieties which have active metabolic functions, appears to be the part of the cell in which the assimilated food is utilized in the production of chemical substances, either fresh cytoplasm or some other product, or in the execution of movements or the liberation of energy in other forms. Most of the active processes that are obvious seem to be carried on in the cytoplasm during the greater part of the life-history of the cell.

2. The nucleus appears to preside over the assimilative processes within the cell. If a cell be subdivided so that the uninjured nucleus is retained in one of the portions, that portion may grow and become a perfect cell. But the portions that are deprived of a nucleus do not grow, and while they may retain life for a considerable time, utilizing the assimilated food they retain, eventually perish.

Aside from this assimilative function, the nucleus appears to be the carrier of hereditary characters from the parent cell to its progeny during the division of the cell. This will become clearer when the process of cell-division is described.

3. The centrosome appears to be the organ presiding over the division of the cell. It inaugurates those activities in nucleus and cytoplasm which result in the production of new cells, and seems to guide them, at least during the greater part of the whole process.

It is evident, from these statements, that the cell has an exceedingly complex organization, which a simple microscopical study cannot wholly reveal. Notwithstanding this fact, obvious microscopical differences are presented by cells which have become specialized in different directions, and we must know something of the visible structure of the primitive cell before we can appreciate these departures from it.

The cytoplasm is not a simple substance. Its constitution is so complex that our present means of research are not adequate to reveal its structure. We know that its solid constituents are chiefly proteids, together with relatively small quantities of carbohydrates, fats, and salts. To these is added a large proportion of water which, while not entering into a definite chemical union with the other constituents, is so intimately associated with them as to form an integral part of the cytoplasm.

The visible structure of cytoplasm differs somewhat in different cells, even among those that appear to be comparatively unspecialized. In the fixed cells of the higher animals and man it appears to consist of a very delicate network or reticulum of minute fibres, termed the spongioplasm. The points of junction of these fibres and their optical cross-sections give a finely granular appearance to the cytoplasm.

In the meshes of the spongioplasm is a clear, homogeneous substance, the hyaloplasm. This may also contain some granules, but they are probably not constituent parts of the cytoplasm and are grouped under the term metaplasm. Some of them are composed of material taken from without, either in their original form or slightly modified; others have been produced within the cell by chemical transformations, and are either useful products, to be subsequently turned to account by the cell itself or to be discharged as a secretion, or they are waste matter destined for elimination from the body.

The relative proportions of the hyaloplasm and the spongioplasm and the arrangement of the fibres of the latter both vary in different cells.[1]

When seen under the microscope the structure of the nucleus, except during the division of the cell, closely resembles that of the cytoplasm. It is traversed by a number of delicate fibres, which branch and give the nucleus a reticulated appearance. At its surface these filaments unite to form a delicate membranous envelope, sharply defining the nucleus from the surrounding cytoplasm, but it is a question whether this membrane is continuous, or whether it is an exceedingly close meshwork with minute apertures permitting a direct communication between the cytoplasm and the interior of the nucleus.

The spaces between the nuclear filaments are occupied by a clear, homogeneous substance, which may be identical and continuous with the hyaloplasm of the rest of the cell.

One or more highly refracting bodies, the nucleoli, may be present in the nucleus, lying freely in the clear substance between the filaments or attached to the latter. Their purpose is not known, but it is thought that they are not essential parts of the cell but correspond more or less closely to the metaplasm in the cell-body.

[1] The reticulated appearance of the cytoplasm may also be explained by assuming it to have an alveolar structure, and the theory that such is its actual structure possesses much plausibility. In that case the visible reticulum would be formed by the walls of the alveoli and their lines and points of intersection, all of which would be included in the spongioplasm, while the contents of the alveoli would constitute the hyaloplasm.

Owing to their affinity for certain coloring matters, the substances composing the nuclear filaments are called chromatin, or chromoplasm. The hyaline substances making up the rest of the nucleus do not receive those coloring matters, and for this reason and in this situation are called achromatin. These terms are only used in a morphological sense and do not specify any definite chemical compounds. The behavior of the nucleoli toward dyes is somewhat different from that of the chromoplasm, which leads to the inference that they are of a different chemical nature.

Except during cell-division, the nucleus usually lies quiescent within the cytoplasm, but some observers have seen it execute apparently spontaneous movements, and it is evidently possible for its position in the cell to vary from time to time.

In marked contrast to this apparently dormant state, as far as visible alterations of structure are concerned, is the *rôle* played by the nucleus during the reproduction of the cell.

There are two modes of cell-division, the "indirect" and the "direct," but they are by no means equivalent to each other. The former, also termed karyokinesis because of the active changes in the nucleus, appears to be the only truly reproductive process. Direct cell-division results in the formation of new cells, but they seem to lack that perfection of organization which would be required for the complete and indefinite transmission of all the characters of the parent cells.

Before entering into a description of karyokinesis, a few words must be said concerning the centrosome. This is an extremely minute granule which is usually situated in the cytoplasm not far from the nucleus. It is often surrounded by a thin zone of hyaloplasm which facilitates its recognition among the fibres and nodal points of union of the spongioplasm. The fibres of the latter are also frequently arranged in a radial manner for a short distance around the centrosome. But in many instances it is extremely difficult to distinguish the centrosome, and its constant presence in cells is largely a matter of inference. Sometimes the centrosome is double, the two granules lying close to each other and often being surrounded by a common clear zone of hyaloplasm.

The first step in the process of cell-division by the indirect method, or karyokinesis, is a division of the centrosome into halves (Fig. 15), which separate and pass to opposite points in the cytoplasm. These points are called the poles of the cell, and when the new centrosomes reach them they are called the polar bodies. In these situations they are surrounded by a more distinct zone of hyaloplasm than that which enclosed the original parent centrosome, and beyond this the spongioplasm is frequently arranged in radiations of unusually thick fibres. The polar bodies with their clear envelopes and the prominent radiations about them are collectively known as the attraction-spheres (Fig. 8).

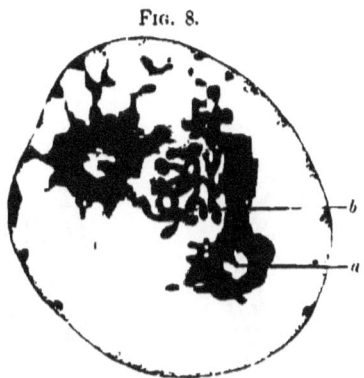

FIG. 8.

Dividing cell from ovum of *ascaris megalocephalus*. (Kostanecki and Siedlecki.) *a*, polar body, centrosome, surrounded by a clear zone; *b*, chromosomes of the dividing nucleus. Between the polar bodies is the achromatic spindle, and radiating from each attraction-sphere are delicate filaments of spongioplasm. The cytoplasm presents indications of vacuolation.

While the polar bodies are separating, or after they have passed into the polar regions of the cell, the nucleus begins to show those changes in structure which constitute karyokinesis. This process may be divided into a number of phases, as follows:

1. **The Formation of the Spirem** (Fig. 9).—This consists in a condensation of the chromoplasm. The branches of the nuclear filaments are withdrawn into the substance of the main fibres, into which the nuclear membrane or peripheral network bounding the nucleus is also absorbed. The vesicular character of the nucleus is lost during these changes in the arrangement of the chromoplasm, which appears as a loose tangle or skein of one or more threads of uniform diameter lying freely in the body of the cell. This skein is called the spirem. The chromoplasm in this condensed condition stains more deeply with nuclear dyes than in the resting condition of the nucleus. The nucleoli in the meantime become faint and seem to ultimately disappear. They play no part in the process

FIG. 9.

FIG. 10.

FIG. 11.

FIG. 12.

FIG. 13.

FIG. 14.

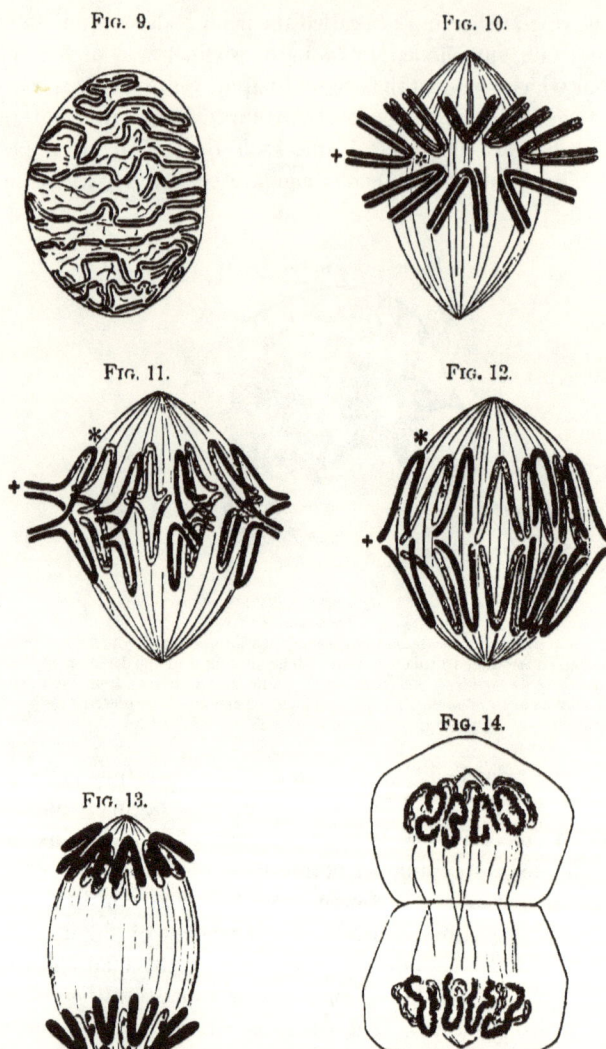

Diagrams illustrating the phases of karyokinesis. (Flemming.)

Fig. 9.—Spirem.
Fig. 10.—Monaster.
Fig. 11.—Metakinesis, early stage.
Fig. 12.—Metakinesis, late stage.
Fig. 13.—Diaster.
Fig. 14.—Dispirem.

The achromatic spindle is represented, but not the centrosomes (polar bodies). The cell-body is also omitted.

of cell-division, unless they participate in the formation of the achromatic spindle.

2. **The Monaster Phase** (Fig. 10).—The threads of the spirem suffer a rearrangement, resulting in the formation of a sort of wreath, situated midway between the poles, in the equatorial plane, i. e., the plane perpendicular to and passing through the centre of a line joining the two polar bodies. This wreath is called the monaster, because of its star-like configuration when seen from above. When viewed in profile it appears as a band of fibres lying in the equator. It is at first made up of a single thread or only a few threads, but subsequently breaks into a number of similar fragments, called chromosomes. The exact number of these varies in different species of animal, but is constant for each species and is always divisible by two. In man it is thought to be sixteen.

The chromosomes are all of nearly, if not quite, the same size, and, in the same kind of cell, closely resemble each other in shape. The most common form appears to be a V-shaped rod lying with its angle directed toward the centre of the wreath or monaster.

3. **Metakinesis** (Figs. 11, 12, 16).—In this phase of karyokinesis

Fig. 15. 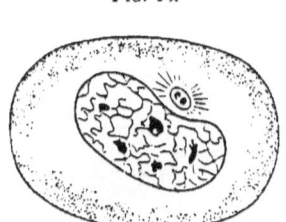   Fig. 16.

Karyokinetic figures in epithelial cells. From a carcinoma removed by operation. (Lustig and Galleotti.)

Fig. 15.—The centrosome has divided, but the nucleus is still in the resting condition. Five nucleoli are represented within the nucleus.
Fig. 16.—Metakinesis. The polar bodies have divided.

the chromosomes split along their axes into two exactly equal parts of similar shape, and these halves separate, each passing toward one of the attraction-spheres.

Meanwhile, the structure known as the achromatic spindle has been formed. This is a system of fibres resembling those that have already been described as radiating from the polar bodies, but of even greater prominence. They are arranged to form a spindle with its apices at the polar bodies and its equator coincident with that of the cell and the plane of the monaster.

It is along the lines of this spindle that the chromosomes travel toward the centres of the attraction-spheres occupied by the polar bodies.

The phases of karyokinesis that follow metakinesis are similar to those that preceded it, but occur in inverse order.

4. **The Diaster Phase** (Fig. 13).—The chromosomes, having reached the attraction-spheres, group themselves around the polar body to form a wreath on a plane perpendicular to the axis joining the poles. These wreaths, with the achromatic spindle, have an appearance somewhat resembling the letter H, with a long cross-piece, formed by the spindle, remaining uncolored or only faintly tinged by nuclear dyes, while the uprights, made up of the chromosomes, are deeply stained.

The ends of the chromosomes now unite to form a thread, and the wreath-like arrangement gradually passes into that of the dispirem.

5. **Dispirem** (Figs. 14 and 17).—The halves of the original chromoplasm of the nucleus are now arranged in two skeins about the poles. From these the two daughter-nuclei of the future cells are formed (Fig. 18).

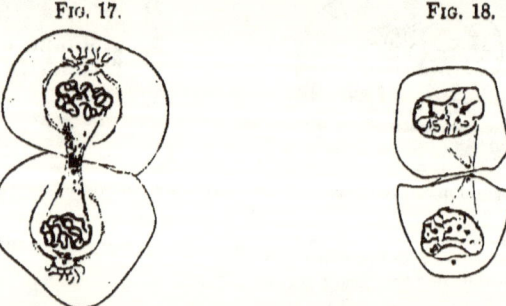

FIG. 17.     FIG. 18.

Fig. 17.—Dispirem. In this case the polar bodies have not divided (compare Fig. 16).
Fig. 18.—Daughter-nuclei which have nearly reached their full development. Centrosomes present in the cytoplasm.
In these figures the structure of the cytoplasm is not given.

During metakinesis the cytoplasm of the cell begins to show signs of division. This may be accomplished through a constriction of the body of the cell, which gradually becomes deeper and finally severs the two portions; or a series of punctiform or short linear enlargements of the lines of the achromatic spindle appear in its equator, and through these a plane of cleavage, dividing the two new cells-from each other, is finally established.

It is rarely that any biological process assumes such mathematical precision as is displayed in karyokinesis. The purpose of that mode of cell-division appears to be an exactly equal partition of all parts of the chromoplasm between the young cells. Whether the amount of cytoplasm given to the daughter-cells is the same or different, the division of the chromoplasm is exactly equal, not only in its whole bulk, but each chromosome, which appears to be the morphological unit of the chromoplasm, is split into exactly equivalent halves, one of which is contributed to the formation of each daughter-nucleus. It is for this reason that the chromoplasm is looked upon as the carrier of hereditary peculiarities.

After the formation of the daughter-nuclei, the centrosome usually passes from it into the cytoplasm. It may divide earlier than has been described, the division taking place while it exists as the polar body, or even earlier (Fig. 16).

A cell nearly always divides to form two new cells, but sometimes three or more cells may be produced, the chromosomes being distributed among them (Fig. 19). Such cases are probably

FIG. 19.

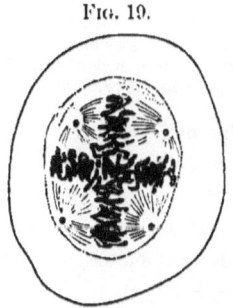

Epithelial cell from a carcinoma. (Galeotti.) The centrosome has divided into four portions, and the chromosomes are arranged with reference to these. The figure represents the metakinetic phase of karyokinesis, which will result in the formation of four imperfect nuclei.

always morbid, and the resulting cells are not wholly the equivalents of the parent cell.

It occasionally happens that the cytoplasm fails to divide after the formation of the daughter-nuclei, and cells with two or more nuclei result. When the nuclei continue to multiply and the cytoplasm increases in amount, but does not suffer division, large multinucleated cells are produced, which have been called "giant-cells." They occur normally in the marrow of bone and are produced in many of the inflammatory processes.

The **direct** or amitotic method of cell-division is inaugurated by an active change in the shape of the nucleus, which may have previously increased in size and become richer in chromoplasm. The nucleus becomes constricted and finally separated into two portions, which are not necessarily equally rich in chromoplasm. The cytoplasm, either at the same time or later, becomes similarly constricted until it is divided into two parts, each containing one of the nuclear divisions (Figs. 20, 21, 22).

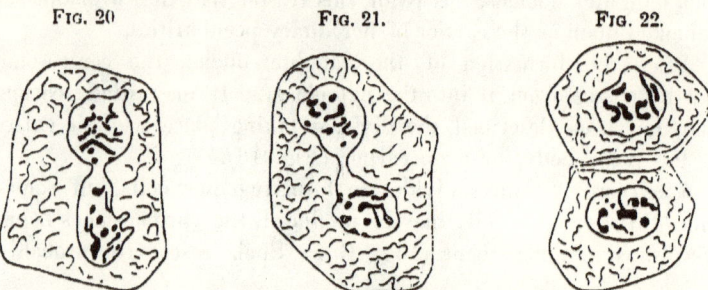

Fig. 20    Fig. 21    Fig. 22.

Amitotic cell-division. (Flemming.) Epithelial cells from the bladder of a salamander.
Figs. 20 and 21 contain nuclei with constrictions dividing them into nearly equal portions.
Fig. 22.—Contiguous cells, each containing a nucleus about half the size of those prevailing in the tissue, and, therefore, probably the result of cell-division by the direct process.

It is believed that this mode of division does not result in the formation of cells that have the complete character of the parent-cell, and that their descendants form a degenerate race that is destined to extinction. It is quite obvious that no such precise partition of the chromatic substance is likely to take place as that which is characteristic of karyokinesis, and if the chromosomes are really the carriers of hereditary peculiarities, this mode of division can hardly favor their perfect transmission.

# CHAPTER II.

## THE ELEMENTARY TISSUES.

The various parts of the body are composed of a small number of "elementary tissues." Each of these elementary tissues has a definite structure, but the details of that structure may vary within certain limits in different parts of the same mass or in different situations within the body. Such variations can usually be referred to differences in the functional activity assigned to the tissue, which is not always exactly the same throughout the body. For example, epithelium is an elementary tissue consisting of cells which are nearly always rich in cytoplasm and are separated from each other by a very small amount of homogeneous intercellular substance. Wherever epithelium is found it has these general peculiarities of structure. But the functions demanded of epithelium are of widely diverse character in different situations, and its structure shows a corresponding diversity in its details. The fact that it is made up almost exclusively of cells leads to the natural inference that the usefulness of epithelium depends upon cellular activities. Inasmuch as these may be of very different character, we should expect the tissue to vary chiefly in the structure and arrangement of its component cells according to the particular activity which was needed and the manner in which it was utilized. Such, as a matter of fact, is the case. These considerations will be made clearer if we follow a little more closely the example offered by epithelium.

In some situations epithelium serves to protect the underlying tissues from injury. But the usual injurious influences which threaten the tissues differ in different parts of the body, and must, therefore, be averted by different means. Upon the surface of the skin they are chiefly of a mechanical or chemical nature, and to resist them the cells of the epithelium forming the epidermis undergo a modification in structure, resulting in the formation of a superficial horny layer which is highly resistant to abrasion and chemical change. Upon the inner surfaces of the respiratory passages the conditions are different. Here the tissues require protection from particles of dust that may be inhaled. For this purpose the epithelial cells lining those passages are provided with minute, hair-like processes, "cilia," which execute lashing movements toward the outlets of the passages and occasion the transportation of substances coming into contact with them toward the outer world. In the digestive tract the conditions are again different. The tissues underlying the epithelial lining need protection from the chemical action of the fluids in the stomach and intestine, as well as from friction with their solid contents. The cells of the epithelium meet these needs by a secretion of mucus, which is discharged upon the inner surfaces of the digestive organs, where it serves as a protective layer and as a lubricant.

In other situations epithelium has an excretory function, which is less clearly of value in protecting its immediate surroundings, but is essential for the protection of the whole organism from substances which would exert an injurious effect if they were permitted to accumulate in the circulating fluids of the body. These substances are absorbed from those fluids by epithelial cells, from which they are discharged from the body either unchanged or after transformation into other chemical compounds. Here the most obvious products of cellular activity are of no use in the economy, and are eliminated from it; but it is not improbable that the cells which separate them or their antecedents from the circulating fluids may also discharge useful substances into those fluids ("internal secretion"). We must not assume that the most obvious function exercised by a tissue is the only service it does to the organism.

The epithelium which carries on this eliminative function is nearly always associated with other elementary tissues to form an organ, called a "gland," in which the epithelium is the functionally active tissue, the other tissues being subservient to it. The glands of the body differ considerably in both structure and function, but in all of them it is epithelium which elaborates the materials essential to the formation of their normal secretions. Mention has already been made of those glands which furnish secretions charged with waste materials to be eliminated from the body. Such glands are called excretory glands, and are exemplified by the kidney. Other glands, distinguished as secretory in a restricted sense, furnish secretions which are of service to the organism. Examples of such glands are those which discharge their secretions into the alimentary tract where, by virtue of the ferments they contain, they prepare the food for absorption. Another example of a secretory gland is furnished by the sebaceous glands of the skin, which produce an oily substance serving to keep the epidermis upon which it is discharged soft and pliable.

In the secretory glands the cells of the functional epithelium elaborate within their bodies the substances necessary to give the glandular secretion its peculiar and useful characters. These substances accumulate within the cells, where they are stored until required, when they are discharged into the secretion. While in the stored condition within the cells these substances may have a different chemical constitution from that which they acquire when they are discharged from the cells. A simple example of this chemical transformation is furnished by the liver, in the epithelial cells of which carbohydrates are stored as glycogen, to be liberated as a closely related chemical substance, glucose. In like manner the ferments stored in the epithelial cells of the digestive glands are not fully formed while in that situation, but exist in states known as "zymogens," from which the potent ferment appears to be readily formed when the cells are called upon to furnish it.

It is apparent, then, that the elementary tissue, epithelium, cannot have the same microscopical structure in all the situations in which it is found; but, notwithstanding these variations, wherever epithelium occurs it presents certain general structural peculiarities which are constant and which distinguish it from the other elementary tissues. Similarly, each of the other elementary tissues presents variations in the details of its structure in different situations, but always retains certain general structural characteristics distinguishing it from all the other elementary tissues. It is the first task of the student of histology to learn to recognize and identify these elementary tissues wherever they occur and however they may vary from the type which is first presented to him for study.

In the following chapters an attempt is made to give the student an idea of the essential structure of the elementary tissues, so that he may recognize them in specimens which he examines with the microscope. For this purpose they have been arranged in the order of their structural simplicity.

When examining a specimen under the microscope with a view to recognizing the elementary tissues it contains, the student should habitually ask himself the following questions: (1) What are the general characters of the cells entering into the structure of the tissue? (2) What kind of intercellular substances separates those cells? (3) How are the cells arranged with reference to each other and the intercellular substances? Correct answers to these three questions will enable him to quickly determine the nature of the tissue he is observing, even if it should vary considerably in structural details from examples of the same tissue with which he has already become familiar.

# CHAPTER III.

## THE EPITHELIAL TISSUES.[1]

### I. ENDOTHELIUM.

**General Characters.**—(1) The cells possess thin membranous bodies, except at the site of the nucleus, to enclose which the cell-body is thickened. (2) The intercellular substance is minimal in amount; clear and homogeneous in character. (3) The cells are arranged, edge to edge, in a single layer. The wavy or denticulate edges of neighboring cells fit into each other, being separated by a mere line of the intercellular substance which in this tissue has received the name of "cement-substance" (Fig. 23).

Endothelium forms a thin membranous tissue composed almost exclusively of cells. It occurs in its most isolated form in the capillary bloodvessels, the walls of which are simply tubes of endothelium, supported externally by the surrounding tissues and fluids and internally by the enclosed blood. It also covers the tissues surrounding the serous cavities of the body, where it serves both as a lining to the cavities and a smooth covering to the organs, diminishing the friction resulting from their movements against each other. It does not occur in any situation where it would be exposed directly to the external world.

The cells of endothelium vary somewhat in size and shape. They may be polygonal, diamond, or stellate in form, and during life are soft and extensible so that their sizes may be modified by stretching or tension in one or more directions. The cell-bodies, or cytoplasm, are usually clear and apparently structureless or only slightly granular, but occasionally some of the cells are smaller and more granular than the majority. This is especially marked in the cells surrounding minute apertures that are found here and there in the endo-

---

[1] The term "epithelial" is used here in its most inclusive sense to designate those tissues which cover surfaces, whether those surfaces are exposed to the outer world, as, for example, the skin and the mucous membranes, or are wholly enclosed, as are the inner surfaces of the bloodvessels, lymphatics, and serous surfaces.

thelial lining of the serous cavities (Fig. 24). These openings are called stomata and furnish a direct communication between the serous cavities and the lymphatic spaces in the tissues surrounding them. These openings virtually convert the serous cavities into enormous lymph-spaces forming a part of the general lymphatic system.

FIG. 23.

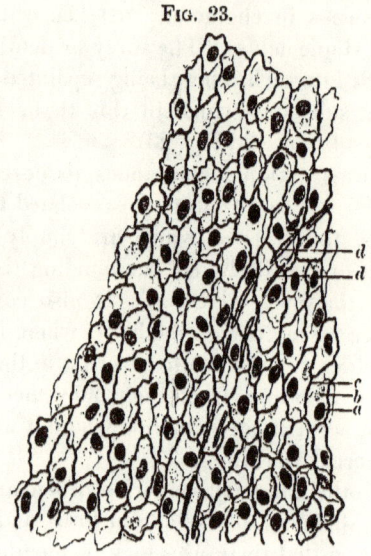

Mesentery of frog treated with silver nitrate. The mesentery is covered on both surfaces with a layer of endothelium. Between these is areolar connective tissue containing bloodvessels, lymphatics, and nerves. In this figure only the two endothelial layers and a capillary bloodvessel are represented: *a*, nucleus of endothelial cell belonging to uppermost layer; *b*, nucleus of cell belonging to deep layer forming the lower surface of the specimen; *c*, intercellular cement between cells of upper layer of endothelium; *d, d*, nuclei of endothelial cells, forming a capillary bloodvessel, seen in profile. The bodies of these cells are not reproduced in the figure. The cement in the deep layer of endothelium is represented by finer lines to distinguish it from that belonging to the upper layer.

The edges of contiguous endothelial cells are not everywhere in equally close approximation to each other (Fig. 25). The occasional points where they are more widely separated than usual are occupied either by an increased amount of the cement-substance, or processes from cells in the underlying tissues are here intercalated between the endothelial cells, reaching the surface of the serous membrane. In either case these points of separation of the endothelial cells are not openings through the tissue, though, as we shall see in a subsequent chapter, they are spots where the tissue is relatively more pervious than elsewhere. They are called pseudostomata, to distinguish them from the stomata already mentioned.

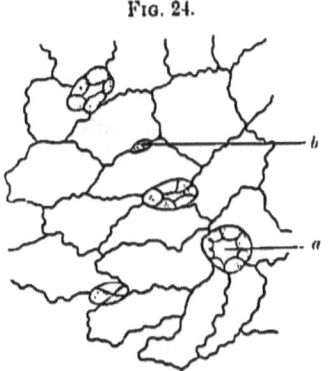

Fig. 24.

Endothelium on a serous surface of the frog. (Klein.) *a*, stoma bounded by endothelial cells with granular cytoplasm; *b*, pseudostoma. The nuclei of the cells are not represented.

The intercellular substance in endothelium is so small in amount and so homogeneous and transparent that it escapes observation

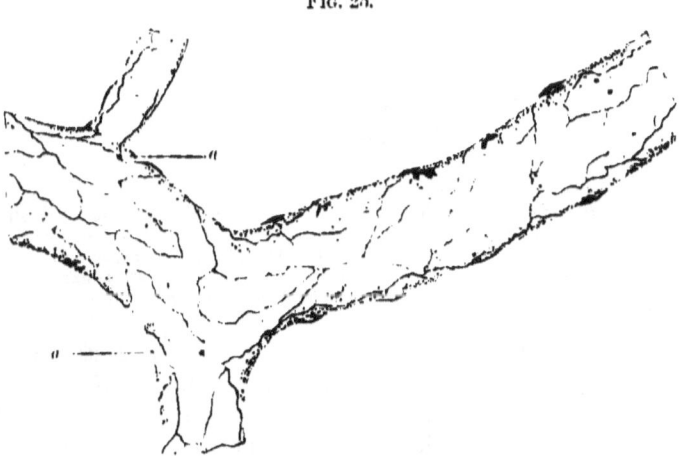

Fig. 25.

Endothelial lining of a small vein treated with silver nitrate; dog. (Engelmann.) The figure represents a tube formed of endothelium the cells of which vary in size and shape. The whole wall of a capillary has essentially the same structure as this venous lining, but its calibre is smaller. The upper branch in this figure may represent a capillary opening into the vein. *a, a*, pseudostomata occupied by cement-substance.

under the microscope unless special means are employed for its demonstration. The simplest of these consists in treating the fresh tissue with a 1 per cent. solution of nitrate of silver for a few moments, washing with distilled water, and then exposing it to the rays of the sun. During this treatment the intercellular substance enters into combination with the silver. Upon exposure to strong light this compound is destroyed, leaving an insoluble black precipitate of silver oxide. When the specimen is examined under the microscope, the site of the cement-substance is marked by the presence of this precipitate. Endothelium so treated shows a network of fine dark lines, the meshes of which are occupied by the cells of the tissue. When no such method has been employed to render the intercellular substance conspicuous, the outlines of the cells cannot be distinguished, and the tissue appears as a continuous, nearly homogeneous membrane containing nuclei at more or less regular intervals. When seen in profile or vertical section, endothelium appears as a delicate line, expanded at intervals to enclose a nucleus (Fig. 26). The nuclei of the endothelial cells are round or

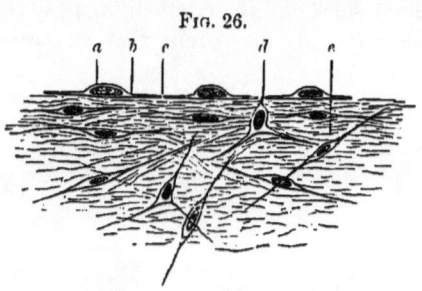

Fig. 26.

Diagram of vertical section through a serous membrane: *a*, nucleus of endothelial cell: *b*, body of cell; *c*, line of junction between two cells occupied by cement-substance; *d*, process of connective-tissue cell occupying a portion of the intercellular space between two endothelial cells, one variety of pseudostoma: *e*, areolar tissue with fusiform and stellate cells. The vessels and nerves in the areolar tissue have been omitted.

oval, and each cell usually possesses but a single nucleus situated near its centre, but occasionally cells with two nuclei are observed.

Functionally, endothelium appears to play only a passive *rôle* in most situations in which it is found. It furnishes a smooth covering for those internal surfaces of the body which are exposed to friction, as, for example, in the serous cavities and the inner surfaces of the vascular systems. In the capillary bloodvessels and lymphatics endothelium forms the entire wall of the vessels, and its thinness permits the passage of fluids through those walls. The fact that the lymph in different parts of the body varies somewhat in composition has led to the inference that the endothelium of the capillary walls exercises an active function in determining what shall pass through it; that the lymph is a sort of endothelial secretion. It is difficult, however, to reconcile this view with the fact that the endothelial cells are so poor in cytoplasm.

Endothelium is developed from the mesoderm.

## II. EPITHELIUM.

**General Characters.**—(1) The cells are nearly always large and rich in granular cytoplasm. They contain distinct round or oval, vesicular nuclei, of which there is usually only one in each cell. (2) The intercellular substance is very small in amount and is clear and homogeneous. (3) The arrangement of the cells and their size and shape all vary greatly, giving rise to a number of varieties of epithelium, which are classified according to the shape and arrangement of the cells. In pavement-epithelium the cells are thin and arranged in a single layer, not unlike endothelium. In cubical epithelium the cells are thicker and also usually arranged in but a single layer. In columnar epithelium the cells are prismatic in form and rest with their bases upon the surface of the tissues beneath. They are usually separated at their bases by pyramidal cells, so that the layer of epithelium cannot be said to consist strictly of but one layer of cells, and in some situations there are several distinct layers. In stratified epithelium the cells are superimposed upon each other to form a layer of cells, the thickness of which is several times the diameter of a single cell. The cells of the variety of epithelium called ciliated epithelium differ from those of the other varieties in possessing delicate, hair-like processes which project from the free surface of the tissue.

Epithelium resembles endothelium in being composed almost exclusively of cells separated by a minimal amount of intercellular substance. Like endothelium, it is nearly always found covering other tissues and having one free surface. The two tissues differ greatly in the character of their cells, with one notable exception. This exception is found in the epithelial lining of the pulmonary alveoli, where the pavement-epithelium contains cells that closely resemble those of endothelium. These cells are, however, directly exposed to the inspired air, while endothelium is only found in situations where it is protected from all contact with the external world.

1. **Cubical Epithelium.**—The cells of this variety of epithelium

are approximately of the same diameter in all directions. They may be almost strictly cubical or spherical, but are usually polyhedral as the result of mutual compression, their contiguous surfaces being flattened. They are usually disposed in a single layer upon a surface furnished by the underlying tissues, as, for example, in tubular or racemose glands, but they may be aggregated to form a solid mass of cells filling a sac, as in the sebaceous glands of the skin, or in strands or columns, variously disposed, as in the liver and suprarenal bodies.

It is this form of epithelium that is chiefly concerned in performing the functions of secretion, and, for this reason, it is frequently designated as "glandular epithelium."

The appearance of the individual cells varies considerably according to the functions that they perform and the stage of functional activity which obtained at the time cellular changes were arrested when the particular specimen was prepared for study. It will suffice for present purposes of description to call attention to the fact that the cytoplasm is usually highly granular, partly because of its own structure, partly because many of the substances elaborated and stored within the cells as the result of their functions appear in the form of granules (metaplasm). The nature of these granules varies. They may be albuminoid, zymogenic granules, or minute drops of fatty substances, which may coalesce to form distinct oily globules, or they may consist of carbohydrates, e. g., glycogen. The granular condition of the cytoplasm may be so marked

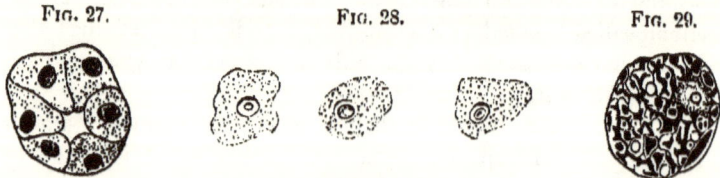

FIG. 27.  FIG. 28.  FIG. 29.

Cubical epithelium.

Fig. 27.—Six cells from the sublingual gland of a man who was executed. (Schiefferdecker.)
Fig. 28.—Three isolated cells from the gastric tubules of the dog and cat. (Trinkler.)
Fig. 29.—Cell with highly granular cytoplasm, the result of stored metaplasm, chiefly glycogen. (Barfurth.)

as to render the detection of the nucleus difficult in unstained specimens (Figs. 27, 28, and 29).

In this form of epithelium the presence of two nuclei in a single cell is more frequent than in the other varieties.

2. **Pavement-epithelium.**—This variety of epithelium consists of thin cells arranged edge to edge to form a single layer. With the exception of certain regions on the surfaces of the pulmonary alveoli, the cells are more cytoplasmic and granular than are those of endothelium which this tissue in other respects closely resembles. During fœtal life the smaller air-passages and alveoli of the lung are lined by a pavement-epithelium, the cells of which are nearly as thick as those of some varieties of cubical epithelium. When, however, the lung is expanded by the respiratory acts following birth, many of the cells lining the alveoli become greatly extended and flattened until their bodies are thin and membranous and their nuclei inconspicuous or even destroyed (Fig. 30). These greatly flattened epithelial cells are found covering those portions of the

FIG. 30.

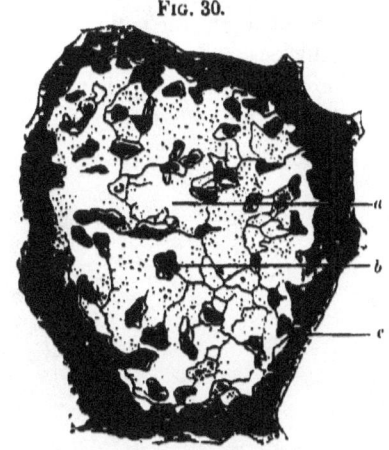

Pavement-epithelium. Surface view of the lining of a pulmonary alveolus; man. (Kölliker.) a, membranous cell without a nucleus; b, nucleated granular cell; c, cut surface of the vertical wall of the alveolus, the structure of which is not represented.

alveolar walls in which the capillary bloodvessels are situated and permit a ready interchange of gases between the air in the alveolar cavities and the blood circulating in their walls. Many of the epithelial cells covering the tissues in the meshes between the capillaries retain the cytoplasmic and granular character possessed before birth and appear capable of multiplying and, perhaps, replacing such of the thinner cells as may be thrown off or destroyed.

It will be evident, from the foregoing descriptions, that there is no sharp structural line separating cubical from pavement-epithelium. Functionally, pavement-epithelium is a much less active tissue than the cubical variety.

3. **Columnar Epithelium** (Figs. 31, 32, 33).—The cells of this

FIG. 31.

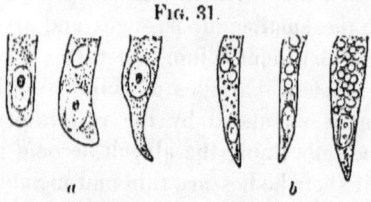

Columnar epithelium. From tongue of *pseudopus*. (Seiler.) *a*, three cells with intact cytoplasm, except the central one, which contains a vacuole; *b*, three cells of which the distal ends contain drops of fluid (vacuoles) or of metaplasm.

form of epithelium are of a general columnar or prismatic shape and possess a single nucleus and a cytoplasm that is usually distinctly granular. They are arranged with their long axes parallel to each other, so that their free ends form the surface of the epithe-

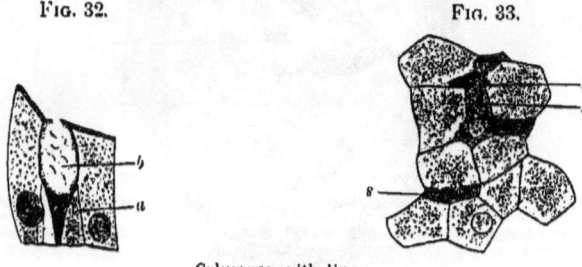

FIG. 32.      FIG. 33.

Columnar epithelium.

Fig. 32.—From small intestine of the mouse. (Paneth.) *a*, pyramidal reserve cell, nucleus not included in section; *b*, "goblet" cell, enclosing a large drop of secretion.
Fig. 33.—From small intestine of the mouse. (Paneth.) Columnar epithelial cells seen from above; *b*, goblet-cell, the mucous contents darkened by the hardening process; *s, s*, highly granular cells which have recently discharged their secretion.

lium, while their deeper ends either rest upon the tissues beneath the epithelium or upon other epithelial cells of different shape which form one or more layers between the columnar cells and the underlying tissues. When they rest directly upon the tissues beneath there are usually other epithelial cells of a pyramidal or oval shape which may be regarded as immature cells ready to take the place of such fully developed cells as may become detached or destroyed. The presence of these cells occasions a narrowing of the deep ends of the columnar cells, so that they are not strictly prismatic in form. In cross-section, or when viewed in a direction parallel to their long axes, the cells have a polygonal form due to the lateral pressure they exert upon each other (Fig. 33).

The nuclei of the columnar cells are oval, situated nearer the base of the cell than its superficial end and with their long axes parallel to those of the cells themselves, and are vesicular in structure with a distinctly reticular arrangement of the chromatin filaments.

Columnar epithelium is found chiefly upon the free surfaces of mucous membranes, but also occurs in some of the secreting glands. The minute structure of the cells varies somewhat in different situations, but the consideration of these minutiae must be deferred until a description of the structure of the different organs is undertaken in a subsequent chapter.

4. **Ciliated Epithelium** (Figs. 34, 35, 36).—Ciliated epithelium

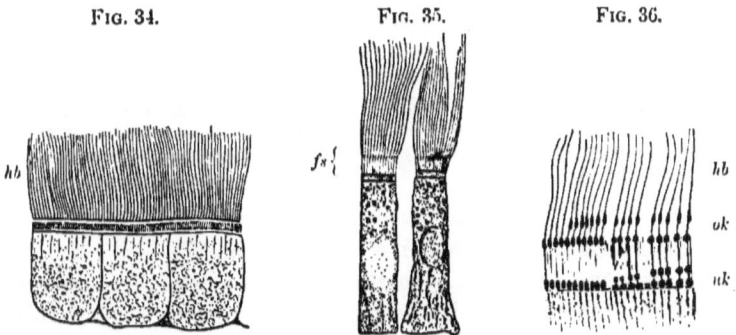

Ciliated epithelium. (Frenzel.)

Fig. 34.—Cubical cells with long cilia (*hb*). The nuclei of the cells are obscured by the granular cytoplasm.
Fig. 35.—Columnar cells. The rodded margin, *fs*, corresponds to the cuticle in Fig. 37.
Fig. 36.—Diagram illustrating variations in the structure of the ciliated ends of cells. The rodded portion, *ok* to *nk*, corresponds to the cuticle of other varieties of epithelium, though the latter do not possess the knobbed ends of the rods represented in this figure; *hb*, cilia.

is merely a variety of either columnar or cubical epithelium in which the free ends of the cells are beset with delicate hair-like processes, which execute lashing movements in some one direction. It is found lining the trachea and bronchi, the cilia here serving to propel toward the larynx such particles of dust as are brought into the respiratory passages by the currents of air during respiration. Ciliated epithelium also occurs on the lining membranes of the nose and the adjoining bony cavities, the mucous membrane of the uterus and the Fallopian tubes, the vasa efferentia of the testis and a part of the epididymus, the ventricles of the brain (except the fifth), the central canal of the spinal cord, and the ducts of some glands.

The possession of cilia, which are very motile organs, presents a marked departure in specialization from the usual metabolic functions of epithelium. Ciliated epithelium rarely exercises a secretory function, its stock of energy being utilized to produce motion instead of chemical change. But there are secreting varieties of epithelium possessing a "cuticle" which appears to be morphologically analogous to the cilia, but in which the fibrils are less highly developed, probably not motile, and, therefore, functionally not the equiva-

FIG. 37.

Cuticularized epithelium, intestine of dog. (Paneth.) Rodded cuticle of the free ends of columnar cells. In most specimens of ciliated epithelium from human tissues, where no special care has been taken to preserve the cilia, the ciliated border presents the appearances shown in Fig. 37.

lents of cilia. This cuticle is highly developed in the cells covering the mucous membrane of the intestine (Fig. 37).

5. **Stratified Epithelium.**—In the varieties of epithelium hitherto considered the cells are, in the main, disposed upon some surface in a single layer, some, at least, of the cells usually extending from the bottom of the layer to its surface.

Stratified epithelium is distinguished from these by being of greater depth and consisting of several layers of cells. The epithelium lining the cheek or the œsophagus may be taken as a typical example of this variety.

The most deeply situated cells are small and nearly filled by the round or oval nucleus. They undergo frequent division, and as they multiply some of them are crowded toward the surface. For a time these increase in size through a growth of their cytoplasm. But as they are pushed nearer to the surface and farther from the sources of nutrition in the vascular tissues underlying the epithelium, they become flattened and their bodies lose their cytoplasmic character, being converted into a dry, horny substance, keratin.

Upon the free surface they are reduced to thin scales, closely adhering to each other and their subjacent neighbors, but entirely devoid of both cytoplasm and nucleus (Fig. 38).

Stratified epithelium is found upon surfaces exposed to friction, which it serves to protect against mechanical injury, and, in some

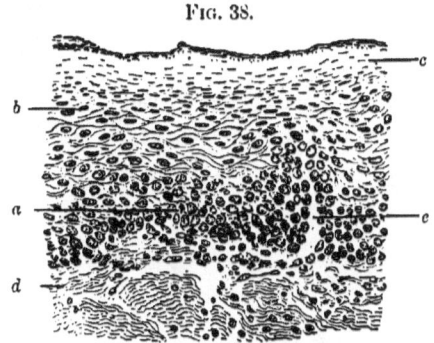

Fig. 38.

Stratified epithelium, œsophagus of the rabbit: $a$, karyokinetic figure in a cell of the deep layer, demonstrating the fact that the cells multiply in this region; $b$, larger flattened cell nearer the surface; $c$, horny layer made up of cells that have undergone keratoid degeneration; $d$, underlying fibrous tissue. In one place, near the centre of the figure, six blood-corpuscles reveal the presence of a small vessel; $e$, tangential section of a small fibrous papilla extending into the epithelium and surrounded by young epithelial cells.

cases, against desiccation. It forms the epidermis of the skin, and lines the mouth, œsophagus, rectum, and vagina. In these situations the scaly or squamous cells of the surface are constantly being removed by the attrition to which they are exposed, but are as constantly replaced by fresh cells from the deeper layers of the epithelium. Pressure and moderate friction stimulate the multiplication of the cells in the deepest layers of the tissue, so that parts, $e.\,g.$ of the skin which are especially subjected to such influences acquire a thicker epidermis (callus).

Where the stratified epithelium consists of many layers of cells, as is the case, for instance, upon the skin, there is a provision for the nourishment of the growing cells which are somewhat removed from the vascularized subjacent tissues. The cells of the deeper layers are somewhat separated from each other, leaving a space between them through which nutrient fluids can circulate. Across this space numerous minute projections or "prickles," springing from neighboring cells, join each other, forming connecting bridges between the cells. When isolated, such cells appear covered with these small spicules ("prickle-cells"), and their presence probably increases the tenacity with which the cell-remains adhere to each other when they become hardened and toughened on the surface of the epithelial layer (Fig. 39).

These delicate bridges connecting neighboring cells are not peculiar to stratified epithelium, though they are more conspicuous in that tissue than elsewhere. They have been observed between the cells of the columnar epithelium of the intestinal mucous membrane, and also between the cells of other elementary tissues; *e. g.*, smooth muscular tissue.

6. **Transitional Epithelium** (Figs. 40 and 41).—This variety resembles stratified epithelium in forming layers several cells in thick-

Fig. 39.

Prickle cells from human stratified epithelium. (Rabl.) Four cells with delicate processes uniting across an intervening space are represented. The lower right-hand cell is just below the upper surface of the section, so that its surface is seen. This is covered with minute spots, which are end views of the prickles directed toward the observer. The nucleus of this cell is not in sharp focus, a fact indicated by the fainter outline in the figure.

ness, but differs in the character of its superficial cells. These do not undergo the horny change peculiar to stratified epithelium, but continue to increase in size, forming a covering of very large cells lying upon those beneath. Under these largest superficial cells are pyriform cells lying with their larger, rounded ends next to the topmost layer, while their deeper and more attenuated ends lie between the oval or round cells that form the one or two deepest layers of the epithelium and rest upon the underlying tissues.

Transitional epithelium is found lining the renal pelves, ureters, and bladder. Its structure permits of a considerable stretching of the tissues beneath without rupture of the epithelial layer over them, the cells of which become flattened to cover the increased surface, to return to their first condition when the viscus which they line is emptied. This is notably the case in the bladder, the epithelial lining of which may be taken as a type of this variety of tissue.

The **functional activities** of epithelium are in marked contrast to the comparatively inert character of endothelium. The cytoplasmic

Fig. 40.

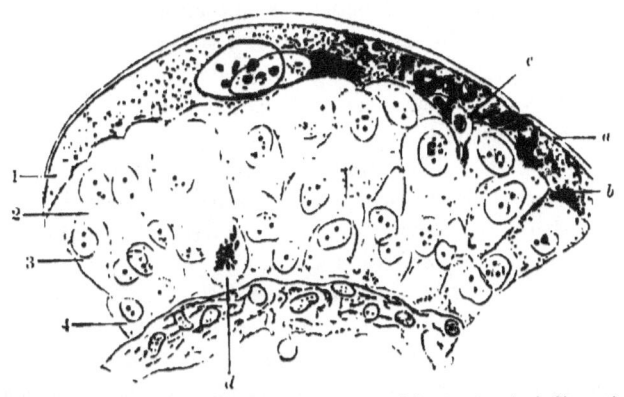

Transitional epithelium from bladder of the mouse. (Dogiel.) *1, 2, 3,* and *4* indicate the layers of cells, not everywhere equally well defined. *a*, hyaloplasmic surface, and, *b*, cytoplasmic body of large superficial cell; *c*, leucocyte—*i. e.*, white blood-corpuscle that has wandered into the epithelium by virtue of its amœboid movements; *d*, karyokinetic figure in a cell belonging to the deepest layer. Beneath this layer is the fibrous tissue, which is covered by the epithelium and forms a part of the wall of the bladder. The superficial cell, which is fully represented, contains two nuclei, a not very infrequent occurrence in these cells.

nature of the epithelial cell, when contrasted with the poverty in cytoplasm of the cell in endothelium, would lead us to expect this difference in the cellular activities of the two tissues. At the beginning of this chapter a sketch of the manifold functions of epithe-

Fig. 41.

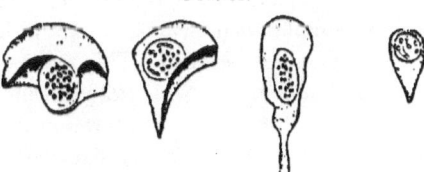

Transitional epithelium. Isolated cells from the bladder of the frog. (List.)

lium was given. It is a fair general statement of its usefulness to say that epithelium is chiefly concerned in bringing about chemical changes in substances brought to it. Sometimes these substances are elaborated into fresh cell-constituents, and the activity of the tissue is displayed chiefly in an active multiplication and growth of its cells. This is especially true in the stratified variety, where protection is provided by a constantly renewed supply of cells. In other cases the substances received by the cells are elaborated into definite compounds destined to form the essential constituents of a secretion. This secretory function of epithelium is an extremely important one, and for its performance that tissue is usually arranged in a special structure or organ, called a gland. A brief statement of the general characters and classification of these organs may here appropriately find a place.

**Secreting Glands.**—The simplest type of secreting structure consists of a surface covered with a layer of epithelium, the cells of which are endowed with the power of elaborating a secretion and discharging it upon their free surfaces (Fig. 32, b). The tissues supporting the epithelium belong to the connective tissues, and are fibrous in character and well provided with bloodvessels, lymphatics, and nerves. These bring to the epithelium the substances necessary for its nourishment and work, and place its activities under the control of the nervous system. Between the epithelium and the fibrous tissue supporting it there is frequently a thin membranous layer of tissue that often appears quite homogeneous, evidently belongs to the connective tissues, and has received the name of "basement-membrane." This appears to offer a smooth surface for the attachment of the epithelial cells, which receive their nourishing fluids through it.

The epithelial surfaces of many of the mucous membranes are examples of the foregoing simple secreting structure. The secretory function is here of use as an adjunct to the protective function assigned to the epithelial covering, and the quantity of secretion is but slight under normal conditions. Where the volume of secretion required is considerable some provision for an increase in the extent of secreting surface is necessary. This may be accomplished by an invagination of that surface, which then forms the lining of one or more tubes or sacs, into which the secretion furnished by the epithelial cells is discharged. Such an arrangement of the tissues constitutes a *gland*, and it is evident that these may be arranged into groups or classes according to whether the secreting surface forms a single tube or sac, or several such tubes or sacs, uniting to form a single gland. Thus, there may be simple or compound tubular glands, or simple or compound saccular glands. Whether the deeper portions of the gland have a tubular or saccular structure, the secretion of the gland is discharged upon some free surface through a tubular outlet, called the duct. This is frequently lined with a non-secreting layer of epithelial cells differing in character from the actively secreting epithelium in the deeper portions of the glandular passages (Figs. 42-47).

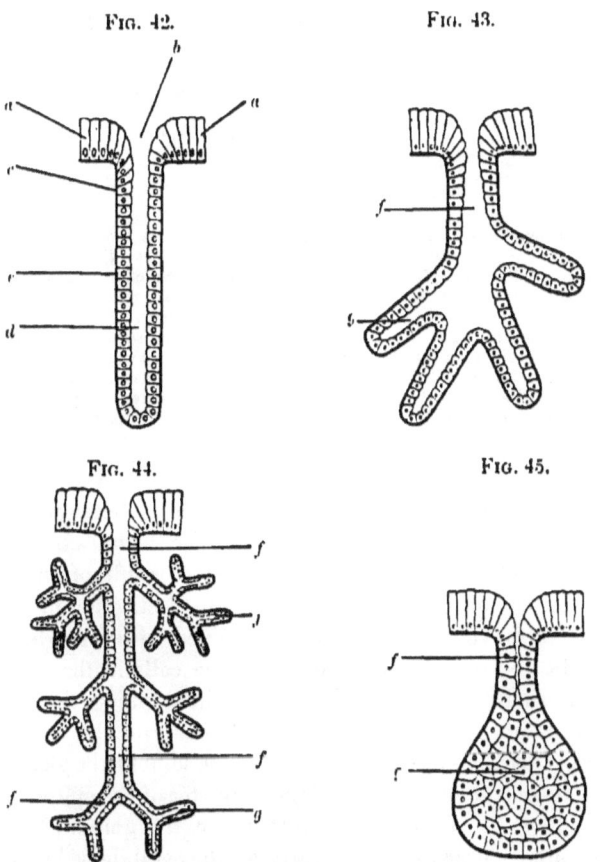

Diagrams representing various types of gland.

Fig. 42.—Simple tubular gland: a, epithelium covering the surface on which the secretion is discharged; b, mouth of gland; c, epithelium lining the duct. This gradually passes into the secreting epithelium. Some simple tubular glands have no such distinction between the cells near the mouth and those nearer the fundus, but all the cells are of the secreting variety—i.e., exercise that function. e, secretory epithelium; d, lumen. The sweat-glands are simple tubular glands which are coiled in their lower part to form a globular mass.

Fig. 43.—Compound tubular gland: f, duct; g, acinus.

Fig. 44.—Racemose tubular gland: f, f, f, ducts; g, g, acini.

Fig. 45.—Simple saccular gland: f, duct; g, acinus.

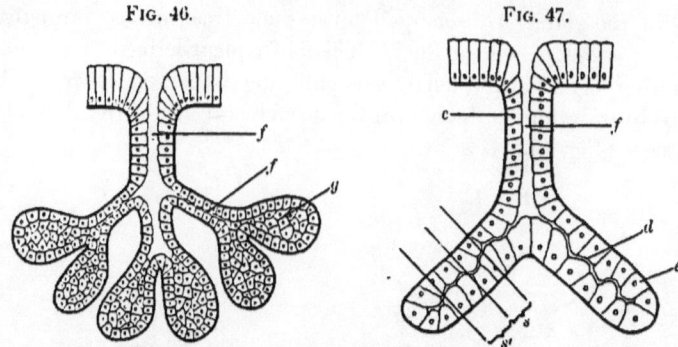

Diagrams representing various types of gland.

Fig. 46.—Racemose saccular gland: *f, f*, ducts; *g*, acinus.

Fig. 47.—Compound tubular gland, with a marked distinction in the character of the epithelium in the duct and acini: *c*, duct epithelium; *f*, duct; *d*, lumen of the acinus; *e*, secreting epithelium. This type of gland is common. This figure is introduced to show how difficult it might be to detect the lumen of the acinus in sections of such a gland. The lumen is of very small diameter (its size is exaggerated in this diagram) and runs such a tortuous course among the epithelial cells that even perfect cross-sections of the acinus might fail to reveal it if it happened at that point to run obliquely to the axis of the acinus. It would then appear merely as a small clear spot upon the granular cytoplasm of the cell that lay immediately beneath it. *s, s'*, represent the way in which two such sections would contain portions of the acinus. The lumen in *s'* would be more easily detected than in *s*, because its general direction is more rectilinear and more nearly coincident with the line of vision.

It is rarely possible to trace the connection between the ducts and other portions of a gland in sections, for the axes of these different parts seldom lie in one plane. As a result of this circumstance, sections of glands usually present a collection of round or oval sections of tubes or sacs, which are lined with a single layer of epithelial cells, surrounding a lumen. The cells in the deeper portions are usually granular and cubical; those lining the ducts are generally more columnar in shape and less granular in character. The deeper portions are called the alveoli or acini of the gland, to distinguish them from the ducts, and the character of the epithelium they contain differs according to the function of the gland. Sometimes the cells are so large that they nearly fill the acini, leaving a scarcely perceptible lumen. In other glands the cells are less voluminous and the lumen of each acinus is distinct. It occasionally happens, *e. g.*, in the submaxillary glands, that the acini contain two sorts of cells which secrete different materials. Both kinds of cell may be present in the same acinus, or each kind may be confined to different acini. In studying sections of glands it must be borne in mind that the tangential section of an acinus would appear as a group of cells surrounded by fibrous tissue, with no trace of a lumen among the epithelial cells (Fig. 48).

Glands develop from surfaces which are covered by epithelium.

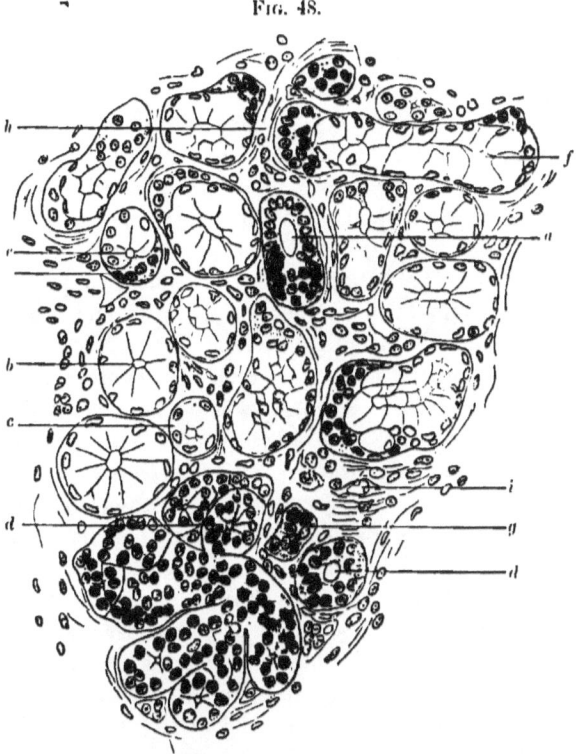

FIG. 48.

Section of gland from human lip. (Nadler.) *a*, duct, cut in slightly oblique direction (lumen oval), and probably near a branch, which would account for the apparent thickness of its epithelial lining in the lower half; *b*, cross-section of acinus secreting mucus; *c*, tangential section of a similar acinus near its extremity and beyond the end of the lumen. Cross-sections of the cells at the fundus occupy the centre. *d*, cross-section of an acinus secreting a serous fluid, revealing a small lumen; *d'*, a similar acinus with a larger lumen, probably cut near its junction with a duct; *e*, acinus with crescentic group of cells with granular cytoplasm (*e'*), and other cells like those in *b*. The granular cells of small size are considered to be cells which have discharged their secretion and are accumulating material for a fresh supply. *f*, nearly axial longitudinal section of a portion of a mucous acinus; *g*, tangential section of a serous acinus; *h*, fibrous connective tissue between the acini; *i*, capillary bloodvessel in the fibrous tissue.

The cells of this epithelium multiply and penetrate into the underlying tissues, forming little solid tongues or columns of cells (Fig. 181). If the gland is destined to be of the simple tubular variety, this column of cells then becomes hollowed to form the lumen, the cells being arranged in a single layer lining the tubule. If the gland is to be compound, the solid column of cells branches within the tissues, and then the lumina of the different portions are formed, the epithelium in the different parts becoming differentiated as specialization of function develops.

The foregoing general description of the structure of secreting glands applies to those glands which have a purely secretory function, discharging the products of their activities upon some free surface, such as the skin or a mucous membrane. There are other glandular organs which perform more complicated functions and the structure of which deviates from that of the simpler glands. Examples of these are furnished by the liver and kidney, the structures of which must be deferred to a subsequent chapter. Other exceptions are exemplified in the thyroid body and other "ductless" glands, which discharge no secretion into a viscus or upon a free surface, but which have an alveolar structure similar to an ordinary secreting gland. These alveoli do not communicate with ducts, which are wanting; but whatever products they may contribute to the whole organism are apparently discharged into the circulating fluids of the body by a process of absorption similar to that through which the glandular epithelium obtains its materials from those fluids, or by a direct discharge into the lymphatics. (See chapter on Ductless glands.) This process is indicated by the term "internal secretion," and is probably of commoner occurrence than is usually supposed. In fact, it but represents a special interpretation of the phenomena of interchange of material that is constantly going on between all the cells of the body and its circulating fluids.

Epithelium is developed from the epiderm or hypoderm; never from the mesoderm. In this respect, as well as in its functional rôle, it differs from endothelium.

# CHAPTER IV.

## THE CONNECTIVE TISSUES.

The two varieties of elementary tissue that have just been considered—namely, endothelium and epithelium—owe their qualities directly to the characters of the cells that enter into their composition. The intercellular substances are insignificant in amount and subordinate in function.

In marked contrast to these are the tissues composing the group known as the "connective tissues." Here the usefulness of the tissues depends upon the character of the intercellular substances which confer upon the tissues their physical properties. The activities of the cells entering into the composition of these tissues appear to be confined to the production of those important intercellular substances and the maintenance of their integrity. The cells may, therefore, be considered as of secondary importance in determining the immediate usefulness of the tissues, the first place being given to the intercellular substances. As was stated in the introductory chapter, these connective tissues are essentially passive—i. e., they are useful because of their physical characters rather than because of any ability to transform either matter or energy. Where the ability to accomplish those transformations is of importance the tissues are found to be essentially cellular in character, as we have already seen to be the case in the epithelial tissues.

The connective tissues may be divided into three main groups: the cartilages, bone, and the fibrous tissues. Each of these groups has certain general structural characters that distinguish it from the other elementary tissues, but within each group there are varieties which differ considerably in the detailed character of their intercellular substances and in the arrangement of these with respect to the cells.

All the elementary tissues belonging to the connective-tissue group are developed from the mesoderm.

## I. THE CARTILAGES.

**General Characters.**—(1) The typical cell of cartilage is round or oval in shape, rich in cytoplasm, and possesses one (rarely two) nucleus of oval form and vesicular and reticulated structure. Within the cytoplasm there are frequently one or more clear spots, which are drops of homogeneous fluid, "vacuoles." The cells frequently depart somewhat from this type. Where the tissue is growing they are usually flattened on the sides turned toward their nearest neighbors. This is because they are the offspring of a cell that has recently divided, and are as yet separated by only a small amount of intercellular substance. Under these circumstances each cell is frequently surrounded by a thin layer of intercellular substance, probably of relatively recent formation, which differs a little from that further from the cell and gives an appearance as though the cell were enclosed in a capsule. In older cartilage this appearance is no longer evident. Where cartilage is being replaced by

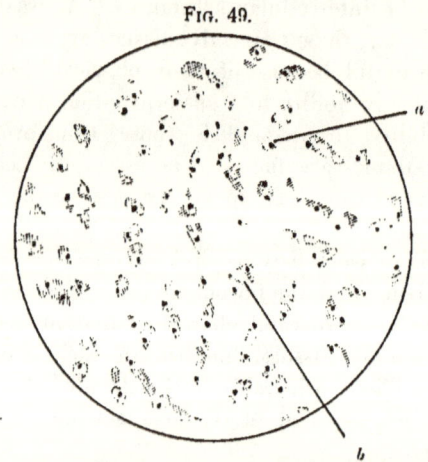

FIG. 49.

Hyaline cartilage. Section of human costal cartilage: *a*, nearly spherical cell containing two vacuoles; *b*, recently formed intercellular substance ("matrix"), separating two cells that have been produced by the division of a single cell. There are several other examples of a similar grouping of cells, due to the same cause, in the figure. Between the cells is the hyaline, nearly structureless "matrix."

bone, "ossification," the cells are arranged in columns, with only a small amount of intervening intercellular substance, and have a general cubical form.

(2) The intercellular substance is abundant in amount and has received the special designation "matrix." According to the character of this matrix, the cartilages have been divided into three varieties: hyaline cartilage, fibro-cartilage, and elastic cartilage. In hyaline cartilage the matrix is clear and homogeneous and has the consistency of gristle. In fibro-cartilage it is traversed by or nearly wholly composed of delicate fibres similar to those of white fibrous tissue, which will be described presently. In elastic cartilage the matrix contains coarse, branching, and anastomosing fibres similar to those of elastic fibrous tissue (*vide infra*).

(3) The arrangement of the cells and intercellular substances varies considerably. Sometimes the cells are pretty uniformly distributed throughout the intercellular substance. Sometimes they

Fig. 50.

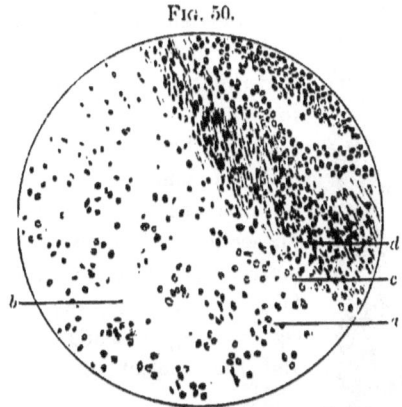

Hyaline cartilage and perichondrium. Human costal cartilage. Same specimen as Fig. 49; *a*, group of cells formed by division, but not yet separated by matrix; *b*, matrix; *c*, cells with a comparatively slight amount of cytoplasm, marking the transition from cartilage to fibrous tissue; *d*, perichondrium, composed of fibrous tissue (spindle-shaped cells with a fibrous intercellular substance).

are arranged in groups of from two to four or even six cells. Toward the surface of a piece of cartilage the cells are apt to be smaller than those nearer the centre, and are frequently flattened. Here, also, they often lose the characters that distinguish them in the body of the tissue, and more and more closely resemble the cells of the fibrous tissue surrounding the cartilage. This fibrous tissue is called the "perichondrium," and is usually not sharply defined from the cartilage itself, the matrix of the latter becoming more and more fibrous in character and the cells less distinctly like those

Fig. 51.

Hyaline cartilage. Section from human thyroid cartilage. (Wolters.) *a*, perichondrium; *b*, peripheral zone of cartilage with flattened cells. In the deeper portions of the cartilage the cells are larger, are arranged in groups, and are surrounded by recently formed matrix. The cells in the deepest portions of the cartilage are vacuolated, and about the groups of cells are fine granules of lime salts. In the matrix are numerous anastomosing lines, which are interpreted as fine canals, serving to carry nourishment to the cells in the cartilage.

typical of cartilage until the distinction between the two tissues is lost. The perichondrium is wanting over the free surfaces of the articular cartilages.

1. **Hyaline Cartilage** (Figs. 49, 50, and 51).—Although under ordinary powers of the microscope and in specimens which have not been specially prepared the matrix of hyaline cartilage appears clear and almost, if not quite, homogeneous, closer study reveals the presence of a fine network within the clear intercellular substance. This network is thought to be a system of minute channels through which the nutrient fluids permeate the tissue and reach its cells. It may be, however, that this reticulum is of fibrous character, in which case the fibres might be more pervious than the surrounding matrix, and bear the same relations to the nutrition of the tissue as a system of minute channels. In sections stained with hæmatoxylin the matrix of hyaline cartilage often acquires a faint bluish tinge, the cytoplasm of the cells a deeper shade of the same color, and the nuclear chromatin a very dark blue.

Hyaline cartilage forms the costal cartilages, the thyroid cartilage, the ensiform process of the sternum, the cartilages of the trachea and bronchi, and the temporary cartilages which are subsequently replaced by bone.

2. **Fibro-cartilage** (Fig. 52).—This variety of cartilage is found in only a few situations: in the interarticular cartilages of joints, in some of the synchondroses, in one region in the heart, and in the intervertebral disks. In the latter situation some of the cells possess branching processes, extending for

some distance between the fibres of the intercellular substance, and giving the whole tissue a character closely resembling that of

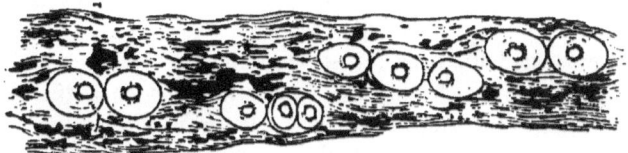

Fig. 52.

Fibro-cartilage. Section from human intervertebral disk. (Schäfer.) The cell to the left presents a branching process extending into the intercellular substance.

white fibrous tissue. The cells are, however, more cytoplasmic than those of ordinary fibrous tissue.

3. **Elastic Cartilage** (Figs. 53 and 58).—This form of cartilage

Fig. 53.

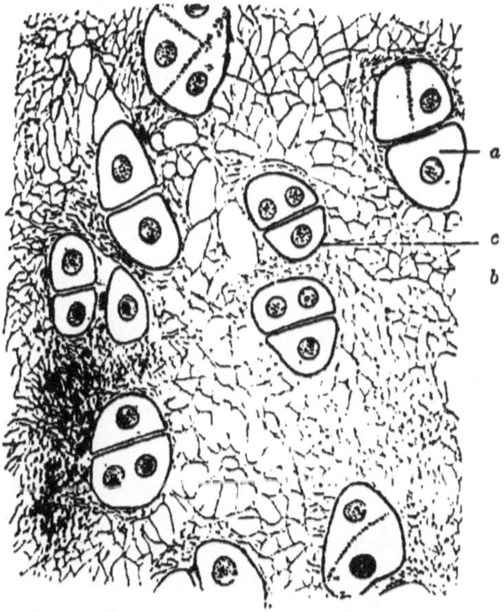

Elastic cartilage. Section from cartilage of human external ear. (Böhm and Davidoff.) a, cartilage-cell; b, c, network of elastic fibres in the intercellular substance; b, with large meshes; c, fine-meshed. Opposite a is a cell showing indications of a division of the cytoplasm following division of the nucleus.

is found in the epiglottis, the cornicula of the larynx, the ear, and the Eustachian tube. The coarseness of the anastomosing fibrous network of the matrix varies in different situations and in different parts of the same piece of cartilage. The reticulum is usually more open and composed of larger fibres toward the centre of the tissue than at the periphery, where it becomes more delicate and finally blends with the fibrous intercellular substance of the perichondrium.

It is evident, both from the structure of the cartilages and from the situations in which they are found, that they constitute elastic tissues suitable for diminishing the effects of mechanical shock. This is obviously the case in the joints, where both the hyaline and the fibrous varieties are found. Their elasticity and moderately firm consistency are also of obvious utility in the larynx and other air-passages and in the ear, nose, and synchondroses.

## II. BONE.

**General Characters.**—(1) The cells of bone, called "bone-corpuscles," have an oval vesicular nucleus, surrounded by a moderate amount of cytoplasm, which is prolonged into delicate branching processes that join those of neighboring cells. (2) The intercellular substance is composed of an intimate association of an organic substance and salts of the earthy metals. (3) The arrangement of these constituents is as follows: the organic basis of the intercellular substance is arranged in laminæ, which are closely applied to each other except at certain points where there are cavities, called "lacunæ," giving lodgement to the bone-corpuscles. Joining these lacunæ with each other are minute channels in the intercellular substances, "canaliculi," which are occupied by the fine processes of the corpuscles. In the compact portions of the long bones, and wherever the osseous tissue is abundant, the laminæ are arranged concentrically around nutrient canals, the "Haversian canals," which traverse the bone, anastomosing with each other and containing the nutrient bloodvessels of the tissue. In cancellated bone these Haversian canals are absent, and the thin plates of bone are made up of parallel laminæ of intercellular substance, between which are the lacunæ, connected with each other by canaliculi. The bone-corpuscles are nourished from the fluids circulating in the marrow, which occupies the large spaces of this spongy variety of bone.

It is not possible in a single preparation to study even these general characters of bone. The earthy salts in the intercellular substance prevent the preparation of sections by means of the knife, and, unless they be removed, specimens of bone must be made by grinding. This can best be accomplished after the bone has been dried. But drying the bone destroys the corpuscles, which appear as little desiccated masses, devoid of structure, within the lacunæ. Ground sections of bone can, therefore, give only an idea of the intercellular substance and the arrangement of the lacunæ, canaliculi, Haversian canals, etc. (Fig. 54). Sections may be cut if

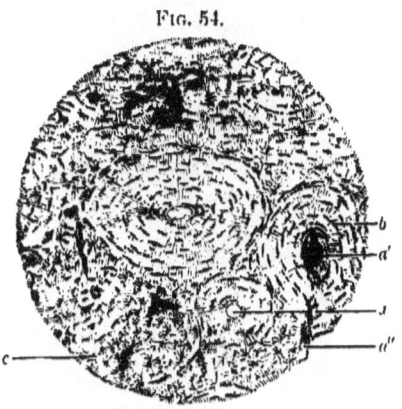

FIG. 54.

Ground section of dried bone. Human femur. *a*, Haversian canal in cross-section; *a'*, Haversian canal occupied by débris; *a''*, anastomosing branch from *a'*, in nearly longitudinal section; *b*, lacuna belonging to the Haversian system, of which *a* occupies the centre; *c*, lacuna in excentric laminæ of bone between the Haversian systems. The delicate lines connecting the lacunæ are the canaliculi.

the bone be first decalcified—*i. e.*, if the earthy salts be dissolved through the action of acids. This treatment not only removes the earthy constituents of the intercellular substance, rendering it soft and pliable, but causes the organic constituents to swell. The effect of this swelling upon the appearance of the bone is very marked. The fine canaliculi are closed and the lacunæ diminished in size, so that the structure of the bone appears much simplified, being reduced to a nearly homogeneous mass of intercellular substance in which there are spaces arranged in definite order and enclosing the somewhat compressed bone-corpuscles. The delicate processes of the latter are not discernible within the canaliculi, but blend with the swollen intercellular substance forming the walls of those minute channels. It is important that the student should learn to recognize these mutilated preparations of bone, since it is in this form that the tissue will most frequently come under his observation (Fig. 55).

Minute study of the structure of the intercellular substance of bone makes it appear that the organic basis is not homogeneous, but is composed of minute interlacing fibres, held together by

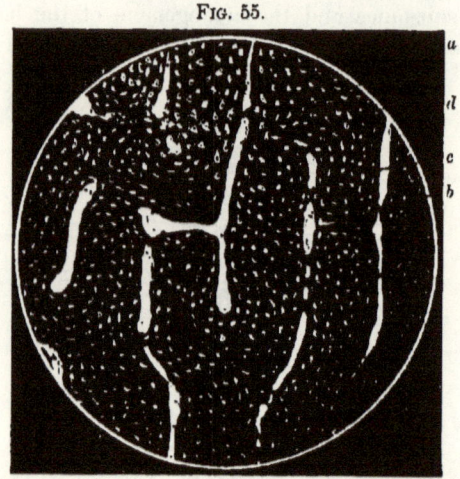

Fig. 55.

Section of decalcified bone, parallel to axis of human femur. *a*, longitudinal section of Haversian canal giving off transverse branch to the left; *b*, tangential section of a transverse branch; *c*, lacuna occupied by bone-corpuscle; *d*, intercellular substance deprived of its earthy salts and so swollen that the canaliculi are obliterated.

a cement or "ground" substance, containing the deposit of earthy salts. To these salts, which are chiefly phosphate and carbonate of calcium, the bone owes its hardness, while the fibres contribute toughness and elasticity to the tissue. The general arrangement of the fibres in the intercellular substance is in laminæ, which have a general parallel direction; but there are occasional fibres of some size which pierce these laminæ in a perpendicular direction and appear to bind them together, very much as a nail would hold a series of thin boards in place, "Sharpey's fibres."

Bone occurs in two forms, the compact and the cancellated. These do not differ in the nature of the tissue itself, but merely in the arrangement of that tissue with respect to its sources of nourishment. Where the bone is massed in compact form, as in the shafts of the long bones, special means for supplying it with nourishment is provided by a series of channels, the Haversian canals, which contain the nutrient bloodvessels, and which anastomose with each other throughout the whole substance of the tissue. The nourishing lymph, derived from the blood, reaches the cells through the canaliculi and lacunae, which connect with each other to form a network of minute channels and spaces pervading the bone, and not only opening into the Haversian canals, but also upon the external and internal surfaces of the tissue.

In the shafts of the long bones the Haversian canals lie for the most part parallel with the axis of the bone, with short transverse branches connecting them with each other. It is around these longitudinal Haversian canals that the laminae of bone are arranged in concentric tubular layers. Each Haversian canal, with the laminae surrounding it, is known as an Haversian system. Between these Haversian systems there are excentric laminae of bone, which do not conform to the concentric arrangement of the Haversian systems.

In the spongy or cancellated variety of bone the thin plates of that tissue derive their nourishment from the lymph of the contiguous marrow filling the spaces between them, and there is no occasion for Haversian canals. The concentric arrangement of the laminae is, therefore, absent.

Except where bounded by cartilage at the joints, the external surfaces of the bones are covered by a fibrous investment, the periosteum, in which the bloodvessels supplying the bone ramify and subdivide before sending their small twigs into the Haversian canals of the compact bone. The deep surface of the periosteum contains connective-tissue cells, " osteoblasts," capable of assuming the functions of bone-corpuscles and producing bone. These facts explain the importance of the periosteum for the nutrition and growth of bone. The tendons and ligaments attached to the bones merge with the periosteum, which has a similar fibrous structure and serves to connect them firmly with the surface of the bone.

The central cavities of the long bones and the spaces of cancellated bone are occupied by marrow, which may be of two kinds, the "red" or the "yellow." A description of the structure of marrow must be deferred until the other varieties of the connective tissues have been considered.

In the embryo the parts which are destined to become bony first consist of some other variety of connective tissue, either cartilage or fibrous tissue. This subsequently "ossifies," during which process it is not really converted into bone, but is gradually absorbed as that tissue develops and replaces it.

## III. THE FIBROUS TISSUES.

**General Characters.**—This group of elementary tissues, which may be said to constitute the connective tissues *par excellence*, includes a number of varieties which are not very sharply defined, because of transitional modifications which bridge over the differences between the more distinct types. It will, therefore, be best to describe these well-marked types of structure, and then to indicate the direction in which they are modified in particular cases so as to simulate in greater or less degree other typical varieties of the same group.

(1) The cells of the fibrous tissues vary considerably in character, three more or less distinct forms being distinguishable. First, flattened, almost membranous cells with oval nuclei and nearly clear and homogeneous bodies, possibly identical with the cells that form endothelium; second, granular cells, rich in cytoplasm and usually ovoid or cubical shape, though sometimes elongated; third, elongated or fusiform cells, with oval nuclei surrounded by a moderate amount of cytoplasm which is frequently prolonged into processes of greater or less length and delicacy, and sometimes dividing into branches. These three sorts of cell are present in varying relative proportions in the different tissues belonging to this group. (2) The intercellular substance is composed of distinct fibres, associated with a homogeneous cement- or "ground-substance," lying between the fibres. The fibres are of two kinds: the "white," non-elastic, and the elastic or "yellow." The relative abundance of these and of the ground-substance associated with them, and also their arrangement, vary greatly in the different members of the group. (3) The arrangement of the constituents of the fibrous tissues in the different varieties is so diverse that a statement of the variations would amount to a description of the tissues themselves. The general characters already enumerated will serve to distinguish the whole group from all the other elementary tissues, and enable the student to recognize the fact that a given form of the tissue which he may have under observation belongs to this group.

Before entering upon a description of the varieties of fibrous tissue, it will be of advantage to note the peculiarities of the two kinds of fibres that are found in their intercellular substance.

The **white**, non-elastic fibres (Fig. 56) are exceedingly delicate, and appear, even under high powers of the microscope, as fine, transparent, homogeneous lines. They are usually aggregated into bundles of greater or less thickness, being held together by a small amount of the cement-substance already referred to. In these bundles the fibres run a somewhat wavy course from one end of the bundle to the other, but lie parallel to each other and never branch. When treated with dilute acetic acid, without previous hardening, they swell and become almost invisible. They are converted into gelatin when boiled in water.

Fig. 56.

Fibres of white fibrous tissue teased apart to show the individual fibrils.

The **yellow**, or elastic, fibres (Figs. 57–59) are coarser than the

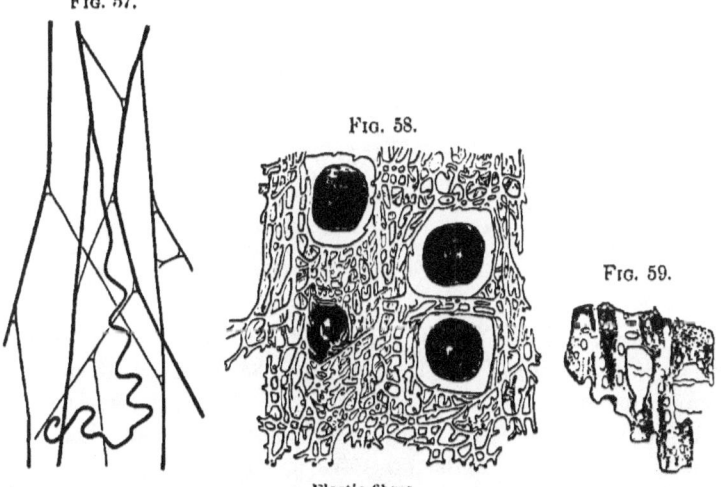

Elastic fibres.

Fig. 57.—From the subcutaneous areolar tissue of the rabbit. (Schäfer.)
Fig. 58.—Section of ear. (Hertwig.) The intercellular substance contains a reticulum of coarse anastomosing elastic fibres. (See Fig. 53.)
Fig. 59.—Fenestrated membrane from a branch of human carotid artery. (Triepel.)

white and more highly refracting, appearing more conspicuous when viewed under the microscope. They may be nearly straight, but more usually run a sinuous course. At intervals they divide, and the branches anastomose with each other to form a fibrous network, the meshes of which may be large, as is the case in areolar tissue, or so small and bounded by such broad fibres that the network resembles a membrane pierced by somewhat elongated apertures, as is exemplified in the fenestrated membranes of the arteries. The formation of such a network is, however, not an essential characteristic of these fibres, for they appear as isolated wavy fibres in some of the fibrous tissues of open and loose structures. Elastic fibres are not affected by acetic acid, nor do they yield gelatin on boiling in water. According to Schwalbe, they have a tubular structure, consisting of a membrane enclosing a substance called "elastin."

We may now turn our attention to the different varieties of the fibrous tissues.

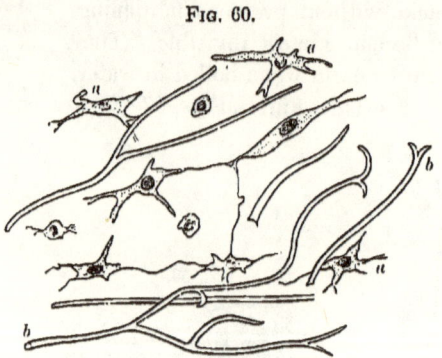

FIG. 60.

Mucous tissue. (Ranvier.) *a*, stellate cells with long and branching processes; *b*, elastic fibres in the homogeneous, mucoid, intercellular substance, which is not visible under the microscope unless artificially colored. Three of the cells are represented in cross-section.

1. **Mucous Tissue** (Fig. 60).—The cells of this elementary tissue are chiefly of the third variety mentioned above. They are spindle-shaped or stellate in form, and many of them possess processes that extend far into the intercellular substance, where they may branch and unite with the processes of neighboring cells. The predominant constituent of the intercellular substance is a gelatinous ground-substance, which contains a variable amount of mucin and appears nearly, if not quite, homogeneous under the microscope. It is this which gives the whole tissue its soft and gelatinous consistency. A variable number of fibres of both the kinds already described run through this ground-substance. The white fibres are

## THE CONNECTIVE TISSUES. 75

FIG. 61.

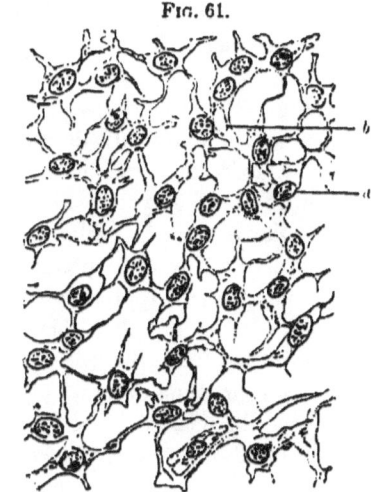

Embryonic connective tissue (mesenchymatous tissue). (Böhm and Davidoff.) a, nucleus of stellate cell; b, cytoplasmic process. The intercellular substance is of gelatinous consistency and optically homogeneous.

arranged in fine bundles, but the elastic fibres appear to be isolated, and, though they may branch, do not appear to form a network.

FIG. 62.

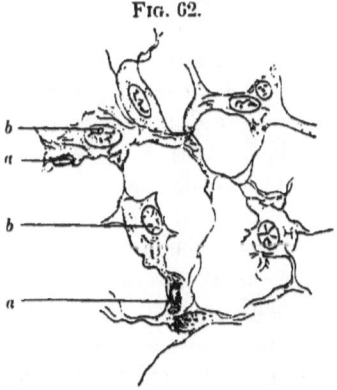

Reticular tissue. Section through a lymph-sinus in a lymph-node of the rabbit. (Ribbert.) a, nuclei of stellate cells of the reticulum; b, endothelial cells which are closely applied to the reticulum. The lymphoid cells, or leucocytes, have been removed from the meshes of the reticulum.

Mucous tissue of a rather highly cellular character is abundant in the embryo, where it constitutes an early stage in the development of the fibrous tissues (Fig. 61). A variety less rich in cells forms the Whartonian jelly of the umbilical cord. It does not occur in the adult under normal conditions, except, perhaps, in the vitreous humor of the eye.

2. **Reticular Tissue** (Fig. 62).—The fibres of this variety of elementary tissue are disposed in extremely delicate bundles, which anastomose with each other to form a fine meshwork. The spaces between the fibrous bundles are filled with lymph, which is usually so crowded with cells similar to the white blood-corpuscles that the structure of the tissue is masked by their presence. The cells of this tissue are flattened and closely applied to the surfaces of the bundles of fibres, which are so fine that they simulate delicate branching processes emanating from the cells. The cement- or ground-substance is reduced to a minimum, only a small amount lying between the fibres and the cells of the reticulum. The tissue is bounded by denser forms of fibrous tissue, with the fibrous bundles of which the reticulum is continuous. It is possible that reticular tissue contains stellate cells of the third variety mentioned as occurring in fibrous tissues, as well as the thin cells already described, which belong to the first variety. Where this is the case it is probable that the branching processes of those cells take part in the formation of the reticulum.

Where the meshes of the reticulum are crowded with lymphoid cells—*i. e.*, cells identical with some of the white corpuscles of the blood—the tissue has received the name "lymphadenoid tissue." This tissue is the chief constituent of lymph-glands and follicles, and is also found in a more diffuse arrangement in many of the mucous membranes (Fig. 107, *L*).

3. **Areolar Tissue.**—This is the most widely distributed variety of fibrous tissue. It contains all three kinds of cells mentioned at the beginning of this section, though not always in the same relative abundance. The intercellular substance consists chiefly of bundles and laminæ of fibres, which interlace in all directions. The white fibres predominate over the elastic, but there are always some of the latter which either form a wide-meshed reticulum, interlacing with the bundles of white fibres, or are applied to the latter in a sort of open spiral, binding them together. In the developing tissue the cement- or ground-substance at first fills all the interspaces between the cells and the fibres; but as development proceeds spaces appear in the tissue, which are occupied by lymph and intercommunicate throughout the tissue. The ground-substance is then restricted to

a mere cement uniting the fibres within the bundles and laminae. The flat or endothelial cells of the tissue lie within these bundles or are applied to their surfaces, forming a more or less perfect lining to the lymph-spaces within the tissue and becoming continuous with the endothelial walls of the lymphatic vessels. It is within these spaces that the lymph accumulates after its passage through the walls of the smaller bloodvessels, to find its way into the lymphatic circulation. The spindle-shaped and cuboidal cells of the tissue lie between or within the bundles of fibres embedded in the cement-substance (Figs. 63 and 64).

FIG. 63.

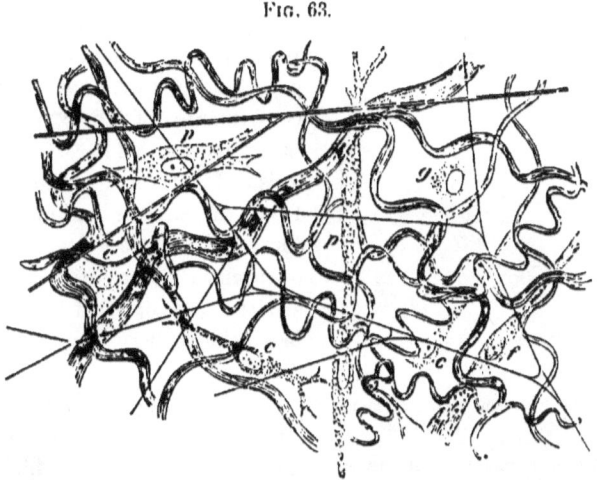

Areolar tissue. Preparation from the subcutaneous tissue of a young rabbit. (Schäfer.) *e'*, endothelioid cell; *p, p*, cells with granular cytoplasm; *c, c, f*, cells of the fusiform or stellate variety not yet fully developed. The white fibres are in bundles pursuing a wavy course; the elastic fibres are delicate and form a very open network; *g*, leucocyte of a coarsely granular variety.

Areolar tissue varies greatly in different situations in the density of its structure—*i. e.*, in the size of the fibrous bundles and their relative abundance, as compared with the number and size of the spaces separating them. The name is derived from that form in which the structure is open and the courses of the fibrous bundles very diverse, so that they interlace, leaving relatively large spaces between them. In this form it occurs in the subcutaneous tissues, between the muscles, forming the loose fasciae in that situation, and in many other parts of the body where adjacent structures are loosely connected with each other. The sinuous course of the interwoven fibrous bundles renders the tissue easily distensible in all directions and permits considerable freedom of motion between the parts which it unites.

In other situations the spaces in the tissue are smaller and the fibrous bundles closer together and less tortuous in their arrangement, so that the parts connected with each other are more firmly held in place. This form of the tissue occurs in all the glandular organs of the body, supporting and holding in place the functionally active tissues of the organs and constituting the chief constituents of their interstitia (see Chapter VII.). To distinguish this form of fibrous tissue from the areolar or more open form it may be designated as connective tissue in a more restricted use of that term than has hitherto been employed (Fig. 65, $b$, $b'$).

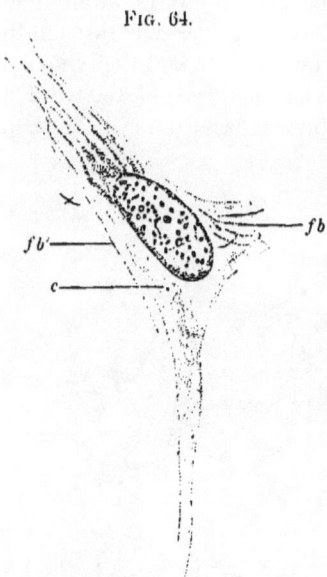

Fig. 64.

Cell from subcutaneous tissue of human embryo. (Spuler.) $c$, centrosome; $fb$, fibrillæ in the cytoplasm of the cell; $fb'$, fibril detached from the cell, but evidently derived from it. This cell corresponds to $c$, $e$, and $f$, in Fig. 63. They are sometimes called fibroblasts because of their activity in the formation of fibres.

A still denser form of the tissue occurs in the fasciæ and aponeuroses, in which the fibres are aggregated in thick bundles and layers that run a comparatively straight course and are firmly held together. Ligaments and tendons differ from these only in the greater density of the fibrous bundles and in their parallel arrangement. These denser varieties of the tissues may be designated by a restricted use of the term, fibrous tissue.

4. **Adipose Tissue** (Fig. 65).—Fat or adipose tissue is a modification of the more open or loosely-textured areolar tissue, caused by the accumulation within the cytoplasm of the cuboidal cells of drops of oil or fat. The cells which have become the seat of this fatty infiltration are enlarged, and their cytoplasm, with the enclosed nucleus, is pressed to one side, the great bulk of the cell being occupied by a single large globule of fat. This globule, together with

the cytoplasm, is enclosed in a delicate cell-membrane. The fatty cells may occur singly in the midst of an apparently normal areolar tissue of the usual type, but they are more frequently grouped to form "lobules," held in position within the tissue by bands and layers of unaltered areolar tissue.

In sections of adipose tissue prepared after hardening the tissue in alcohol the fatty globules can no longer be seen, since the alcohol dissolves the fat from the tissues. The partially collapsed

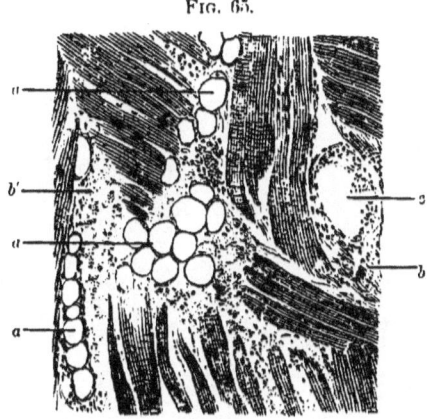

Fig. 65.

Section from the tongue of a rabbit: a, a, a, groups of fat-cells forming small masses of adipose tissue in the connective tissue; b, b', connective tissue, b in longitudinal, and b' in cross-section; c, small vein containing a few red blood-corpuscles. Near the centre of the figure is another bloodvessel filled with corpuscles. The remainder of the figure represents striated muscle-fibres in nearly longitudinal section. In the upper left hand corner these show a tendency to split into longitudinal fibres (sarcostyles).

membranes of the cells, with the cytoplasm and contained nucleus forming an apparent thickening at one side, are all that remain to distinguish the tissue (Fig. 65, a).

Adipose tissue is widely distributed in the body. It serves as a store of fatty materials which can be drawn upon as a reserve stock of food when the nutrient supply of the body falls below its needs.

The usefulness of the fibrous tissues can be readily inferred from their structure. The more open varieties of areolar tissue serve to give support to the structures they unite and to the bloodvessels, lymphatics, and nerves supplied to them. They also afford spaces and channels for the return of the lymph, which transudes through the walls of the capillary bloodvessels, carries nourishment to the tissue-elements it bathes, and then returns to the blood in the veins through the interstices and lymphatic vessels contained in the areolar tissue. In pursuance of these functions, areolar tissue pervades nearly all parts of the body. Wherever bloodvessels are found, there more or less areolar tissue is present, surrounding them, giving them support, and furnishing channels for the lymphatic circulation. As has already been stated, this areolar tissue varies in the closeness of its texture in different parts of the body.

The fibrous tissues of tendons and ligaments form inextensible

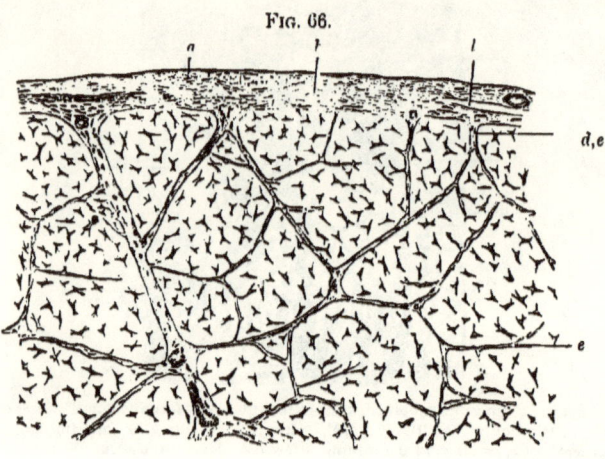

Fig. 66.

Portion of a large tendon in transverse section. (Schäfer.) *a*, sheath of areolar tissue surrounding the tendon; *b*, longitudinal fasciculus of fibres within that sheath; *l*, lymphatic space; *c*, section of a broad extension of the ensheathing areolar tissue, dividing the tendon into larger bundles; *d*, *e*, more delicate layers of areolar tissue subdividing the larger bundles of fibres. Between these areolar septa are the bundles of fibres constituting the tendon. The cells which lie between the smallest fasciculi of fibres appear in stellate form; the cross-sections of the individual fibres, among which these cells lie, are not represented. They would appear as minute dots.

bands or cords highly resistant to tensile stress, but very pliable. They consist of bundles of fibres lying parallel to each other and to the direction in which they are to resist pulling forces. Layers of loose areolar tissue penetrate the ligaments and tendons, dividing them into fasciculi, which in turn are united into larger bundles by thicker layers of areolar tissue (Fig. 66). These sheaths of areolar tissue support the vessels and nerves supplied to the denser forms of the fibrous tissue making up the ligaments or tendons. The thicker aponeuroses of the body may be regarded as broad and flat ligaments, in which the bundles of fibres run in various directions. They present a structural transition between the fibrous arrangement in ligaments and tendons and that in the more open varieties of areolar tissue. The fibres of these tissues are mostly of the white variety, but in some situations, notably in the ligamentum nuchae, they are chiefly of the elastic variety.

Reticular tissue may be regarded as a special modification of areolar tissue, in which the main bulk of the tissue consists of a series of freely intercommunicating lymph-spaces. These are often densely crowded with lymphoid cells, among which the lymph slowly circulates, thereby being subjected to the modifying influences of their activities.

# CHAPTER V.

## TISSUES OF SPECIAL FUNCTION.

The elementary tissues included in this group are highly differentiated in structure so as to adapt them for the performance of some special function of a high order. The constituent of the tissues which appears most highly specialized is the cell, which is often so greatly modified in structure as to have lost many of the general characters of the cells hitherto studied. Thus, for example, the cells of striated muscle are multinucleated, and the cytoplasm has become transformed into a substance known as contractile substance, which occupies nearly the whole bulk of the cell, leaving only a small amount of relatively undifferentiated cytoplasm immediately surrounding the nuclei.

In like manner the intercellular substances of some of these tissues show a complexity of structure in great contrast to those with which we have become familiar in the preceding tissues. In fact, it is stretching a point to regard the tissues lying between the cells of striated muscle as forming an intercellular substance belonging to that tissue. In this case those tissues are identical in structure with the loose areolar tissue that was described in the preceding chapter. We may, therefore, with propriety, regard the striated muscles as *organs* in which the muscle-cells constitute the parenchyma and this areolar tissue the interstitium (see Chapter VII.). But in other tissues of the group there is either an intercellular substance resembling those of the preceding tissues, or some special form of sustentacular tissue—*e. g.*, the neuroglia of the central nervous system.

The tissues of special function are arranged in two groups: the muscular tissues and the nervous tissues. As is implied in the title, these tissues are grouped together because of their functional powers, and not with regard to peculiarities of structure, so that it is impossible to give concise statements of any common general structural characters possessed by all the members of each of these two groups. Thus, the individual muscular tissues differ considerably from each other in structure, but are closely related in function, each variety being specialized so as to execute a particular kind of contraction when functionally active. We must also assume that the variations in structure met with in the nervous system have reference to the translation of various impressions into nervous impulses, or the liberation of such impulses under different conditions, as well as to their transmission and application to the functional activities of other tissues.

The complex functions exercised by the nervous system appear to necessitate a great variety of nervous structures, and it would be a matter for surprise to find the visible structure of the nervous system as simple as it is, were it not for the fact, already learned, that cells apparently similar in structure may have widely different, though related, functional powers.

## I. THE MUSCULAR TISSUES.

There are three varieties of muscular tissue, which differ from each other both in structure and in the character of their functional activities. One variety is that found in the walls of the hollow viscera and larger bloodvessels. Its activities are not under the control of the will, and the cells are devoid of marked cross-striation of the contractile substance. It has, therefore, received the names, "involuntary" or "smooth" muscular tissue. The other two varieties present distinct and rather coarse cross-striation of the contractile substance, but differ in other structural details. One of these is called "voluntary" or "striated" muscle; the other is found only in the heart, is not under the control of the will except in rare instances, and is known as "cardiac" muscular tissue.

1. **Smooth Muscular Tissue.**—This elementary tissue is composed of elongated or fusiform cells, which gradually taper to a sharp point. The body of the cell, except close to the ends of the nucleus, consists of a modified cytoplasm, called "contractile substance," which stains a coppery red with eosin, and presents fine, indistinct, longitudinal and transverse markings, possibly the optical expression of certain ridges that are in contact with similar ridges on neighboring cells. Each cell has a single, greatly elongated, rod-shaped nucleus situated in its centre, with the long axis coincident with that of the cell (Fig. 67). The nuclei are vesicular and possess

a distinct intranuclear reticulum of chromatin. The intercellular substance is a mere cement of homogeneous character. Fig. 67. The cells are arranged with their long axes parallel to each other and with the tops of their minute ridges in contact, so that fine channels exist between the contiguous cells. This is apparently a provision for the circulation of nutrient fluids between the cells (Fig. 68).

Smooth muscular tissue occurs in the form of bundles or layers, in each of which the cells or fibres run in the same direction. The tapering ends of the individual cells interdigitate with each other, masking the intercellular substance, so that the tissue appears as though wholly composed of cells. Surrounding the muscular bundles or between the layers of that tissue is vascularized areolar tissue, giving it support and containing its nerve-supply.

The microscopical appearances of sections of smooth muscular tissue depend upon the direction in which the individual cells have been cut. A brief analysis of the different appearances that may result will be useful as an

Fig. 68.

Smooth muscular tissue.

Fig. 67.—An isolated fibre from the muscular coat of the small intestine. (Schäfer.) The nucleus is somewhat contracted, so as to appear broader and shorter than when in the extended state.

Fig. 68.—Cross-section of smooth muscular tissue; human sigmoid flexure. (Barfurth.) Two of the muscle-cells have been cut in the region occupied by the nucleus, which appears in round cross-section. The other cells have been cut between the site of the nucleus and the end of the cell. The structural details of the cytoplasm or contractile substance are not represented, but the connecting ridges of the cells, with the channels between them, are shown. These minute ridges can, however, only be seen when the tissue has been exceptionally well preserved and is studied under a high power of the microscope.

illustration of the way in which microscopical appearances must be interpreted in order to gain a correct conception of the structure of an object under observation. It is rarely that sections happen to be made in such a direction that they reveal the complete structure of an object. It is nearly always necessary to study the appearances presented by the section, and to infer what the structure of the object must be in order to yield the appearances seen. This is sometimes a matter of considerable difficulty.

If the plane of the section lie parallel with the long axes of the cells, the nuclei of the latter will appear as rod-like or long, oval bodies lying parallel to each other and distributed at regular intervals throughout the tissue. The outlines of the cells will be distinctly visible in some places, but in most of the section the boundaries of the deeper cells will be obscured by the bodies of the cells at the surface of the section, and the borders of the latter will be difficult of detection, because in many places the knife has left only a portion of the cell with a very thin and transparent edge (Figs. 69 and 70). For the practical recognition of the tissue, when cut in this direction, we must, therefore, in many cases, depend solely upon the shape and distribution of the nuclei and the color of the material between them after the section has been treated with certain stains (*e. g.*, eosin).

If the cells of the tissue have been cut perpendicular to their long axes, the section will contain true cross-sections of the individual fibres. These appear as round, oval, or, more usually, polygonal areas of various size, according to the part of the cell included in the section. If the cell has been cut near one of its ends, the cross-section will be small; if near the middle, it will be large, and will contain a cross-section of the nucleus, situated near its centre and appearing as a round dot (Fig. 71). It is in such sections that one may sometimes see the minute prickles or ridges, already referred to, projecting from the cell-bodies and joining with those of the contiguous cells to form delicate bridges across the narrow intercellular spaces. The only tissue with which this aspect of smooth muscular tissue is liable to be confounded is dense fibrous tissue, as seen in the cross-sections of tendons or ligaments (Fig. 66). There we also see polygonal areas of various sizes, separated for the most part by only a thin layer of cement. But these areas never contain nuclei, because they are composed, not of cell-bodies, but of intercellular substance. The nuclei of the flattened connective-tissue cells may be seen here and there apparently lying within the cement, the body of the cell being

## NORMAL HISTOLOGY.

### Fig. 69.

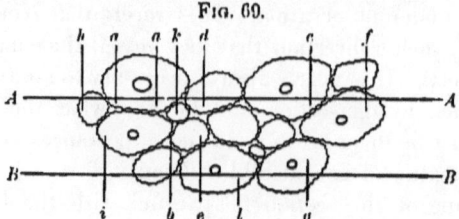

### Fig. 70.

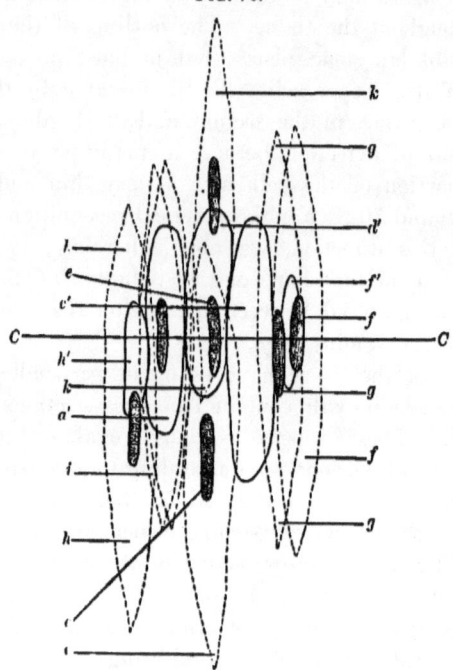

Diagrams illustrating the appearance of a longitudinal section of smooth muscular tissue. The distance between the lines $AA$ and $BB$ in the upper figure represents the thickness of the section, the line $AA$ being in the plane of its upper surface. The line $CC$ in the lower figure is in the plane of the transverse section represented in the upper figure.

It will be noticed that only portions of the cells, $h$, $a$, $d$, $c$ and $f$, will be contained in the longitudinal section (lower figure). The upper cut surfaces of those cells will appear as oval areas when seen from above, $h'$, $a'$, $d'$, $c'$, $f'$. Where the edges of those sections are thin—e. g., $a$—the outlines of the corresponding oval ($a'$) will be difficult of detection. At the same time those portions of cells which lie at the top of the section will obscure the outlines of the cells beneath. Thus, at the point $b$ the outlines of the cells $i$ and $e$ will be difficult of detection because covered by the cells $k$ and $a$, and also because the cell $i$ overlaps the cell $e$. If the plane of junction were perpendicular to the surface of the section, the outlines of $i$ and $e$ would be much more clearly defined.

This brief analysis will serve to show that the outlines of the cells will rarely be seen with distinctness in longitudinal sections of smooth muscular tissue. On the other hand, the nuclei of the cells will be prominently visible in stained sections because of the color they have received. For the recognition of this tissue, when so cut, we must, therefore, depend chiefly upon the character and distribution of the nuclei.

In order to avoid an unnecessarily complicated diagram, many of the cells represented in the upper cut have been omitted from the lower figure.

# TISSUES OF SPECIAL FUNCTION.

so thin as often to escape observation. These differences render it easy to distinguish the two kinds of tissue in spite of their general similarity when seen in cross-section.

It is, of course, rarely that sections contain smooth muscular

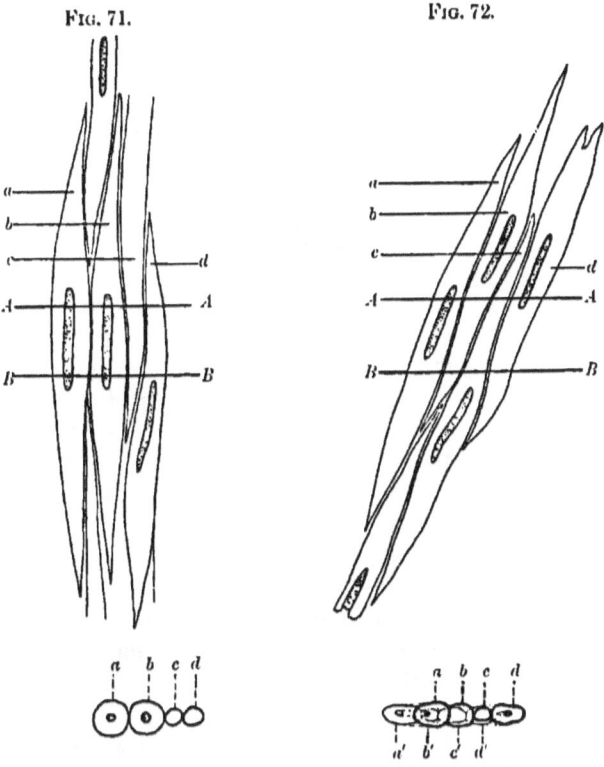

Fig. 71.
Fig. 72.

Diagrams of smooth muscular fibres cut in various directions.

Fig. 71.—Fibres cut exactly perpendicular to their long axes. The lines $A\,A$ and $B\,B$ in the upper figure indicate the portions of the fibres included in the section, which is viewed from above in the lower figure. The cross-sections of the fibres $a$ and $b$ contain cross-sections of the nuclei; those of $c$ and $d$ are smaller and devoid of nuclei.

Fig. 72.—Fibres cut obliquely to their long axes. When the upper surface of the section, marked by the line $A\,A$ in the upper figure, is in sharp focus, the sections of fibres appear as in in $a$, $b$, $c$, and $d$ of the lower figure. When the bottom of the section, indicated by the line $B\,B$ in the upper diagram, is in clear view, the sections of the fibres appear as shown at $a'$, $b'$, $c'$, and $d'$. It will be noticed that the optical section of fibre $a$ in the upper diagram has moved from $a$ to $a'$ in the lower diagram. As the focus was changed, the nucleus in fibre $a$ was constantly present and its optical section appeared of uniform size. This could only be the case when the nucleus was rod-shaped and of the same diameter throughout that portion contained in the section. In the fibre $d$ the nucleus was visible when the upper surface of the section was in focus, but disappeared when the focal plane was depressed.

### Fig. 73.

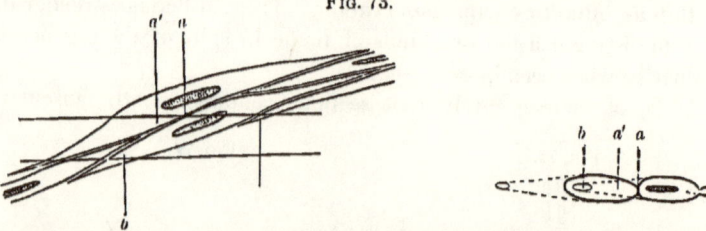

Diagrams of smooth muscular fibres cut very obliquely. The explanation of Fig. 72, already given, will make this one clear. In this case the outlines of the fibres in section will be less sharply defined than in the preceding case, because, for instance, at the point $a$ the fibre $a'$ is cut so as to leave only a thin edge, difficult of detection, and the fibre $b$ has had such a thin slice removed from it that the loss would be hardly perceptible. The appearance of the section would, therefore, be much less easy of interpretation than is represented in the lower figure, where the outlines of the sections are equally distinct throughout.

tissue cut exactly in either of the directions just considered. In the majority of sections that come under observation the muscle-fibres are cut obliquely, and the oval or polygonal areas which result are, therefore, elongated. The nuclei of the cells lie at an angle with the line of vision, and, in consequence, appear foreshortened. If now we focus the instrument so as to get a sharp image of the upper surface of the section, and then rapidly turn the fine adjustment so as to bring the lower surface into focus, we shall notice an apparent lateral motion of the nuclei and cell-sections. This apparent lateral movement, due to change of focus, is an evidence of the elongated shape and oblique position of the objects exhibiting it; and a little reflection will convince the student that such oblique sections, when carefully studied, are better calculated to reveal the shapes and relative positions of the tissue-elements than either perfect longitudinal or cross-sections. He should seek to train his powers of observation so that he may readily interpret the instructive, though at first confusing, images presented by such sections (Figs. 72 and 73).

Smooth muscular tissue is not under the direct control of the will. For this reason it is frequently called "involuntary muscle." It is also sometimes designated as non-striated or unstriped muscle, in contradistinction to the other two varieties of muscular tissue, the fibres of which present distinct cross-striations.

The functional contractions of smooth muscular fibres are sluggish. The fibres are slow in responding to stimulation, contract leisurely, maintain the contracted condition for a long time, and then gradually relax. These properties render the tissue of value in conferring "tone" to certain structures in which it is found, notably the walls of the arteries and veins. They also render it of service in producing the vermicular movements that are essential for the functional activity of such organs as the stomach and intestine.

Smooth muscular tissue has a wide distribution in the body. It is found in greatest bulk in the uterus, middle coats of the arteries

FIG. 74.

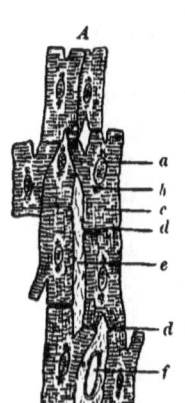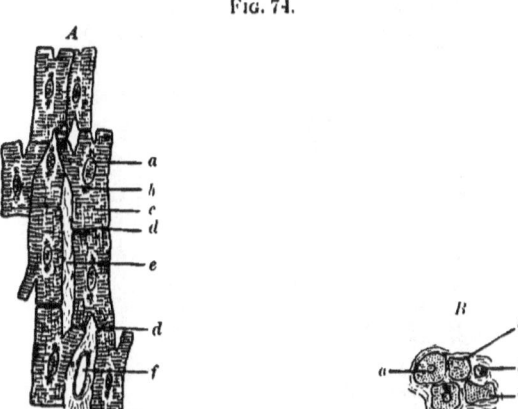

Diagrams of cardiac muscular tissue.

*A*, longitudinal section: *a*, nucleus of muscle-cell; *b*, unmodified cytoplasm; *c*, contractile substance with longitudinal and transverse striations; *d*, cement-substance uniting contiguous cells; *e*, areolar tissue (vessels omitted) between the muscle-fibres formed by the union of the individual cells; *f*, small bloodvessel within the areolar tissue. If the lines of junction between the cells were not visible, the tissue would appear as though composed of interlacing and anastomosing fibres, none of which could be traced for any considerable distance. Such is the usual appearance of longitudinal sections of cardiac muscle.

*B*, transverse section: *a*, section of a cell, including the nucleus; *c*, section above the nucleus and just below a crotch formed by the divergence of a branch; *b*, section above the nucleus and the point where the branch *b'* is given off.

and veins, the muscular coats of the stomach and intestine, and the wall of the bladder; but we shall find it present in greater or less amount in many of the organs, the structure of which we shall presently have to study.

2. **Cardiac Muscular Tissue** (Figs. 74 and 75).—The heart-muscle is composed of cells having a general cylindrical form and containing a single (occasionally, two) nucleus. The nucleus is vesicular, has a distinct reticulum of chromatin, and is usually oval. It is situated near the centre of the cell, and is surrounded by a small amount of cytoplasm, which is a little more abundant at the ends

of the nucleus. The rest of the cell-body is composed of contractile substance, a modification of the cytoplasm of which the cell was first composed, which presents a fine longitudinal and a somewhat coarser transverse striation. The proper intercellular substance is a homogeneous cement, which lies between the ends of the cells. These are arranged end to end so as to form fibres, the lines of

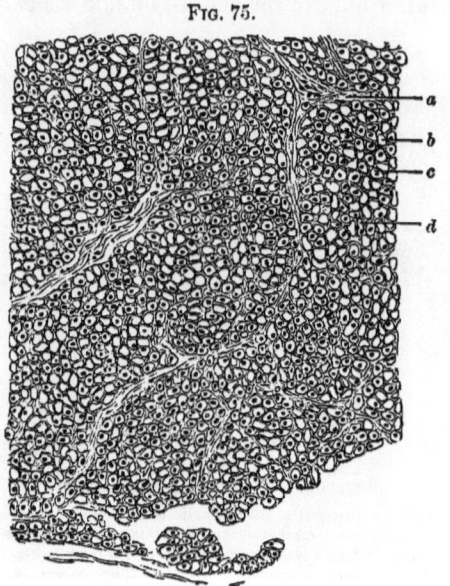

Fig. 75.

Section of human heart. The direction of the section is such that the muscular cells are cut exactly perpendicular to their long axes. *a*, intermuscular areolar tissue. From this, more delicate fibrous tissue penetrates between the muscle-fibres forming the muscular bundles, which are imperfectly separated from each other by the broader septa of fibrous tissue. *b*, muscle-cell cut beyond the nucleus; *c*, cell cut so as to include the nucleus; *d*, cell cut just below a branch. The index line *d* points to that part of the cell which passes into the branch. The granular character of the contractile substance when seen in cross-section has been omitted from the figure. At the lower edge of the figure the section has been torn, but a small amount of the subpericardial areolar tissue is represented.

junction between the cells, which are occupied by the cement-substance, being usually invisible. The cells give off branches which unite with each other in such a way as to convert the heart-muscle into a reticulum of muscular fibres. The meshes of this reticulum are occupied by areolar tissue, in which the vascular and nervous supply of the tissue is situated. Where this tissue is abundant it may also contain a few fat-cells. The cardiac muscle-cells are destitute of a cell-membrane, in which respect they differ from the voluntary striated muscle-fibres.

When seen in longitudinal section it is difficult to trace a given muscle-fibre for any considerable distance, because the occasional anastomosing branches of the cells cause a blending of the neighboring fibres with each other. In cross-section the cells have a round, oval, or polygonal shape, and vary considerably in size, owing to the branching. Their cut surfaces are dotted with the minute polygonal cross-sections of the elements of the contractile substance, which give the cell its appearance of longitudinal striation. These elements are called the "sarcostyles."

Cardiac muscle occurs only in the heart. It is not under the control of the will, but differs from the other involuntary muscles in the force and rapidity of its contractions, which resemble those of the voluntary muscles.

3. **Striated Muscular Tissue** (Figs. 76–79).—The voluntary muscles have for their characteristic tissue-element greatly elongated, multi-

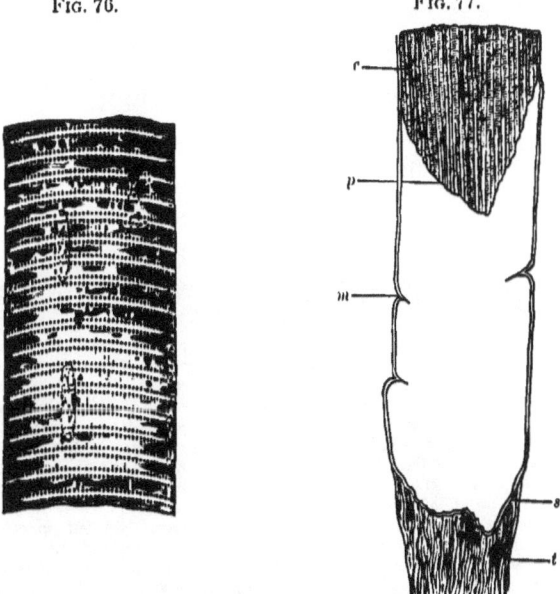

Fig. 76. Fig. 77.

Striated muscular tissue.

Fig. 76.—Portion of a muscle-fibre from a mammal. (Schäfer.) This figure represents the appearances of the fibre when the surface is in sharp focus.

Fig. 77.—Termination of a muscle-fibre in tendon. (Ranvier.) c, contractile substance; p, retracted end of contractile substance, separated from the sarcolemma during the preparation of the specimen; m, sarcolemma, slightly wrinkled; s, sarcolemma in contact with fibrous tissue of tendon; t, tendon.

# NORMAL HISTOLOGY.

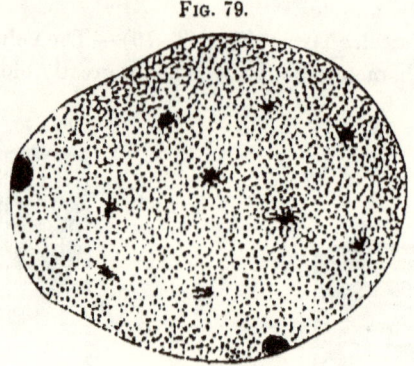

Fig. 78.

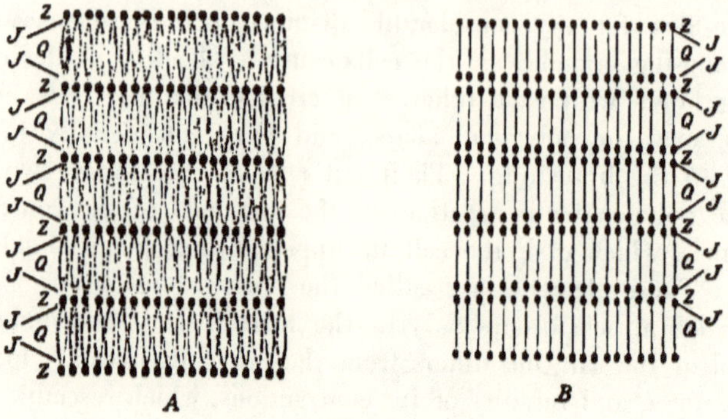

Fig. 79.

Striated muscular tissue.

Fig. 78.—Diagrams of the structure of the contractile substance. (Rollet.) $Q$, sarcous elements, appearing dark in $A$, light in $B$; $Z$ and $J$, sarcoplasm. The sarcoplasm also lies between the sarcous elements in $Q$, appearing as light bands in $A$ and as dark lines in $B$. $A$ is the appearance of the fibre when the focal plane is deep; $B$, the appearance when the focal plane is superficial (see Fig. 76). The dots $Z$ in $A$ and $J$ in $B$ are optical expressions of differences in the refraction of the sarcoplasm and sarcous elements, and do not represent actual structures. A complete explanation of the way in which a microscopical image may contain apparent objects which have no actual existence cannot be entered into here. It is due to the fact that regularly alternating structures of different powers of refraction affect rays of light very much as they are affected by a fine grating, producing diffraction spectra. These spectra may interfere with each other, occasioning an alternation of light and dark bands or areas above the specimen. When the focal plane is changed the light areas become dark and the dark areas light, but sometimes with an alteration in their outline and relative sizes, as exemplified in the cuts.

Fig. 79.—Cross-section of a muscle-fibre. (Rollet.) The fine reticulum, collected into larger masses at a few points in the midst of the contractile substance, is composed of sarcoplasm. The clear areas within this reticulum are the cross-sections of the sarcous elements. These cross-sections are sometimes called "Cohnheim's areas." Immediately beneath the sarcolemma are cross-sections of two nuclei.

nucleated, cylindrical cells. The body of these cells is almost exclusively composed of a very complex, contractile substance which presents both longitudinal and transverse striations, the latter much coarser and prominent than the former. It must suffice us to consider this contractile substance as made up of a number of prismatic bodies, "sarcous elements," which are arranged end to end to form columns, sarcostyles, extending parallel to each other, from one end of the cell to the other. The sarcous elements of all the sarcostyles lie in planes perpendicular to the long axis of the cell. It is, therefore, possible to separate the contractile substance into a number of fibre-like columns (sarcostyles, Fig. 65), made up of sarcous elements attached at their ends, or to split it transversely into disks composed of sarcous elements lying side by side. Between the sarcous elements is a substance which has received the name "sarcoplasm."

The contractile substance is enclosed in a thin, homogeneous membranous envelope, called the "sarcolemma." The nuclei of the cell lie immediately beneath the sarcolemma, between it and the contractile substance, and are surrounded by a small amount of unmodified cytoplasm.

The muscle-fibres lie parallel to each other and to the general direction of the muscle which they compose, and are separated by loose areolar tissue, containing their vascular and nervous supplies. When seen in cross-section they are circular or polygonal in form, and the cut surface of the contractile substance appears crowded with small polygonal areas, the sections of the sarcous elements, between which is the sarcoplasm. Where the nuclei are included in the section they appear somewhat flattened and lie at the edge of the contractile substance, where a thin zone of cytoplasm may sometimes be detected around them. The sarcolemma which lies outside of these constituents of the cell is so thin that it can rarely be distinctly seen.

The muscle-fibres are in close contact at both ends with the dense fibrous tissue of the tendons attached to the muscle.

# CHAPTER VI.

## TISSUES OF SPECIAL FUNCTION (CONTINUED).

### II. THE NERVOUS TISSUES.

THE nervous tissues, like the muscles, are tissues of special function, and are composed of highly specialized structures. Of these, only the ganglion-cells, the nerve-fibres, the neuroglia, and a few of

FIG. 80.

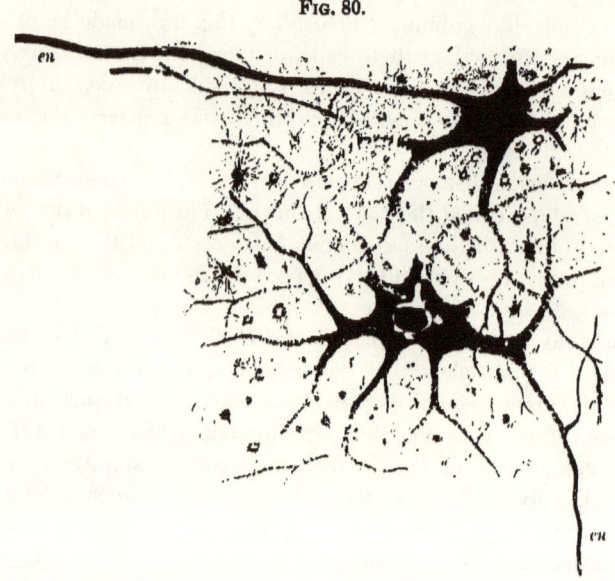

Nerve- and neuroglia-cells from gray matter of spinal cord; calf. (Lavdowsky.) The figure represents two isolated ganglion-cells, with branching protoplasmic processes, and each with a single axis-cylinder process, *cn*. The axis-cylinder process of the lower cell gives off a branch a short distance from the cell. Between the ganglion-cells are those of the neuroglia. The protoplasmic processes of the nerve-cells subdivide into very delicate fibres, which lie among those of the neuroglia-cells.

the modes of terminal distribution of the nerves will be considered here.

1. **Ganglion- or Nerve-cells** (Figs. 80 and 81).—Nerve-cells vary greatly both in shape and size. They are rich in cytoplasm, and contain an unusually large nucleus, generally spherical in shape, within the reticulum of which there is nearly always at least one conspicu-

Fig. 81.

Section of unipolar nerve-cell from gray matter of spinal cord. (Flemming.) This figure shows the fibrillation of the axis-cylinder process and the cytoplasm of the cell, as well as the prominent chromophilic granules in the latter.

ous nucleolus. The cell-bodies may be spherical, ovoid, polyhedral, or stellate in form, and are prolonged into one or more long processes. Some of these taper and branch repeatedly, the ultimate delicate fibrils terminating in free extremities lying in the intercellular substance, "dendritic processes." At least one of the processes emanating from each cell is coarser than these dendritic processes, and is prolonged into a nerve-fibre, forming the essential constituent of that structure. This process is called the "axis-cylinder process." It does not branch as freely as the other processes, but may give off one or more lateral twigs near its origin.

It is customary to divide the nerve-cells into unipolar, bipolar, and multipolar cells, according to the number of processes proceeding from them. The unipolar cells are connected by their single processes with nerve-fibres, and many of the bipolar cells, which have a fusiform shape, lie in the course of a fibre with which the two processes are continuous. In such cases one of the processes is an axis-cylinder process. The multipolar cells have one axis-cylinder process, the rest being of the dendritic type already mentioned, which are distinguished as "protoplasmic" processes.

Nerve-cells are, as a rule, larger than the other cytoplasmic cells of the body, with the exception of the larger epithelial cells. Their cytoplasm is so finely granular that the cells look much more transparent than those of epithelium. With a high power the cytoplasm frequently exhibits fine striations, which are prolonged into the processes, giving them an appearance of longitudinal fibrillation. These appearances are due to the arrangement of the fibrils of spongioplasm. Considerable attention has of late been given to certain granules, which become evident in the cytoplasm when nerve-cells have been fixed in alcohol or in acid solutions. These granules have an affinity for dyes, "chromophilic granules," and usually occur in groups in the neighborhood of the nucleus. Their significance is not yet understood.

The protoplasmic processes of the nerve-cells diminish in diameter as they branch, and they also present occasional varicosities, which give them an irregular contour. They terminate either in fine-pointed extremities or in little, knobbed ends, and do not unite with those of neighboring cells, but form with them an intricate interlacement of delicate nervous twigs.

The axis-cylinder processes arise in conical extensions of the cell, and then become uniform in diameter and of a smooth contour without varicosities. When they branch the two divisions retain their size throughout their course until they enter into the formation of some terminal structure.

The average size of the nuclei of nerve-cells is greater than that of the other nuclei in the body, but they appear to contain less chromatin, and therefore stain less deeply and present a less distinct intranuclear reticulum.

Nerve- or ganglion-cells are found in the gray matter of the central nervous system, in the ganglia, and sometimes in the course of nerves and in their peripheral terminations (Fig. 82).

2. **Nerve-fibres.**—There are two varieties of nerve-fibres: the white, or medullated, and the gray, or non-medullated. These differ both in their appearance when seen by the unaided eye and in their microscopical structure.

(*a*) Medullated nerve-fibres consist of a central cylindrical struct-

are running a continuous course from the cell giving it origin to the peripheral termination of the nerve, called the "axis-cylinder"; an external membranous envelope, the "neurilemma"; and a semisolid material, the "myelin," "white substance of Schwann," or "medullary sheath," lying within the neurilemma and surrounding the axis-cylinder.

The axis-cylinder is a greatly elongated process (axis-cylinder process) springing from a nerve-cell. It is marked by longitudinal

Fig. 82.

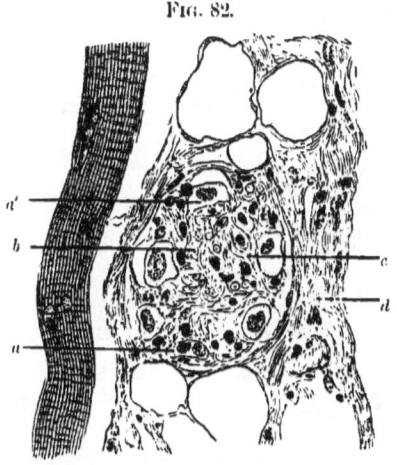

Small ganglion in the tongue of a rabbit; *a, a'*, ganglion-cells; *a'*, cell, with the beginning of its axis-cylinder process; *b*, medullated nerve-fibre in cross-section; *c*, fibrous tissue within the ganglion (part of this fibrous structure may be composed of non-medullated nerve-fibres); *d*, areolar tissue surrounding the ganglion and containing adipose tissue in the upper and lower parts of the figure. To the left is a striated muscle-fibre. The ganglion is seen in cross-section, so that its connection with the nerves, in the course of which it lies, is not visible.

striations, which appear to represent exceedingly delicate fibrils composing the axis-cylinder. These fibrils frequently separate at the distal extremity of the nerve and take part in the construction of the various forms of nerve-endings. A more minute study of the axis-cylinder leads to the inference that it is composed of spongioplasm, continuous with that of the body of the cell, and that the appearance of longitudinal striation is due to the elongated shape of the spongioplasmic meshwork and the greater thickness of its longitudinal threads, the transverse threads uniting them being much less conspicuous.

7

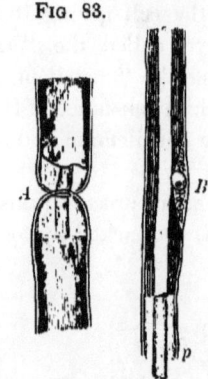

FIG. 83.

Medullated nerve-fibre. (Key and Retzius.) *A*, node of Ranvier; *B*, nucleus belonging to the neurilemma; *c*, axis-cylinder; *p*, neurilemma, rendered distinct by the retraction of the myelin of the medullary sheath. In the left-hand figure the clefts of Lantermann are shown as white lines in the dark myelin. These figures are taken from specimens treated with osmic acid, which colors the fatty constituent of the myelin a dark brown or black.

The neurilemma, or external investment of the nerve-fibre, called also the "primitive sheath," or "sheath of Schwann," is a thin, homogeneous membrane enclosing the medullary substance or myelin. At regular intervals, upon the inner surface of the neurilemma, and surrounded by a small amount of cytoplasm, are flattened, oval nuclei, which appear to belong to the neurilemma. About midway between these nuclei the nerve-fibre is constricted, forming the "nodes" of Ranvier. The neurilemma appears to pass through these nodes without interruption, so that the neurilemma of one internode is continuous with that of the adjacent internodes. At the nodes, and apparently within the neurilemma, is a disk, perforated for the passage of the axis-cylinder, called the "constricting band" of Ranvier. It may be that this band is of the nature of a cement-substance, joining the neurilemma of neighboring internodes; for the latter appear to be developed from cells, probably of mesoblastic origin, which surround the nerve-fibres after their formation, becoming flattened to form membranous investments of the nerve-fibre. If this view be correct, the neurilemma of each internode, with its single nucleus, is to be regarded as a single, specialized cell, derived from the surrounding connective tissues, and serving to protect the nerve-fibre. In perfect harmony with this conception of its nature are the facts that the nerves within the brain and spinal cord are destitute of neurilemma, and that when a nerve-fibre branches in its course the point of division is always at one of the nodes of Ranvier (Fig. 83).

The medullary sheath, or myelin, is a soft material interposed between the neurilemma and axis-cylinder. It is not a simple substance, but contains at least one constituent closely resembling fat or oil in its chemical nature; also a substance chemically allied to the keratin of horns and the superficial cells of the epidermis, called neurokeratin; and a homogeneous, clear fluid. The way in which these constituents are combined is a matter of doubt, the apparent structure of the medullary sheath varying greatly when different modes of preparing the nerve for microscopical study have been employed. But the neurokeratin appears to exist as a delicate reticulum pervading the medullary substance. The medullary sheath appears to be interrupted at irregular intervals by oblique clefts, which surround the axis-cylinder like the flaring portion of a funnel. These "Lantermann's" clefts are occupied by a soft material, probably similar to that composing the constricting bands (Figs. 84 and 85).

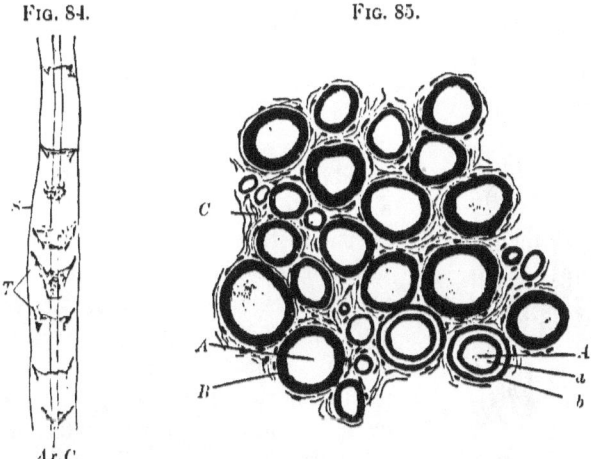

Fig. 84. — Longitudinal view of portion of nerve-fibre from sciatic of dog. (Schiefferdecker.) S, neurilemma; T, stained substance within the clefts of Lantermann.

Fig. 85. — Cross-section from sciatic nerve of frog. (Böhm and Davidoff.) A, axis-cylinder, showing punctate sections of the fibrillæ; B, medullary sheath stained with osmic acid; a, b, apparent duplication of the medullary sheath, due to the presence of a Lantermann cleft; C, areolar tissue between the fibres.

The medullary sheath is developed after the formation of the axis-cylinder, and is, at first, continuous along the course of the latter. Subsequently it becomes interrupted at the nodes of Ranvier by the constricting disk. It seems to be derived from the axis-cylinder, and may, therefore, be regarded as a product of that greatly extended arm of the cytoplasm of the nerve-cell.

The amount of medullary substance present in different nerves varies greatly. Sometimes it is so slight as to be hardly distinguishable. In other cases its thickness considerably exceeds the diameter of the axis-cylinder. It is present within the spinal cord and brain, although not enclosed in neurilemma in those situations. At the peripheral ends of the nerves, on the contrary, it usually disappears before the neurilemma.

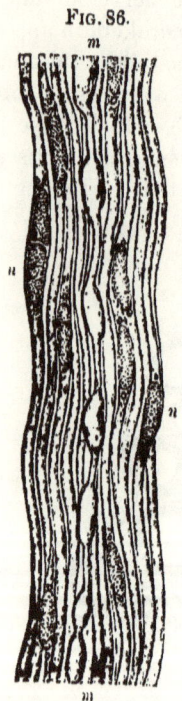

FIG. 86.

Nerve-fibres from the sympathetic system. (Key and Retzius.) All the fibres except that marked *m* are non-medullated. The fibre *m* has an incomplete medullary sheath. *n, n*, nuclei of the neurilemma. These are surrounded by a small amount of cytoplasm, which is not clearly represented in the figure.

The individual nerve-fibres are isolated only at their extremities. Throughout most of their course they are collected into bundles, forming the "nerves" of the body. Within these bundles the nerve-fibres are held together by fibrous tissue in the following manner: a delicate areolar tissue containing their vascular supply lies between the individual fibres. This fibrous tissue is called the "endoneurium." The nerve-fibres, thus held together, are aggregated into bundles, called "funiculi," which are surrounded by sheaths of still denser fibrous tissue, rich in lymphatic spaces, which are called the "perineurium." This perineurium on its inner surface becomes continuous with the endoneurium just described. The funiculi, enclosed by their perineurium, are, in turn, held together by an areolar sheath, which has received the name, "epineurium," and forms the outer covering of the nerve.

The funiculi do not run a distinct course throughout the length of the nerve, but give off nerve-bundles, enclosed in perineurium, which join other funiculi; the nerve-fibres themselves do not, however, anastomose with each other.

(*b*) The gray, or non-medullated, nerve-fibres are, as their name implies, destitute of medullary substance. They consist of an axis-cylinder, which at intervals appears to be nucleated. These nuclei are presumably constituents of a membranous investment or neurilemma; but the latter is difficult of demonstration because of its thinness and transparency, and its constant presence is not definitely established (Fig. 86).

Unlike the medullated variety, the gray nerve-fibres frequently give off branches, which join other fibres and constitute true anastomoses.

Non-medullated fibres are most abundant in the sympathetic nervous system, but occur also in the nerves derived directly from the brain and spinal cord.

3. **Neuroglia.**—The nerve-cells and fibres of the central nervous system are surrounded and supported by a tissue which is derived from the epiderm, and is called the "neuroglia." It must be regarded as a variety of elementary tissue having functions similar to the connective tissues, although its origin makes its relations to the epithelial tissues very close.

Neuroglia consists of cells, the "glia-cells," which vary consider-

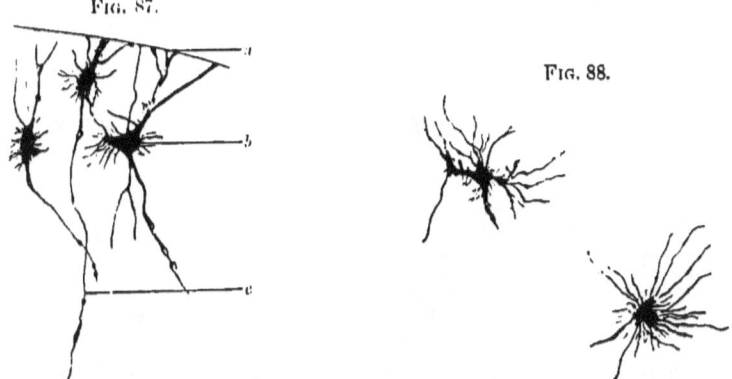

Glia-cells from the neuroglia of the human spinal cord. (Retzius.)
Fig. 87.—Three cells from the anterior portion of the white matter: a, processes extending to the surface of the cord; b, cell-body; c, long, delicate process extending far into the white matter.
Fig. 88.—Two cells from the deep portion of the white matter.

ably in character, and an intercellular substance, which is for the most part soft and homogeneous, resembling in this respect the cement-substance found in epithelium, but which may, here and there, contain a few delicate fibres, possibly derived from the processes of some of the cells, or possibly of mesodermic origin, and, in consequence, belonging to the connective tissues.

The glia-cells possess delicate processes, which lie in the cement- or ground-substance and form a felt-like mass of interlacing filaments, but do not unite with each other. Two types of cell may be distinguished, but they are not sharply defined, because intermediate forms are met with. In the first type the cells have relatively large

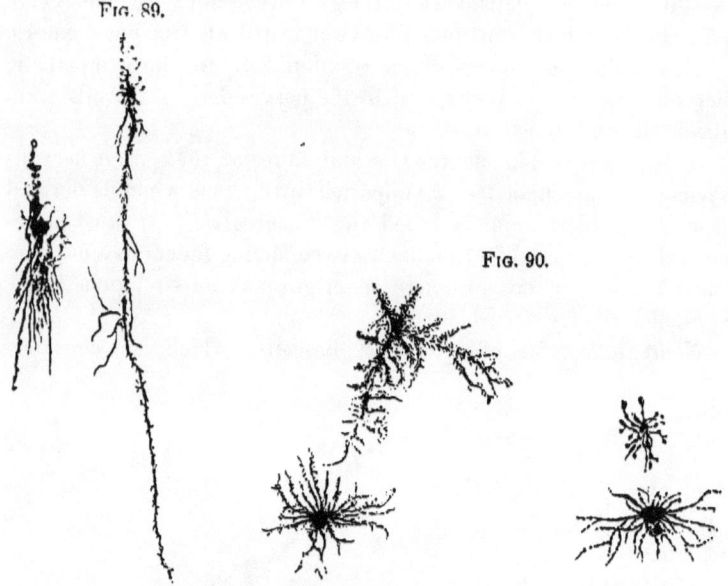

Glia-cells from the human spinal cord. (Retzius.)

Fig. 89.—Cells from the *substantia gelatinosa Rolandi* of the posterior horn. The cell to the right has a long process beset with fine, bluish branches.
Fig. 90.—Four cells from the gray matter.
Figs. 87-90 are taken from specimens stained by Golgi's method, which fails to reveal the internal structure of the cells, but is extremely well adapted to show the shapes of the cells and their extension into fine processes.

bodies, beset with a multitude of comparatively short, very fine, and frequently branching processes (Figs. 89 and 90). This type is most frequently met with in the gray matter. The second type of glia-cell is represented by cells with smaller bodies and longer and somewhat coarser processes that branch much less freely (Figs. 87 and 88). They also often possess one particularly large and prominent process of greater length than the others. The small bodies of these cells serve to distinguish them from nerve-cells, with which they might otherwise be easily confounded. This type predominates in the white matter.

Aside from the processes of the glia-cells already mentioned, the

## TISSUES OF SPECIAL FUNCTION. 103

central nervous system contains fibrous prolongations of the epithelial cells of the ependyma and central canal of the spinal cord (Fig. 91). Fibrous constituents are also derived from the areolar tissue which extends into the organs of the central nervous system from their fibrous investments, the pia mater, in company with the vascular supply.

The central nervous system, then, consists of a small amount of a ground-substance and a great number of cells, most of which possess numerous delicate fibrillar processes which interlace in all directions. Some of these cells are the functionally active elements of the organs, the nerve-cells. Others belong to the sustentacular tissue, and are probably functionally passive, constituting the interstitium. Both kinds of cell are developed from the epiderm, and are therefore genetically closely related to each other.

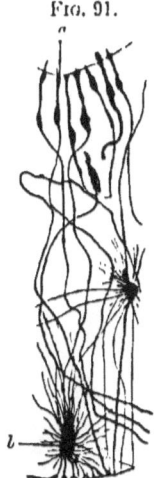

FIG. 91.

Ependyma and glia-cells from the spinal cord. (Retzius.) *a*, ependyma in the wall of the central canal; *b*, neuroglia-cell near the anterior fissure of the cord.

**4. Nerve-endings.**—Nerve-fibres terminate in two ways: first, in free ends lying among the elements of the tissues to which the nerve is distributed; second, in terminal organs, containing not only nerve-filaments, but cells which are associated with them to form a special structure. The simplest mode of termination consists in a separation of the minute fibrillæ

FIG. 92.

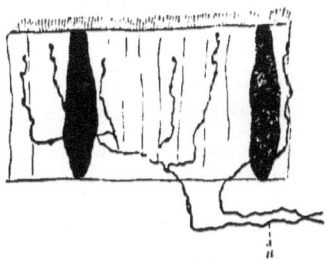

Termination of nerves by free ends. (Retzius.) Nerve-endings among the ciliated columnar epithelium on the frog's tongue. Two goblet-cells, the whole bodies of which are colored black, are represented. The other cells are merely indicated.

composing the axis-cylinders of the medullated fibres, or the chief bulk of the non-medullated fibres, into a number of delicate fila-

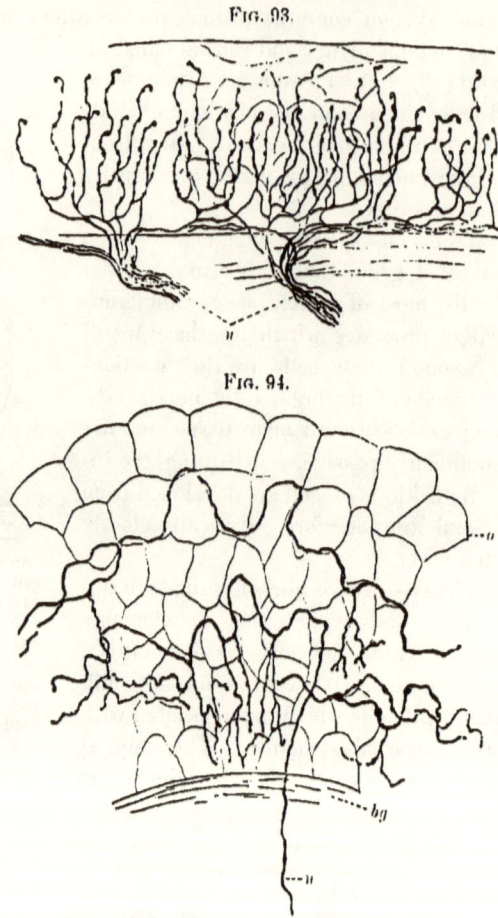

Termination of nerves by free ends. (Retzius.)

Fig. 93.—Two nerves terminating in the stratified epithelium covering the vocal cords of the cat.

Fig. 94.—Nerve-fibres distributed among the cells lining the bladder of the rabbit: *o*, superficial layer of the transitional epithelium; *bg*, fibrous tissue underlying the epithelium.

ments, which branch and finally end among the tissue-elements to which the nerve is supplied. The filaments often present small varicosities, and sometimes end in slight enlargements corresponding to one of those swellings. In other cases the terminations are filiform (Figs. 92-94).

A more complex mode of termination is that exemplified in the "motor-plates" of the striated muscle-fibre. Here the axis-cylinder divides into coarse extensions, which form a network of broad varicose fibres, lying in a finely granular material containing two sorts of nuclei. This whole structure lies in close relations to the contractile substance of the muscle-fibre, but whether it is covered by the sarcolemma or not is a matter of doubt. The nuclei in the motor-plate are derived in part from the muscle-fibre, from the cytoplasm of which the granular material surrounding the nerve-

FIG. 95.

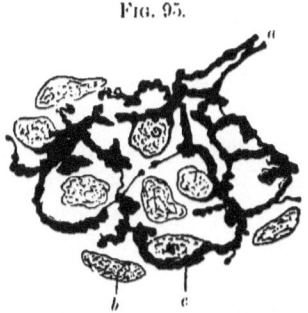

Motor-plate. Tail of a squirrel. (Galeotti and Levi.) *a*, two branches of axis-cylinder terminating in a plexus of varicose filaments; *b*, muscle-nucleus; *c*, nucleus derived from neurilemma. The finely granular substance surrounding these structures has been omitted.

endings appears to be derived, in part from cells similar to those forming the neurilemma, which participate in the production of the motor-plate (Fig. 95).

The nerves of sensation, like those supplying the striated muscles, end in bodies in which the nervous terminations are associated with cellular structures of peculiar form. Their consideration will be postponed until the structure of the nervous system is described.

# CHAPTER VII.

## THE ORGANS.

In the lowest order of animals, the protozoa, the single cell, which constitutes the whole individual, performs all the functions necessary to the life of the animal; but in the higher multicellular animals, the metazoa, those functions are distributed among a number of different but definite structures, called *organs*, each of which is composed of certain of the elementary tissues arranged according to a definite and characteristic plan peculiar to the organ.

Within each organ certain of the elementary tissues are charged with the immediate performance of the function assigned to that organ. These tissues are collectively termed the **parenchyma** of the organ. Thus, for example, the epithelium entering into the composition of the liver and doing the work peculiar to that organ, constitutes its parenchyma. The parenchyma of the heart is its muscular tissue, through the activity of which it is enabled to contract upon its contents.

Functionally ancillary to its parenchyma, each organ possesses a variety of elementary tissues, some of which belong to the connective-tissue group, which serve to hold the tissue-elements of the parenchyma in position, to bring to them the nutrient fluids necessary for their work, and to convey to them the nervous stimuli which excite and control their functional activities. These subsidiary tissues are collectively known as the **interstitium** of the organ. For example, the fibrous tissue and the elementary tissues forming the bloodvessels, lymphatics, and nerves of the liver, or of the heart, form the interstitia of those organs.

Two sets of structures entering into the formation of the interstitia of the organs—namely, the nerves and the vessels, including those which convey blood and those through which the lymph circulates—have a similar general structure in all the organs, and are connected with each other throughout the body, forming "systems." These systems serve to bring the various parts of the body, so diverse in structure and function and yet so interdependent upon each other, into that intimate correlation that makes them subordinate parts of a single organism.

Through the medium of the circulatory system the exchanges of material essential to the well-being of each organ and of the whole body are made possible, and through the nervous system the activities of the different parts of the body are so regulated that they work in harmony with each other and respond to their collective needs.

Because of their wide distribution throughout the body, we can hardly study any structures which are not in intimate relations with both vessels and nerves. It will, therefore, be well to consider the structure of the circulatory system before proceeding to a study of other organs. The study of the nervous system must, because of its complexity, be deferred.

# CHAPTER VIII.

## THE CIRCULATORY SYSTEM.

The circulatory system is made up of organs which serve to propel and convey to the various parts of the body the fluids through the medium of which those parts make the exchanges of material incident to their nutrition and functional activities.

For some of these exchanges it appears necessary for the circulating fluids to come into the most intimate contact with the tissue-elements; to penetrate the interstices of the tissues and bathe their structures. For mechanical reasons these fluids must circulate slowly and consume a considerable time in traversing a relatively short distance. Such a sluggish current could not avail for the transportation of oxygen from the lungs to the tissues, and we find that the circulatory system is divided into two closely related portions: the hæmatic circulation and the lymphatic circulation. The former is rapid, and the circulating fluid is the blood, the red corpuscles of which serve as carriers of oxygen. The latter is slow, and the circulating fluid, called "lymph," is derived from the liquid portion of the blood ("the plasma"). The blood is confined within a system of closed tubes, the bloodvessels; but the lymph, when first produced by transudation through the walls of the bloodvessels, is not enclosed within vessels, but permeates the tissues or enters minute interstices between the tissue-elements surrounding the bloodvessels. Thence it gradually makes its way into larger spaces—lymph-spaces—which open into the thin-walled vessels constituting the radicles of the lymphatic vascular system. These smaller lymphatic vessels join each other to form larger tubes, which finally open into the venous portion of the hæmatic circulation, thus returning to the blood the lymph which has made its way through the tissues.

The circulating fluids are kept in motion chiefly by the pumping action of the heart, which forces blood into the arteries, whence it passes through the capillaries into the veins, and thence back to the heart. During its passage through the smaller arteries, the capillaries, and the smaller veins, a part of the plasma of the blood, somewhat modified in composition, makes its way through the vascular walls, partly by osmosis, partly by a sort of filtration, and becomes the nutrient lymph of the tissues. The composition of this lymph varies a little in the different parts of the body, and this variation is attributed to some kind of activity, allied to secretion, on the part of the cells lining the vessels.

The larger veins are provided with pocket-like valves, which collapse when the blood-current is toward the heart, but which fill and occlude the veins when, for any reason, the current is reversed. When, therefore, the muscles contiguous to the larger veins thicken during contraction and press upon the veins the effect is to urge the blood within them in the direction of the heart. This accessory mode of propulsion materially aids the heart, especially during active exercise, when the muscles are in need of an abundant supply of oxygen.

The large lymphatic vessels are similarly provided with valves, and valves guard the orifices by which the lymphatic trunks open into the veins. But the chief reason for the flow of the lymph appears to be the continuous formation of fresh lymph, which drives the older fluid before it—the so-called *vis a tergo*.

For convenient description we may divide the vascular organs into the heart, arteries, veins, capillaries, and lymphatics.

1. **The heart** is covered externally by a nearly complete investment of serous membrane, the epicardium, which is a part of the wall of the pericardial serous cavity. Its free surface is covered with a layer of endothelium resting upon areolar fibrous tissue, and containing a variable amount of fat.

The substance of the heart is made up of a series of interlacing and connected layers of cardiac muscular tissue, separated by layers of areolar tissue, which extends into the meshes of the muscle, forming the interstitial tissue of the heart. The fibres in the different layers of muscle run in different directions, so that sections of the wall of the heart show the individual muscle-cells cut in various ways.

The areolar tissue is more abundant and denser near the orifices of the heart, and at the bases of the valves merges into dense fibrous rings, which send extensions into the curtains of the valves, increasing their strength and giving them a firm connection with the substance of the organ. In the centre of the heart, between the auriculo-ventricular orifices and the aortic orifice, this fibrous tissue is reinforced by a mass of fibro-cartilage.

The cavities of the heart are lined by the endocardium, consisting of endothelium resting on areolar tissue. The deeper portions of the epi- and endocardium merge with the areolar tissue of the body of the heart. Smooth muscle-fibres are of occasional occurrence in the deeper layers of the endocardium.

The auricles and the basal third of the ventricles contain ganglia, connected on the one hand with the nerves received by the heart from the cerebro-spinal and sympathetic systems, and on the other hand with a nervous plexus which penetrates the substance of the heart and gives off minute nervous fibrillæ to the individual cells of the cardiac muscle. These fibrillæ end in minute enlargements connected with the surfaces of the muscle-cells. Many of the ganglia lie beneath the epicardium or in the areolar or adipose tissue situated in its deeper portions.

The valves of the heart are composed of fibrous tissue, continuous with that forming the rings around the orifices. Their surfaces are covered by extensions of the endocardium, except the outer surfaces of the pulmonary and aortic valves, which are covered by extensions of the not dissimilar inner coats of the pulmonary artery or aorta. The fibrous substance of the valvular pockets of those two valves are further strengthened by tendinous strips of fibrous tissue at their lines of contact when the valves are closed. The curtains of the auriculo-ventricular valves are also reinforced by fibrous tissue derived from fan-like expansions of the chordæ tendineæ.

2. **The Arteries.**—It will be best to consider first the structure of the smaller arteries, because the individual coats are less complex in these than in the larger arteries.

The arterial wall consists of three coats: the intima, or internal coat; the media; and the adventitia, or external coat (Fig. 96).

The **intima** consists of three more or less well-defined layers. These are, from within outward: 1, a single layer of endothelium; 2, a layer of delicate fibrous tissue containing branching cells; 3, a layer of elastic fibrous tissue. The endothelial layer consists of cells, usually of a general diamond shape, with their long diagonals parallel to the axis of the vessel they line. When the vessel expands these cells broaden somewhat and appear very thin. When the vessel is contracted they are thicker and the portion containing the nucleus projects slightly into the lumen of the vessel.

The subendothelial fibrous tissue forming the second layer of the intima is composed of very delicate fibrils, closely packed together, with a little cement between them, and enclosing irregular spaces in which the branching cells of the tissue lie. Elastic fibres, spring-

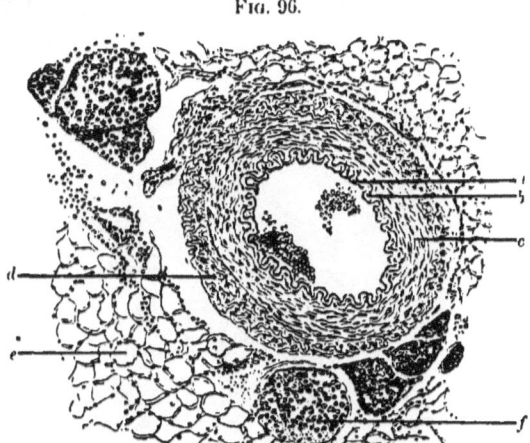

Fig. 96.

Branch of splenic artery of a rabbit: *a*, internal endothelial surface of the intima; *b*, elastic lamina of the intima (fenestrated membrane, see Fig. 59); *c*, media composed of smooth muscular tissue encircling the vessel and therefore appearing in longitudinal section with elongated nuclei; *d*, adventitia of fibrous tissue blending above and to the left with the surrounding areolar tissue; *e*, adipose tissue, between the cells of which a few lines of red corpuscles reveal the presence of capillary bloodvessels; *f*, small nerve, containing both medullated and pale or non-medullated nerve-fibres. There are other similar sections of nerves in the figure. To the left of the artery the section is slightly torn, the adipose tissue being separated from the adventitia of the artery. A few red blood-corpuscles have been extravasated near the nerve at the upper left corner of the figure. There are also a few corpuscles within the lumen of the artery.

ing from the external layer of the intima, may here and there, especially in the larger arteries, make their way into the subendothelial layer.

The elastic lamina of the intima is formed by a network of anastomosing elastic fibres, having a general longitudinal disposition with respect to the axis of the vessel. The spaces left between the fibres of this network vary considerably in size. Where they are small and the fibres between them are correspondingly broad this layer has the appearance of a perforated membrane (the fenestrated membrane of Henle). Even where this membranous character of the elastic layer is well developed, elastic fibres are given off from its surfaces and enter the subendothelial layer on the one side and the median coat of the artery on the other.

The **tunica media,** or middle coat of the arteries, consists essentially of smooth muscular tissue, with the cells arranged transversely to the long axis of the vessels, so that by their contraction they serve to diminish the calibre of the arteries.

The **adventitia** is an external sheath or layer of fibrous tissue

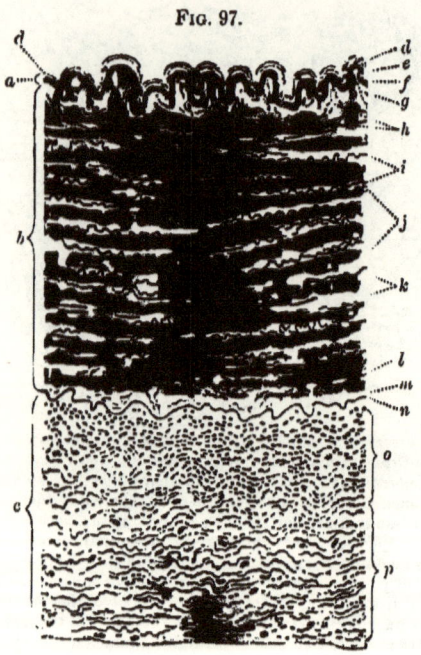

Fig. 97.

Portion of a transverse section of a human lingual artery from an adult. (Grünstein.) *a*, intima; *b*, media; *c*, adventitia; *d*, endothelium; *e*, subendothelial stratum (delicate areolar tissue); *f*, tunica elastica interna (fenestrated membrane belonging to the intima); *g*, stratum subelasticum containing elastic fibres (*h*) that pass from the fenestrated membrane into the media; *i*, concentric elastic fibres within the media; *j*, smooth muscular fibres of media with elongated nuclei; *k*, white fibrous tissue in media; *l*, elastic fibres radiating from the media into the external elastic tunic; *m*, stratum submusculare (areolar fibrous tissue); *n*, tunica elastica externa; *o*, stratum elasticum longitudinale (fibrous tissue containing elastic fibres running parallel with the axis of the vessel); *p*, stratum elasticum concentricum (fibrous tissue containing elastic fibres encircling the vessel). The vasa vasorum supplying the tissues of the vascular wall are not represented.

which merges with the areolar tissue of the parts surrounding the arteries and serves to support the latter without restricting the mobility necessary for their functional activity.

In the larger arteries the muscle-fibres of the media are grouped in bundles, which are separated by white and elastic fibrous tissue (Fig. 97). The muscle-fibres themselves are less highly developed than in the smaller arteries, so that the vessels are less capable of contracting, but are more highly elastic, because of the greater abundance of elastic fibres. In these larger arteries the boundary between the media and the intima is less sharply defined than in the smaller arteries, the elastic tissues of the two coats being more or less continuous. In cross-sections of the smaller arteries this boundary is very clearly seen, the elastic lamina of the intima appearing as a prominent line of highly refracting material, which assumes a wavy course around the artery when the latter is in a contracted state. In such sections the nuclei of the endothelial layer of the intima appear as dots at the very surface of the intima.

3. **The Capillaries** (Fig. 25).—As the arteries divide into progressively smaller branches the walls of the latter and their individual coats become thinner. In the smallest arterioles the elastic tissue of the wall entirely disappears, and the muscular coat becomes so attenuated that it is represented by only a few transverse fibres partially encircling the vessel. These in turn disappear, and the branches of the vessel then consist of a single layer of endothelium continuous with that lining the intima of the larger vessels. These thinnest and smallest vessels are the capillaries. They form a network or plexus within the tissues, and finally discharge into the smallest veins the blood they have received from the arteries. It is chiefly through the walls of the capillaries that the transudation giving rise to the lymph takes place, but some transudation probably also occurs through the walls of the smaller arteries and veins.

4. **The veins** closely resemble the arteries in the structure of their walls, but relative to the size of the vessel the wall of a vein is thinner than that of an artery. This is chiefly because the media is less highly developed. The elastic lamina of the intima is also thinner in veins than in arteries of the same diameter.

The valves of the veins are transverse, semilunar, pocket-like folds of the intima, which are strengthened by bands of white fibrous tissue lying between the two layers of intima that form the surfaces of the valves. The valves usually occur in pairs, the edges of the two coming into contact with each other when the valvular pockets are filled by a reversal of the blood-current.

Behind each valve the wall of the vein bulges slightly. Single valves of similar structure not infrequently guard the orifices by which the smaller veins discharge into those of larger size.

5. **The Lymphatics.**—The lymph at first lies in the minute interstices of the tissues surrounding the bloodvessels from which it has transuded. In most parts of the body those tissues are varieties of fibrous connective tissue, and contain not only the small crevices between their tissue-elements, but larger spaces also, which have a more or less complete lining of flat endothelial cells, but permit the access of lymph to the intercellular interstices of neighboring tissues. The lymph finds its way into these "lymph-spaces," and thence into the lymphatic vessels. These begin either as a network of tubes with endothelial walls, or as vessels with blind ends, and have a structure similar to that of the blood-capillaries. They are larger, however, and are provided with valves. By their union larger vessels are formed, resembling large veins with very thin and transparent walls, consisting of intima, media, and adventitia. These finally unite into two main trunks, the thoracic duct and the right lymphatic trunk, which open into the subclavian veins. Valves are of much more frequent occurrence in the lymphatic vessels than in the veins, but their structure is the same.

In its passage through the lymphatic circulatory system the lymph has occasionally to traverse masses of reticular tissue containing large numbers of lymphoid cells, called "lymph-glands."

That portion of the lymphatic system which has its origin in the walls of the intestine not only receives the lymph which transudes through the bloodvessels supplying that organ, but takes up also a considerable part of the fluids absorbed from the contents of the intestine during digestion. Mixed with this fluid is a variable amount of fat, in the form of minute globules. These globules give the contents of these lymphatics a milky appearance, and the vessels of this part of the lymphatic system have, therefore, received the name "lacteals." They do not differ essentially from the lymphatics in other parts of the body.

**Lymph-glands.**—It is a misnomer to call these structures glands, for they produce no secretion. A better term is "lymph-nodes."

The lymph-nodes are bodies interposed in the course of the lymphatic vessels through which the lymph-current passes. Their essential constituent is lymphadenoid tissue.

Each node has a spherical, ovoid, or reniform shape, with a depression at one point, called the "hilus." It is invested by a fibrous capsule, which is of areolar character externally, where it connects the node with surrounding structures, but is denser, and frequently reinforced by a few smooth muscular fibres internally. Extensions from this capsule penetrate into the substance of the node, forming "trabeculæ," which support the structures making up the body of the node.

The lymphadenoid tissue occurs in two forms: first, as spherical masses, "follicles," lying toward the periphery of the node, except at the hilus, and constituting the "cortex" (Fig. 98); second, in the

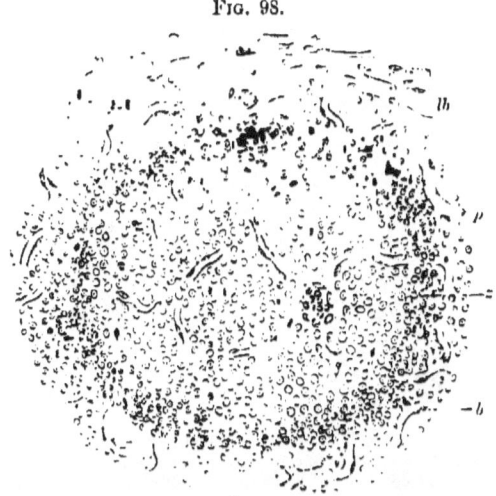

Fig. 98.

Single lymph-follicle from a mesenteric node of the ox. (Flemming.) *lb*, wide-meshed lymphatic sinus at periphery of the follicle. Between this and the peripheral zone of the follicle *z*, and within the follicle, the reticulum of the sinus and that supporting the cells and vessels of the follicle are not represented. The cells are merely indicated by their nuclei, the cytoplasm being omitted. *z*, peripheral zone of the follicle, marked by a close aggregation of small lymphoid cells; *p*, more scattered cells outside of the peripheral zone and at the edge of the lymph-sinus. Within the zone *z* is the germinal centre of the follicle, in which numerous karyokinetic figures are present, demonstrating the active proliferation of the cells in that region. Two such figures are also represented within the lymph-sinus at the upper left corner. *b*, blood vessels.

form of anastomosing strands, which make a coarse meshwork of lymphadenoid tissue in the medullary portion of the node (Fig. 99). The trabeculæ springing from the capsule penetrate the substance of the node between the follicles in the cortex, and then form a network of fibrous tissue lying in the meshes of the medullary lymphadenoid tissue, after which they become continuous with the mass of fibrous tissue at the hilus and, through it, with the capsule at that point.

The lymphatic vessel connected with the node divides into a number of branches, the "afferent vessels," which penetrate the capsule at the periphery and open into a wide-meshed reticular tissue lying between the trabeculæ and the lymphadenoid tissue of the follicles and the medullary strands. This more open reticular tissue, through which the lymph circulates most freely, forms the

FIG. 99.

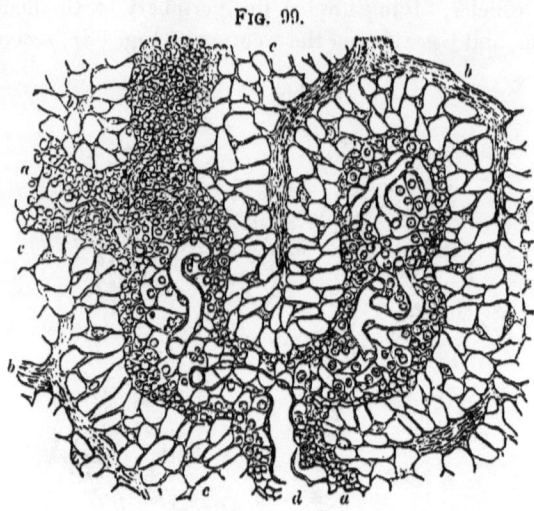

Portion of the medulla of a lymph-node. (Recklinghausen.) *a, a, a,* anastomosing columns of lymphadenoid tissue; *b,* anastomosing extensions of the cortical trabeculæ; *c,* lymph-sinus; *d,* capillary bloodvessels. The lymphoid cells in the sinus are not shown.

"lymph-sinuses" of the node, and is less densely crowded with lymphoid cells than the reticular tissue of the follicles and medullary lymphoid tissue. The walls of these sinuses, which are turned toward the fibrous tissue of the trabeculæ and their extensions in the medulla, are lined with endothelium, and a somewhat similar, but probably much less complete, lining may partially separate the sinuses from the lymphadenoid tissue. However this may be, it is certain that lymphoid cells can freely pass from the lymphoid tissue into the sinuses, or in the reverse direction, and that there is a ready interchange of fluids between the two.

From the sinuses the lymph passes into a single vessel, the "efferent vessel," through which it is conveyed from the node at the hilus.

The arteries supplied to the lymph-node may be divided into two groups: first, small twigs which enter at the periphery and are distributed in the capsule and fibrous tissues of the trabeculæ and the medulla; and, second, arteries which enter at the hilus, pass through the sinuses, and are distributed in the lymphadenoid tissue of the medulla and cortex. The veins follow the courses of the corresponding arteries. The nerve-supply is meagre, and consists of both medullated and non-medullated fibres. Their mode of termination is not known.

In the centre of the follicles the reticular tissue is more open and the lymphoid cells less abundant than toward the periphery. Mitotic figures are of frequent occurrence in lymphoid cells in this region, and it is evidently a situation in which those cells actively multiply. Further toward the periphery the reticular tissue is closer and very densely packed with small lymphoid cells, to become more open again and freer of cells as it passes into the reticulum of the sinus (Fig. 100). This last reticu-

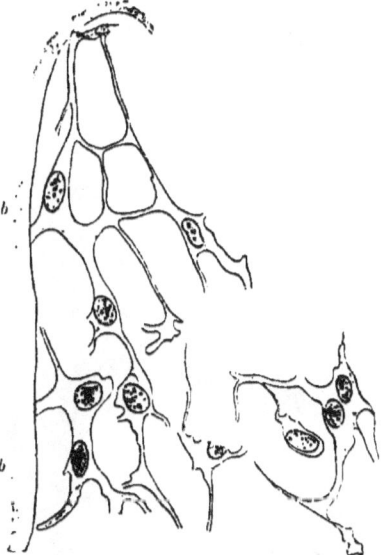

Fig. 101.

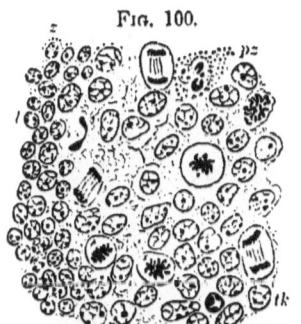

Fig. 100.

Fig. 100.—Portion of lymph-follicle from mesentery of ox. (Flemming.) z, peripheral zone of small, closely aggregated lymphoid cells. To the right of these is a portion of the germinal centre of the follicle, with larger cells, many of which are dividing. Opposite l is a cell executing amœboid locomotion. pz, pigmented cell, which has taken up colored granules from outside; tk, dark chromophilic body, the nature of which has not been determined. Such bodies occasionally occur in lymph-nodes, but their origin and significance are unknown.

Fig. 101.—Section of a small portion of the reticulum of the sinus in a human mesenteric node. (Saxer.) b, b, diagrammatic representation of a portion of the neighboring trabecula.

lum becomes continuous with delicate fibres given off from the tissues of the capsule and trabeculæ (Fig. 101). The distribution of the lymphoid cells gives the follicles a general concentric appearance.

The lymph-follicles of the cortex not infrequently blend with each other, and the activity of the cellular reproduction in their centres varies considerably and is sometimes entirely wanting, when the concentric arrangement of the cells disappears.

The structure of the lymph-nodes causes the lymph entering them to traverse a series of channels, the "sinuses," which, in the aggregate, are much larger than the combined lumina of the vessels supplying them. The velocity of its current is, therefore, greatly reduced, and it remains for a considerable time subjected to the action of the lymphoid cells in and near the sinuses. Small particles which may have gained access to the lymph in its course through the tissues are arrested in the lymph-nodes, and are either consumed by phagocytes—*i. e.*, cells possessing the power of amœboid movement and capable of incorporating foreign substances—or are con-

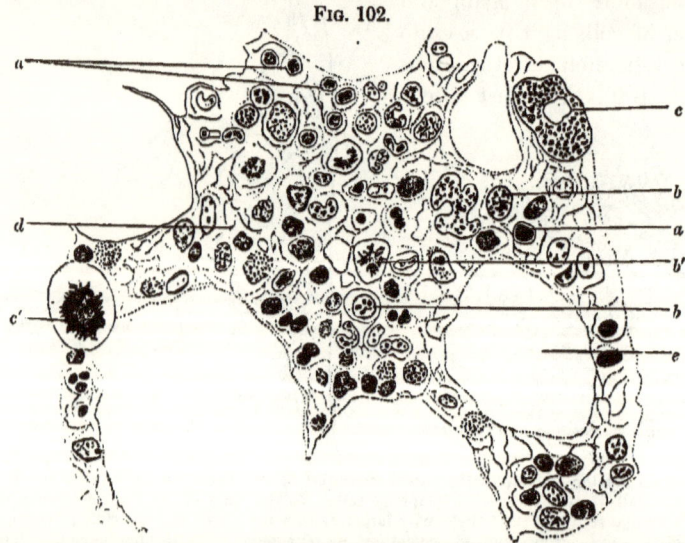

Fig. 102.

Section of red marrow; human. (Böhm and Davidoff.) *a, a*, erythroblasts; *b, b*, myelocytes; *b′*, myelocyte undergoing division; *c*, giant-cell with a single nucleus; *c′*, giant-cell with dividing nucleus; *d*, reticulum; *e*, space occupied by a fat-cell (not represented); *f*, granules in a portion of an acidophilic cell.

veyed into the marginal portions of the follicles, where, if insusceptible of destruction, they remain. It is in consequence of this process that the lymph-nodes connected with the bronchial system of lymphatics are blackened as the result of an accumulation of particles of carbon that have been inhaled and then absorbed into the lymphatics.

The lymph-nodes may, therefore, be considered as filters which remove suspended foreign particles from the lymph; but it is probable that the dissolved substances in the lymph are also affected in its passage through the nodes, and that a purification of that fluid is thereby occasioned. A fresh access of leucocytes further alters the character of the lymph during its transit through the lymph-nodes.

**Bone-marrow** (Fig. 102).—In early life the medullary cavities of the long bones, as well as the cancellæ of the spongy bones, are all occupied by that form of marrow known as "red" bone-marrow. This is functionally the most important variety. In after-life the marrow in the medullary cavities of the long bones becomes fatty through infiltration of its cells with fat, which converts them into cells quite similar to those of adipose tissue. Marrow so modified is called "yellow" marrow. It may subsequently undergo a species of atrophy, during which the fat is absorbed from the cells and the marrow becomes serous, fluid taking the place of the materials that have been removed. This process results in the production of a "mucoid" marrow.

The marrow of bones possesses a supporting network of reticular tissue not unlike that of the lymph-nodes. In the meshes of this tissue are five different varieties of cell (Fig. 103): First, myelocytes, cells resembling the leucocytes of the blood, but somewhat larger in size and possessing distinctly vesicular nuclei. They are capable of amœboid movements, and not infrequently contain granules of pigment which they have taken into their cytoplasm. Second, erythroblasts, or nucleated red blood-corpuscles, which divide by karyokinesis and eventually lose their nuclei, becoming converted into the red corpuscles of the circulating blood. Third, acidophilic cells, containing relatively coarse granules having an affinity for "acid" anilin-dyes, such as eosin. These cells are larger than the majority of the leucocytes circulating in the blood. Their nuclei are spherical or polymorphic and vesicular. Fourth, giant-cells with unusually large bodies and generally several nuclei, though occasionally only one nucleus is present. They possess the power of executing amœboid movements and appear to act as phagocytes. Where absorption of bone is taking place they are found closely applied to the bone that is being removed, and have in this situation been called "osteoclasts." Fifth, basophilic cells, or plasma-cells, the cytoplasm of which contains granules having an affinity

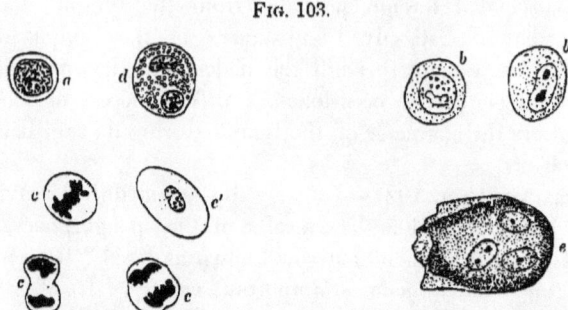

Fig. 103.

Cells from bone-marrow: *a*, small leucocyte from circulating blood, with highly chromatic nucleus and slight amount of cytoplasm, a "lymphocyte" probably derived from a lymph-node; *b, b*, myelocytes, larger than *a*, with vesicular nuclei; *c, c, c*, erythroblasts, with nuclei in karyokinesis; *c'*, mature red corpuscle (erythrocyte); *d*, acidophile (eosinophile) leucocyte. The basophilic leucocytes, or plasma cells, resemble this, but have smaller and less abundant granules of different chemical nature; *e*, giant-cell (myeloplax) with three nuclei; *a, b, c*, and *d*, from the marrow of the fowl (Bizzozero), the red corpuscles of which are oval and nucleated, *c'*; *e*, from the marrow of the guinea-pig. (Schäfer.)

for "basic" anilin-dyes, such as dahlia. These cells are relatively large, and possess vesicular and frequently polymorphic nuclei. Aside from these cells, which may be regarded as forming a part of the marrow, it contains red blood-corpuscles and leucocytes, either formed within the marrow or brought to it by the circulating blood.

The functions of the various cells in bone-marrow have not been finally determined, but it is certain that the erythroblasts, by their multiplication and transformation, maintain the supply of red corpuscles circulating in the blood.

The arteries supplied to the marrow divide freely and open into small capillaries, which appear subsequently to dilate, and either to blend with the endothelial elements of the reticular tissue or to become pervious through a separation of the cells forming their walls. In either case the blood passes into the meshes of the reticular tissue, where it slowly circulates among the constituents of the marrow. It then passes into venous radicles devoid of valves, and is thence conveyed from the bone. In some animals—*e. g.*, birds—the production of red corpuscles appears to be confined to the venous radicles (Fig. 104). The veins leaving the bones are abundantly supplied with valves.

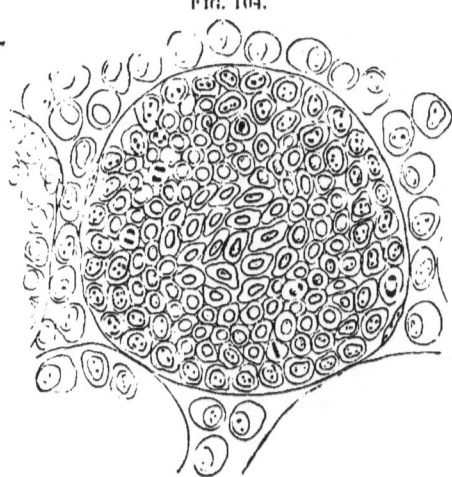

Fig. 104.

Section of small venous radicle in marrow of the fowl. (Bizzozero.) Just within the vascular wall is a zone of leucocytes, one of which contains a karyokinetic figure. Within this zone is a second zone of erythroblasts, four undergoing division, and in the centre of the lumen are a number of matured red blood-corpuscles (containing nuclei in the case of birds). The cytoplasm of the leucocytes contains no hæmoglobin, while that of the erythroblasts does. In birds and, probably, in other classes of animals the marrow of the bones is one of the sites for the production of leucocytes as well as red corpuscles. The latter are not produced from the former, but only from the erythroblasts, which constitute a distinct variety of cell.

Throughout life the cancellated portions of the flat bones and of the bodies of the vertebræ contain red marrow, but the shafts of the long bones are occupied by the yellow variety, which has lost its power of producing red blood-corpuscles and leucocytes, and has, therefore, become functionally passive.

# CHAPTER IX.

## THE BLOOD AND LYMPH.

THE blood consists of a fluid, the plasma, in which three sorts of bodies are suspended: the red corpuscles, the leucocytes or white corpuscles, and the blood-plates.

The plasma is a solution in water of albuminous and other substances. Some of these are of nutritive value to the tissues of the body. Others have been received from those tissues, and are on their way toward elimination from the body. Still other constituents have passed into the blood from one part of the body, and are destined to be of use to other parts.

In the smaller vessels, while on its course through the circulatory system, portions of the plasma make their way through the vascular walls and form the fluid of the lymph. This passage appears to be, in part, a simple filtration through the walls of the vessel, or the result of osmosis; in part, the result of a species of secretion

FIG. 105.

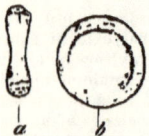 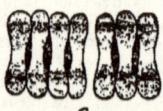

Red corpuscles from human blood. (Böhm and Davidoff.) *a*, optical section of a red blood-corpuscle, seen from the edge; *b*, surface view. (The bounds of the central depression are made a little too distinct in this figure, as is evident from an inspection of *a*.) *c*, rouleau of red corpuscles. When undiluted blood has remained quiescent for a few moments the red corpuscles arrange themselves in such rows, probably because of the attraction which they, in common with other bodies suspended in a fluid having a nearly identical specific gravity, have for each other.

effected by the endothelial cells lining the bloodvessels, these cells promoting the escape of certain constituents of the plasma and restraining or preventing that of others. In the exercise of this secretory function the endothelia in different parts of the vascular system appear to act differently, the composition of the fluid passing through the walls of the vessels not being exactly the same in all parts of the body. It is still a question, however, in what degree the endothelial cells are active in bringing about these differences. Their character is not such as would be expected of cells carrying on active processes.

The **red corpuscles** are soft, elastic discs, with a concave impression in both surfaces (Fig. 105). They are slightly colored by a solution of haemoglobin, and are so abundant that their presence gives the blood an intense red color; but when viewed singly under the microscope each corpuscle has but a moderately pronounced reddish-yellow tinge. The haemoglobin solution is either intimately associated with the substance composing the body of the corpuscles, called the "stroma," or it occupies the centre of the corpuscle and is surrounded by a pellicle of stroma.

Under normal conditions the red corpuscles, in man and most of the mammalia, are not cells, for they possess no nuclei, nor are they capable of spontaneous movement or multiplication. They are, rather, cell-products, being formed either within the cytoplasm of cells of mesoblastic origin, or by the division of cells derived from the mesoblast, and called erythroblasts, the descendants of which become converted into red corpuscles through an atrophy and disappearance (probably expulsion) of the nuclei and a transformation of the cytoplasm into the stroma, which take place after the elaboration of the haemoglobin within the cell. The former, or intracellular, mode of production occurs in the embryo, even before the complete development of the bloodvessels; the latter mode of production seems to be the only one occurring in the adult, the chief location of the erythroblasts appearing to be in the red marrow of the bones, where they are situated either in the tissues of the marrow itself, whence their descendants, while still cellular, pass into the vessels, or in the large venous channels of the marrow, where the blood-current is sluggish and the erythroblasts remain close to the vascular walls. In some anaemic conditions the erythroblasts appear in the circulating blood, where they may be distinguished from the normal red corpuscles by the presence of their nuclei and, frequently, also by a difference in size (see Fig. 103, c).

In the reptilia and birds the red corpuscles are normally nucleated; but, though morphologically resembling cells, they are incapable of multiplication or spontaneous movement, and have undergone such modifications that they are not cells in a physiological sense.

The functional value of the red corpuscles is dependent upon the haemoglobin they contain, which is said to constitute 90 per cent. of their solid matter. It is readily oxidized and reduced again, and serves to carry the oxygen of the air, obtained during the passage of the blood through the pulmonary capillaries, to all parts of the body. The red corpuscles, therefore, subserve the respiratory function of the blood, as the plasma subserves its nutritive function.

The **leucocytes**, or white blood-corpuscles, are cellular elements closely resembling the amoeba in their structure, which are present in the blood in much smaller number than the red corpuscles, the usual proportion being about one to six hundred. They vary somewhat in size and structure, either because of differences in their origin, or because they are in different stages of development. The majority of them are capable of amoeboid movements; but while they are circulating in the more rapid currents of the blood the constant shocks they receive through contact with other corpuscles or with the vascular walls keep their cytoplasm in a contracted state and they maintain a globular form. If, however, through any chance they remain for some time in contact with the wall of a vessel, they are able to make their way between the endothelial cells and pass out of the circulation into the surrounding tissues. Here they creep about, and for this reason have been called the migratory or wandering cells of the tissues. They ultimately either suffer degenerative changes and disappear, or find their way back into the circulation through the lymphatic channels. During these excursions they may incorporate stray particles in the tissues, and thus act as scavengers. This activity has been called their phagocytic function, and may play an important part in the removal of material that should be absorbed or of particles that would otherwise be injurious to the tissues; *e. g.*, bacteria. (See statements regarding the nature of colostrum-corpuscles.)

The emigration of leucocytes from the bloodvessels is pronounced in many of the inflammatory processes, and their phagocytic function may have a marked influence on the result.

The leucocytes are produced in the lymphadenoid tissues of the body, the lymphatic glands, thymus, spleen, and the more diffusely arranged tissues of like structure, but probably most abundantly in the red marrow of the bones.

A close study of the leucocytes has resulted in their subdivision into a number of groups according to their morphological differences or to peculiarities in their behavior toward coloring-matters. The best defined of these groups are:

1. **The polynuclear neutrophilic leucocytes**, in which the nucleus has a very irregular form, often presenting the appearance of two or more nuclei, and the cytoplasm contains granules that have an affinity for neutral anilin-dyes (Fig. 106, $f$ and $g$). This variety constitutes about 72 per cent. of the total number of leucocytes, and is probably produced chiefly in the red marrow of the bones. They

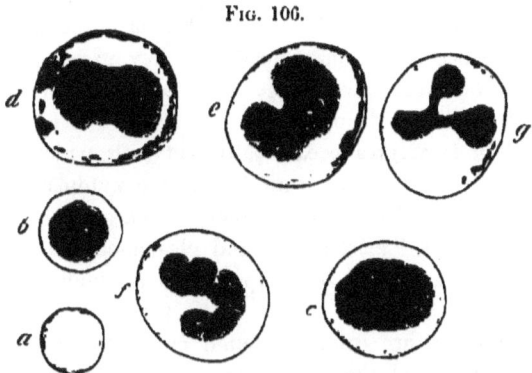

Fig. 106.

Leucocytes from normal human blood. (Böhm and Davidoff.) $a$, red blood-corpuscle, introduced for comparison; $b$, small mononuclear leucocyte (lymphocyte); $c$, large mononuclear leucocyte; $g$, polynuclear leucocyte. These differ in the character of the granules they contain (not represented in the figure). In normal blood these granules are neutrophilic in the vast majority of the polynucleated leucocytes. Occasionally they are acidophilic, "esinophile leucocytes"; sometimes basophilic, "mast-cells" or "plasma-cells." $d$, $e$, $f$, intermediate and probably transitional forms between the large mononuclear leucocytes $c$, and the polynucleated leucocytes, or leucocytes with polymorphic nuclei, $g$.

possess the power of executing amœboid movements and incorporating foreign particles.

2. **The lymphocytes**, with a single round nucleus and a little clear cytoplasm around it. These leucocytes are of about the same size as the red blood-corpuscles (Fig. 106, $b$). They are derived from the lymphadenoid tissue in the lymph-nodes and other situations, and appear to be incapable of amœboid movement. They constitute about 23 per cent. of the total number of leucocytes in normal blood.

3. **The large mononuclear leucocytes**, which are larger than the red corpuscles and have oval nuclei surrounded by clear cytoplasm (Fig. 106, $c$). This variety has also received the name "myelocyte," on the probably correct assumption that they are derived from the red marrow of the bones. They are capable of passing through

transitional forms until they acquire the characters of the polynuclear neutrophilic leucocytes described above. The large mononuclear leucocytes, together with the transitional forms, make up about 3 per cent. of the normal number of leucocytes.

4. **The eosinophilic leucocytes** (Fig. 103, *d*), also larger than the red corpuscles, with irregular, polymorphic nuclei, and a cytoplasm containing relatively large granules which have an affinity for acid dyes; *e. g.*, eosin. These are frequently seen in unusual numbers around inflammatory foci or in tissues undergoing involution; *e. g.*, in the connective tissue of the breast when lactation is suspended. Their significance is not understood, but they appear to be derived from the red bone-marrow. They constitute from 1 to 2 per cent. of the total number of leucocytes.

5. **Basophilic leucocytes**, occasionally met with, which are characterized by the presence of granules in the cytoplasm having a special affinity for basic anilin-colors. These cells have also received the names "mast-cells" and plasma-cells, but the latter term is indefinite, having been applied to a number of cells of different nature.

The **blood-plates** are colorless round or oval discs, about one-fourth the diameter of the red corpuscles. Their function has not been definitely determined, but it is thought that they may play a *rôle* in the production of fibrin, perhaps by the liberation of fibrin-ferment.

Minute globules of fat are occasionally present in the blood, especially during digestion.

The **lymph**, like the blood, consists of a fluid portion, the plasma, and corpuscles held in suspension.

The **plasma**, as would be anticipated from its origin, is very similar in composition to that of the blood.

The **corpuscles** are, for the most part, identical with the small leucocytes (lymphocytes) of the blood, which derives its supply of those cells from the lymph flowing into it.

The **chyle** is the lymph found in the lacteal lymphatics during digestion. When absorption of the products of digestion is in progress this lymph contains a great number of globules of fat, some so minute as to be barely visible under the microscope. In the intervals between absorption this lymph does not differ from that found in the other lymphatics of the body.

**Fibrin** may present the appearance of a delicate network of extremely fine fibres, somewhat resembling a cobweb (Fig. 268), or these fibrils may be aggregated into larger threads variously interwoven, or they may be still further condensed to form masses of a hyaline character. The fibres may undergo a disintegration into granules, when their fibrinous nature is not readily revealed. Fibrin is not found in the body under normal conditions, but separates from the blood if the circulation be arrested for any considerable length of time. It appears to be the result of the interaction of four substances: fibrinogen, fibrinoplastin, fibrin-ferment, and salts of lime. The latter are always present in the tissues; fibrinogen exists in the plasma of the blood and lymph, and is, therefore, very widely distributed. The fibrinoplastin is believed to be derived from the bodies of cells that have undergone some destructive change; and the ferment may be derived from the same source. These four substances are present when the flow of blood through the vessels has been seriously checked for a considerable period; fibrin is then formed, causing a coagulation of the blood. Such a clot, within a vessel during life, is called a "thrombus." Coagulation takes place more rapidly if there be a destruction of tissue; *e. g.*, a break in the wall of the vessel. It may also be occasioned by a roughness on the internal surface of the vessel, if the flow of blood over that obstruction is seriously retarded. In such a case the fibrin-forming elements may be liberated from the bodies of leucocytes that find lodgement behind the obstruction and suffer injury, or they may be derived from blood-plates that have been arrested and undergone similar changes. In a like manner, fibrin may be formed in the lymphatic vessels or the interstices of the tissues.[1]

[1] An explanation of fibrin-formation, offered by Lilienfeld, would serve to elucidate many cases of coagulation under morbid circumstances. According to this observer, fibrin is formed by the union of "thrombosin" with calcium, and is, therefore, a calcium-thrombosin compound. The thrombosin is produced from fibrinogen by the action of nuclein, which in turn is formed from the nucleohiston contained in the nuclei of cells. Coagulation, then, would be the result of the following process: the nucleohiston in the nuclei, during "karyolysis" or disintegration of the nucleus, is decomposed into "histon" and nuclein. The latter, acting on fibrinogen, produces thrombosin, which unites with calcium to produce fibrin.

## CHAPTER X.

### THE DIGESTIVE ORGANS.

THE digestive tract consists of six hollow, and for the most part, tubular organs, which successively open into each other and extend from the pharynx to the anus. The food, after mastication and admixture with saliva in the mouth, passes through (1) the œsophagus into (2) the stomach. Here it undergoes digestive changes under the influence of the gastric secretions. Thence it passes into (3) the duodenum, where the secretions of the liver and pancreas and other glands are mixed with it and still further fit it for absorption. From the duodenum it enters (4) the small intestine, the walls of which take up the available products of digestion, and thence passes into (5) the colon. In the latter the fluid portions are gradually absorbed and the relatively dry residue, the fæces, passes out of the body through (6) the rectum and the anal orifice.

The walls of the digestive organs have a general similarity throughout the whole of the digestive tract. They consist of four coats: 1, an internal mucous membrane; 2, a submucous coat; 3, a muscular coat; and, 4, either a serous or a fibrous external coat. These coats are, respectively, continuous with each other throughout the whole tract. The internal coat, or mucous membrane, varies in both structure and function in the different organs, and will, therefore, require closer study than the other coats. The latter have nearly the same structure in all the organs. The submucous coat is made up of areolar fibrous tissue, which permits some freedom of motion between the mucous and muscular coats, and contains the larger bloodvessels and lymphatics that supply all the coats. The muscular coat consists, in general, of two layers of smooth muscular tissue: an internal circular layer and an external longitudinal layer. Its function is to produce those vermicular or peristaltic movements which mix and gradually propel the food along the digestive tract. The external coat is smooth and serous over those portions of the tract which require the greatest freedom of motion. It is nowhere complete, but, where present, is really a portion of the peritoneum which partially envelops the organs that are contained in the abdominal cavity. Where this serous covering is wanting the external coat consists of areolar fibrous tissue, which serves to connect the organs of the digestive tract with neighboring structures, and thus becomes continuous with the areolar-tissue system pervading the whole body. It supports the vessels and nerves which make their way through it to the different organs.

In addition to the organs above enumerated, it is appropriate to consider here the structure of the tongue, pharynx, salivary glands, and pancreas.

1. **The tongue** consists chiefly of voluntary muscles, the fibres of which are grouped in bundles running in various directions through the substance of the organ. Between the individual striated muscle-fibres, and also between the bundles into which they are collected, there is a variable amount of areolar fibrous tissue containing fat, nerves, and bloodvessels (**Fig. 65**). This areolar tissue

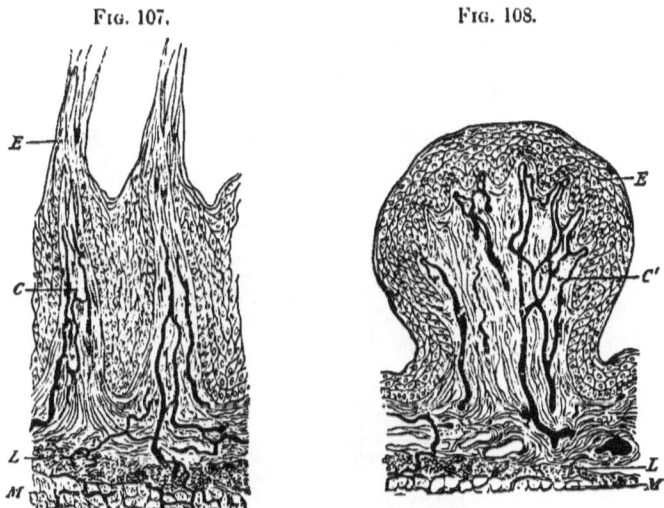

Fig. 107.      Fig. 108.

Sections of papillæ of tongue.

Fig. 107.—Filiform papillæ; human. Heitzmann.
Fig. 108.—Fungiform papillæ; human. (Heitzmann.)
*E*, stratified epithelium; *C*, injected capillaries within the fibrous tissue of the papillæ; *L*, lymphadenoid tissue in lower portion of mucous membrane; *M*, muscular tissue of the tongue.

is more abundant near the surface of the tongue, and is covered with a layer of stratified epithelium, thicker at the sides and on the dorsum of the tongue than on its under surface, where it becomes continuous with the stratified epithelium covering the gums and lining the buccal cavity.

Fig. 109.

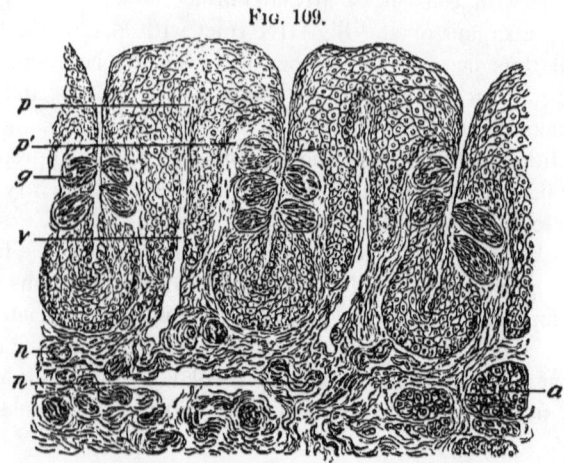

Two circumvallate papillæ; rabbit. (Ranvier.) *p, p'*, fibrous tissue extending into the papilla; *p'*, that containing the nerves passing to the taste-buds; *g*, taste-buds; *v*, small vein; *n, n*, nerves; *a*, acini of a serous gland.

The upper surface and the edges of the tongue are covered with papillæ, some of which are pointed (filiform papillæ), others rounded

Fig. 110.

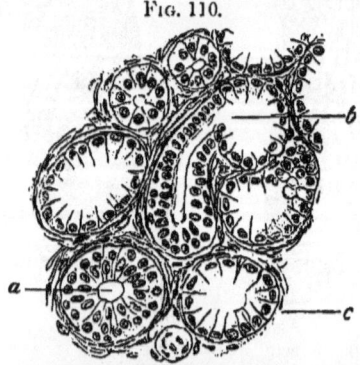

Portion of a section of a mucous gland in the human tongue. (Benda and Guenther's *Atlas*.) *a*, duct; *b*, acinus opening into a duct-radicle; *c*, acinus lined with mucigenous cells, similar to *b*. Between and below *a* and *c*, cross-section of a small artery, recognizable by the elongated nuclei of its muscular coat.

(fungiform papillæ), and still others surrounded by a sulcus (circumvallate papillæ) (Figs. 107-109). Within the epithelium lining this sulcus are peculiar groups of cells, called taste-buds, which will be described in a subsequent chapter. At the junction of the middle and posterior thirds of the upper surface of the tongue there are several of these circumvallate papillae which are of unusual size.

Within the subepithelial areolar tissue, and often extending for some distance between the muscles, there are, here and there, small racemose glands, which secrete a serous or mucous fluid (Figs. 109, a and 110). They are most abundant on the back and sides of the posterior part of the tongue, and their ducts frequently open into the sulci of the circumvallate papillae. Within the subepithelial areolar tissue small collections of lymphadenoid tissue (lymph-follicles) are also of not infrequent occurrence. The papillae covering these are low and inconspicuous, so that the surface of the tongue appears unusually smooth at those points.

2. **The salivary glands** belong to the racemose variety of secreting glands. The secretions which they furnish are of two kinds: 1, a thin, serous fluid, containing albuminoid materials, among which are the specific ferments elaborated by the gland; and, 2, a viscid fluid containing mucin. These two secretions are furnished by acini lined with different varieties of epithelium. The parotid gland secretes only the serous fluid, and is composed of serous alveoli. The sublingual gland secretes only the mucous fluid; but the submaxillary gland secretes both, and, therefore, contains both serous- and mucous-secreting cells.

The cells which line the mucous acini have clear bodies, as the result of a storage of transparent globules of mucin or mucigen within the cytoplasm. Where these globules are abundant the nuclei of the cells are crowded toward the attached ends of the cells. When the mucin is discharged from the cells they become smaller, less clear, and more granular in appearance.

At the periphery of the acini, and especially well marked at or near their blind extremities, are, here and there, crescentic, granular epithelial cells, which may reach the lumen of the acinus or be crowded back by the enlarged cells adjoining them. These cells form the "crescents of Gianuzzi." In the submaxillary gland, at least, many of these crescents secrete the serous or albuminoid fluid mentioned above. This secretion reaches the lumen of the gland through minute intracellular channels (Fig. 111).

The serous alveoli of the salivary glands are lined with cells that, at certain stages of their activity, are so crowded with granules that the nuclei are obscured. These granules are the accumulated material from which the secretion is formed, and when the gland has been functionally active for some time they diminish in number,

Fig. 111.

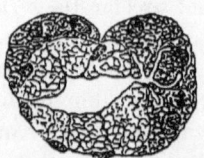

Section of an acinus of the human submaxillary gland. (Krause.) The lumen is surrounded by mucous cells, containing globules of mucigen. Two groups of Gianuzzi's crescents are represented, with the intracellular channels conveying the serous secretion to the lumen.

and the nuclei then come into view. At the same time the cells become smaller, and the lumen within the acinus, which at first was barely distinguishable, becomes more obvious.

The epithelium lining the acini of all the salivary glands rests

Fig. 112.

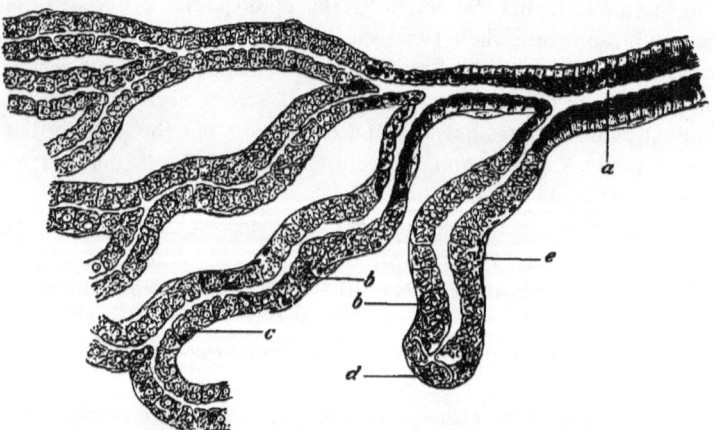

Diagrammatic representation of a portion of a human submaxillary gland. (Krause.) a, duct, lined with columnar cells, striated at their bases and passing into a more cubical epithelium without such striation; b, mucous cells; c, serous cells; d, crescent; e, basement membrane. In this figure the convoluted course of the ducts and tubular acini has been ignored, and they have been represented as though lying in a single plane.

upon a modified connective tissue, called the "basement-membrane," which consists of flattened cells arranged to form a broad, membranous reticulum, the meshes of which are filled with cement. Outside of this basement-membrane there is a small amount of vascular areolar tissue, and broader bands of that tissue divide the whole gland into small lobes and these again into still smaller lobules (Fig. 25).

The ducts of the salivary glands are lined with columnar or pyramidal epithelial cells, the attached ends of which often show a stria-

Fig. 113.

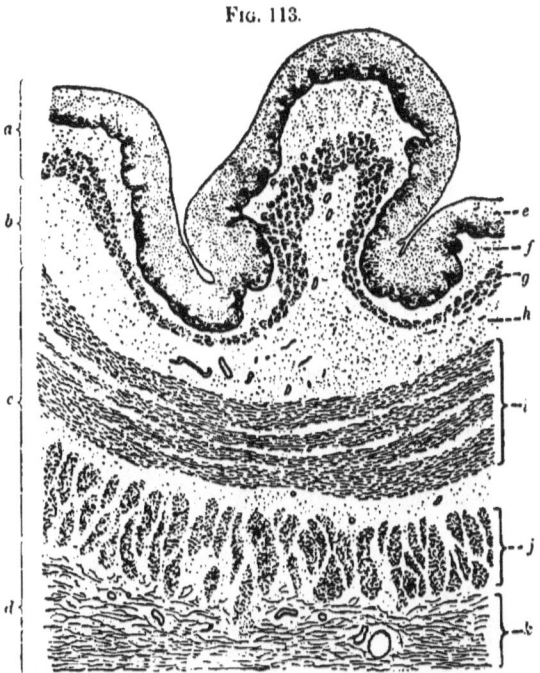

Part of a cross-section of the œsophagus of a dog. (Böhm and Davidoff.) *a*, mucous membrane; *b*, submucous coat; *c*, muscular coat; *d*, fibrous coat; *e*, stratified epithelium; *f*, subepithelial areolar tissue (sometimes called the "tunica propria" of the mucous membrane); *g*, muscularis mucosæ; *h*, areolar tissue of the submucosa, containing the chief branches of the arterial and venous vessels; *i*, internal, encircling layer of the muscular coat. It is the contraction of this coat that has caused a longitudinal wrinkling of the mucous membrane. One of those folds is completely and two are partially shown. *j*, external, longitudinal layer of the muscular coat; *k*, areolar tissue forming the external coat and connecting the œsophagus with neighboring structures. A few large vessels entering the œsophagus are represented in this coat.

tion perpendicular to the surface of the basement-membrane (Fig. 112).

The nerves ramify in the interlobular areolar tissue and send delicate, non-medullated fibres through the basement-membrane to be distributed upon and between the epithelial cells. Occasionally small ganglia are seen upon the larger nerves.

3. **The Œsophagus** (Fig. 113).—The mucous membrane of the œsophagus is composed of three layers. The innermost layer is made up of stratified epithelium. Beneath this is a layer of fibrous tissue, with small papillæ extending into the deeper portions of the epithelium (see Fig. 38). Outside of this is a layer of longitudinal smooth muscular tissue, the "muscularis mucosæ." This is but imperfectly represented at the upper part of the œsophagus, but at the lower end forms a continuous layer separating the "tunica propria" (Fig. 113, *f*) of the mucous membrane from the submucous coat, and becoming continuous with a similar layer of smooth muscular tissue in the mucous membrane of the stomach and intestine. Occasionally small, imperfectly defined lymph-follicles are met with in the mucous membrane.

The submucous coat of loose areolar tissue contains small racemose glands sparsely distributed through it, the ducts of which penetrate the mucous membrane and open upon the internal surface of the œsophagus.

The muscular coat consists of an internal circular and an external longitudinal layer, which at the upper end of the œsophagus are composed of striated muscle. This is gradually replaced by smooth muscular tissue further down the œsophagus, and at its lower end only the latter tissue is found.

The external coat of the œsophagus is represented by a variable amount of areolar tissue which loosely connects it with the surrounding structures.

4. **The Stomach.**—Nearly the whole thickness of the mucous membrane of the stomach is made up of straight tubular glands (gastric tubules), which lie perpendicular to the surface, and are separated from each other by only a small amount of a delicate, highly vascular areolar tissue. This is a little denser and more abundant below the deep ends of the glands, where it separates them from the muscularis mucosæ forming the deepest layer of the mucous membrane.

The mouths of the gastric tubules open into shallow, polygonal depressions or crypts on the surface of the mucous membrane, several glands opening into each depression. These depressions give the internal surface of the stomach a reticular appearance when viewed with a low power. They, and the ridges which separate them, are covered with a rather tall columnar epithelium. The glands which open into them are of two kinds: the "pyloric"

# THE DIGESTIVE ORGANS. 135

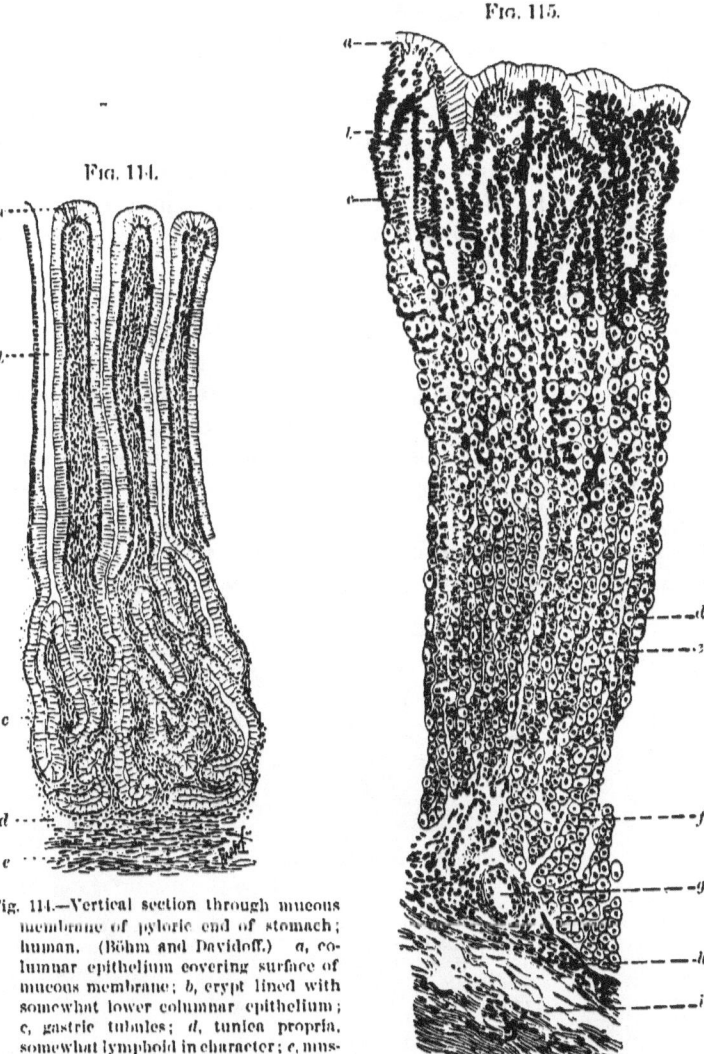

Fig. 114.
Fig. 115.

Fig. 114.—Vertical section through mucous membrane of pyloric end of stomach; human. (Böhm and Davidoff.) *a*, columnar epithelium covering surface of mucous membrane; *b*, crypt lined with somewhat lower columnar epithelium; *c*, gastric tubules; *d*, tunica propria, somewhat lymphoid in character; *e*, muscularis mucosæ, of smooth muscular tissue.

Fig. 115.—Nearly vertical section of the mucous membrane near the cardiac end of the stomach; rabbit: *a*, columnar epithelium covering the surface of the mucous membrane; *b*, that lining a crypt; *c*, duct; *d*, parietal cell extending to the lumen of the gland; *e*, lumen, readily traced for only a short distance; *f*, central or chief cells; *g*, small artery, to the left and above it, a small vein; *h*, muscularis mucosæ, consisting of three thin layers of smooth muscular tissue, the middle layer in transverse, the others in longitudinal section; *i*, portion of submucosa. The specimen was taken from an animal some time after the ingestion of food, and the chief cells are, in consequence, relatively small in comparison with the size of the parietal cells.

variety, so-called because more abundant at the pyloric end of the stomach, and the "cardiac" variety, which preponderate near the cardiac end.

The **pyloric glands** (Fig. 114) have the simpler structure. They possess a comparatively deep and open mouth, lined with columnar epithelial cells similar to and continuous with those lining the depressions already mentioned, and, like them, mucigenous. Into these mouths one or more straight tubular glands, lined with low, granular columnar cells, discharge their secretion.

The **cardiac glands** (Fig. 115) have shallower mouths than the pyloric glands, and the tubes that open into them contain two sorts of epithelial cells: 1, the "chief" or central cells, which line and nearly fill the whole tubule, leaving only a very small and somewhat tortuous lumen in the centre; and, 2, the parietal cells, lying at intervals between the central cells and the surrounding connective tissue, but sometimes projecting between two central cells nearly or quite to the lumen of the gland. Very fine channels run from that lumen to and around these parietal cells, which are believed to produce the free acid of the gastric juice (Figs. 116–118).

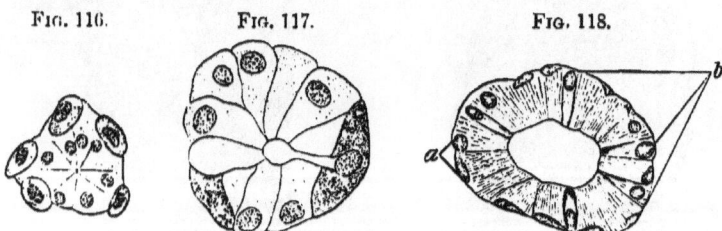

Fig. 116.   Fig. 117.   Fig. 118.

Cross-sections of gastric glands; dog. (Hamburger.)

Figs. 116 and 117.—From the cardiac end of the stomach, showing the chief or central cells and the parietal cells. 116, from a dog killed during the second hour of digestion. The central cells are relatively large, and the lumen is reduced to a mere line, appearing as a dot in the centre of the cross-section. 117, from a dog killed during the seventh hour of digestion. The parietal cells are relatively large, and the lumen more distinct than in 116, owing to loss of material on the part of the central cells and a gain on the part of the parietal cells. One of the latter is in communication with the lumen through a small channel between the central cells.

Fig. 118.—From the pyloric end of the stomach during the fifth hour of digestion. The cells *b* have parted with their secretion and are compressed by the cells *a*, which still retain the materials stored for secretion. The lumen of the gland is much larger than that of the glands at the cardiac end of the stomach.

Besides the secreting glands, the mucous membrane of the stomach sometimes contains small lymph-follicles. Its blood- and lymph-supplies are abundant, and nerves are distributed to its various tissue-elements.

The deepest layer of the mucous membrane is the muscularis mucosae, made up of two or three strata of smooth muscular tissue in which the fibres run in different directions.

The submucous coat of the stomach consists of loose areolar tissue, which allows considerable freedom of motion between the mucous membrane and the muscular coat. When, therefore, the organ is empty the contraction of the muscular coat throws the mucous membrane into coarse folds (rugae). The large arteries, veins, and lymphatics course in this submucous tissue, and thence send branches into both the mucous and muscular coats. The nerves also form a ganglionated plexus in this coat.

The muscular coat consists of an external longitudinal layer, inside of which is another layer encircling the organ. The external layer is continuous with the outer muscular layer of the œsophagus. The internal muscular layer of the latter organ is continued into the wall of the stomach as a scattered set of oblique fibres lying internal to the encircling fibres already mentioned. The muscular coat of the stomach may, therefore, be considered as composed of three layers, the innermost of which is incomplete. At the pylorus the encircling muscular layer is thickened.

Aside from the fibrous tissue that more or less completely separates its layers, the muscular coat contains ganglionated nerve-plexuses.

The external surface of the stomach is covered with a serous investment of peritoneum, except along the curvatures, where the peritoneum is reflected from the organ, permitting the passage of its vessels and nerves.

5. **The Duodenum.**—The structures characteristic of the small intestine first make their appearance in the duodenum. We shall first consider those features which are found throughout the small intestine, and then describe those which are peculiar to the duodenum (Fig. 119).

The mucous membrane presents thin, transverse folds, the valvulae conniventes, which are not obliterated when the intestinal wall is stretched. They are made up of a thin layer of areolar tissue, extending from the submucous coat of the intestine, which is covered on both surfaces with mucous membrane. This arrangement serves greatly to increase the surface of mucous membrane coming in contact with the contents of the intestine, a provision facilitating absorption of the products of digestion.

The valvulæ conniventes begin a short distance below the pylorus, and are very numerous and prominent in the duodenum, but become progressively less frequent and pronounced in the lower portions of the small intestine.

The absorbent surface of the small intestine is still further in-

FIG. 119.

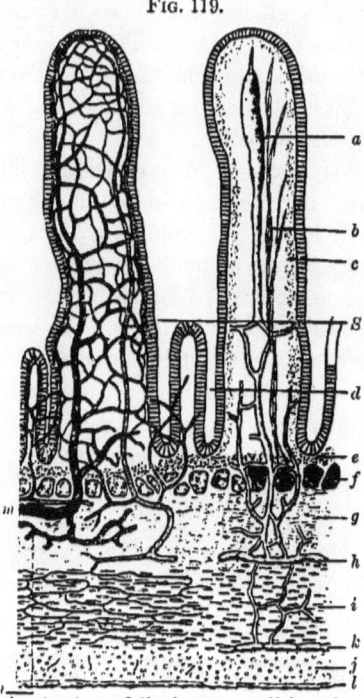

Diagram representing the structure of the human small intestine. (Böhm, Davidoff, and Mall, slightly modified.) Two villi are represented. In the one on the left the bloodvessels are shown; in the one on the right, the lymphatics. The line $S$ indicates the surface of the mucous membrane between the villi. $a$, central lacteal vessel; $b$, smooth muscular fibres extending into the villus from the muscularis mucosæ; $c$, lymphadenoid tissue beneath the epithelial covering of the villus; $d$, crypt of Lieberkühn; $e$, tunica propria of lymphadenoid tissue, and continuous with that of the villus; $f$, muscularis mucosæ, forming the deepest portion of the mucous membrane; $g$, submucosa containing the larger bloodvessels and the lymphatic plexus $h$; $i$, encircling layer of the muscular coat; $j$, longitudinal layer; $k$, lymphatic plexus within the muscular coat; $l$, serous coat; $m$, vein. The crypts are lined, and the villi covered, with columnar epithelium.

creased by the presence of innumerable minute, finger-like projections from the surface of the mucous membrane, the villi. These are just discernible by the unaided eye, and give the internal surface of the intestine a velvety appearance.

Between the attached ends of the villi, and opening upon the surface of the mucous membrane, are tubular depressions extending nearly to the muscularis mucosae. These are the "crypts of Lieberkühn," and have the appearance of simple tubular glands; but it is doubtful if they elaborate any peculiar secretion. These crypts are present, not only in the whole extent of the small intestine, but also throughout that of the colon.

The crypts of Lieberkühn are lined with columnar epithelium, which also covers the surface of the mucous membrane and the villi springing from it. The cells composing this epithelium multiply in the crypts, and, as they mature, are gradually moved toward their orifices, whence they replace those that have been destroyed upon the surfaces of the villi. The cells possess a granular cytoplasm, which becomes infiltrated with fat during digestion; an oval, vesicular nucleus; and a delicate cell-membrane. The free ends of the cells are formed by a well-marked cuticle, which may be either homogeneous in appearance, or present very fine vertical striations (Fig. 37). Many of the cells are mucigenous and contain globules of mucus near their free ends, or appear as goblet-cells after the discharge of that secretion. These cells are more abundant on the villi, where they are older, than in the crypts lined with less mature cells.

The epithelium rests upon a basement-membrane, which contains nuclei, and is therefore composed, in part at least, of cells. Beneath this basement-membrane is a layer of reticular and areolar tissues, containing a variable number of lymphoid cells and numerous capillary bloodvessels. The rest of the mucous membrane, down to the muscularis mucosae, and the axes of the villi are occupied by areolar fibrous tissue.

The thin muscularis mucosae, which forms the deepest layer of the mucous membrane, is made up of two layers of smooth muscular tissue: an internal layer, in which the fibres run transversely to the axis of the intestine, and an external longitudinal layer. From the upper surface of this muscular layer of the mucous membrane muscular fibres extend into the villi, in the areolar tissue in their axes, and serve to shorten the villi by their contraction, so that the villi are moved about in the intestinal contents during the process of absorption. In the centre of each villus is a capillary lymphatic vessel arising in a blind extremity near the apex of the villus. These lymphatics open into a lymphatic plexus situated between the muscularis mucosae and the ends of the crypts of Lieberkühn, and thence discharge their contents into the lymphatics in the submucosa. The muscular fibres in the villi probably aid in the propulsion of the chyle in these lymphatics (Fig. 120).

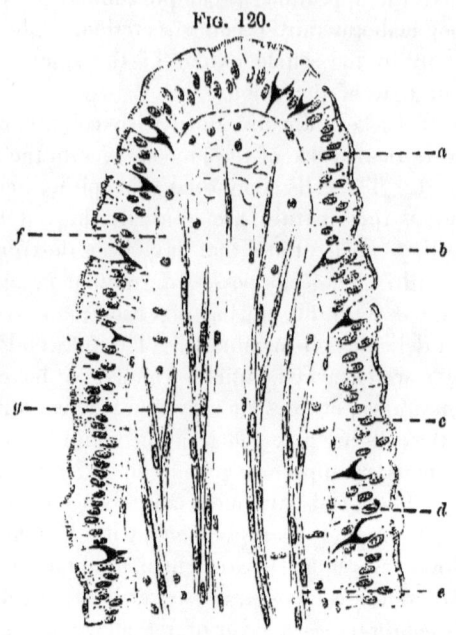

Fig. 120.

Axial section of villus of the dog. (Kultschitzky.) *a*, epithelial covering with cuticle ; *b*, goblet-cell ; *c*, space between tapering ends of the epithelial cells ; *d*, cell of the basement-membrane ; *e*, smooth muscular fibres ; *f*, reticulum of the tunica propria (the lymphoid cells have been, for the most part, removed) ; *g*, lumen of the central lymphatic. The bloodvessels are not represented.

The submucous coat of the intestine is composed of areolar fibrous tissue. Outside of this coat is the muscular coat, divisible into two layers, which is covered throughout the whole circumference of the intestine, except at the line of mesenteric attachment, with a serous investment of the peritoneum.

In the duodenum the submucous coat contains compound tubular glands, the glands of Brunner, the ducts from which penetrate the muscularis mucosæ and open upon the surface of the mucous membrane, between the crypts of Lieberkühn. Here and there, in the duodenum, are little collections of lymphadenoid tissue, occupying an enlarged villus and often extending through the muscularis mucosæ into the submucous areolar tissue (Fig. 121). These lymph-follicles may be regarded as the result of an increase in the amount of reticular tissue of the villus, which has replaced the other structures usually present. In the lower portions of the small intestine there are collections of these solitary follicles, which have received the name "Peyer's patches."

6. **The small intestine** below the duodenum resembles the latter

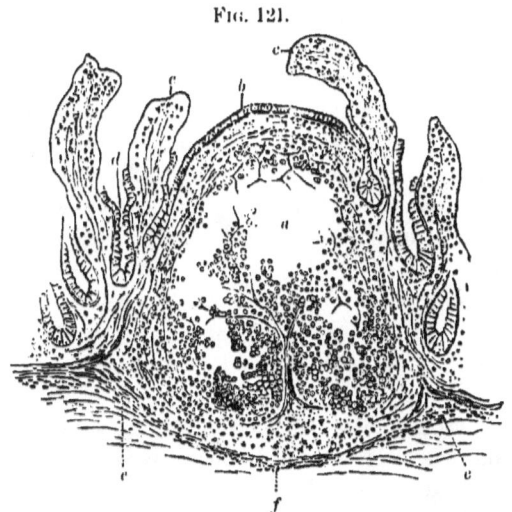

Fig. 121.

Section of solitary follicle from the ileum. (Cadiat.) *a*, space left by the disintegration of the central, delicate lymphadenoid tissue of the follicle during the preparation of the section; *b*, columnar epithelium of intestinal surface; *c, c*, villi, partially denuded of epithelium; *d*, crypt; *e, f*, muscularis mucosae; above *f*, the point where the vessels enter the follicle. The Peyer's patches are collections of such solitary follicles, placed side by side and destitute of villi at their upper surfaces.

in structure, with a few modifications, which become progressively more marked as the distance from the stomach increases.

The glands of Brunner are most abundant near the upper part of the duodenum, more sparsely distributed further down, and usually disappear entirely before the beginning of the jejunum.

The valvulae conniventes, which are most highly developed a little below the entrance of the gall and pancreatic ducts, also become lower and less frequent along the course of the intestine, and finally disappear about the middle of the ileum.

The crypts of Lieberkühn are deepest in the upper part of the intestinal tract, but persist in shallower form throughout its whole extent, as well as along the whole length of the colon.

The Peyer's patches are most abundant in the lower part of the ileum, where they lie in the intestinal wall opposite the line of mesenteric attachment, and form oval areas with their long axes parallel to the axis of the intestine.

7. **The Colon.**—The mucous membrane of the colon is destitute of villi, but contains crypts of Lieberkühn closely arranged side by side and lined with columnar epithelium rich in mucigenous cells. The muscularis mucosæ is similar to that of the small intestine, and gives off occasional fibres that penetrate between the crypts.

The submucous coat resembles that of the small intestine, and, in common with the mucous membrane, contains solitary lymph-follicles, most abundantly in the cæcum and vermiform appendix.

The muscular coat has its outer or longitudinal layer most highly developed in three bands, which are situated about equidistantly around the circumference of the bowel and occasion a pouching of the intervening wall.

The serous coat is similar to that of the small intestine, but is occasionally extended over small pendulous projections of the subserous fibrous tissue, which contain adipose tissue, appendices epiploicæ.

8. **The rectum** resembles the colon in its structure, except that the three muscular bands present in the latter are wanting. The mucous membrane as it passes into the anal canal loses its tubular glands, and subsequently becomes covered, not with columnar, but with stratified epithelium, continuous with the epidermis of the skin around the anus.

9. **The pancreas** (Fig. 122) has a structure similar to that of the salivary glands, but its lobules are separated and held in place by a rather more considerable amount of loose areolar tissue, in which there are occasional groups of cells of uncertain nature, but certainly distinct from those lining the glandular acini. They are called the "interalveolar cell-islets," and may, perhaps, be of the nature of ductless glands (*q. v.*).

As the pancreas exercises its secretory function the granules within its cells move toward the lumina of the acini and successively disappear, the attached ends of the cells becoming clearer and the whole cell diminishing somewhat in size during the process.

The nerves of the stomach and intestinal tract form two ganglionated plexuses, the plexus of Auerbach, which lies between the two layers of the muscular coat, and the plexus of Meissner, situated in the submucous coat. From these plexuses fibres are distributed to the muscles and other structural elements. These fibres are of the non-medullated variety.

The nerves of the pancreas are also non-medullated, possess a few ganglia within the organ, and are finally distributed among the epithelial cells.

**The Tonsils, Lymph-follicles, and Peyer's Patches.**—These collections of lymphadenoid tissue in the alimentary tract have special

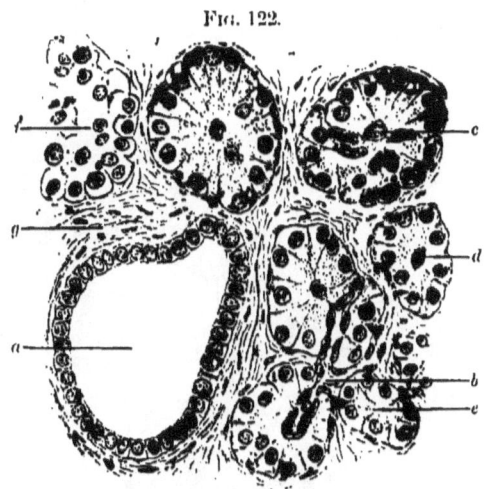

Fig. 122.

Section of human pancreas. (Böhm and Davidoff.) *a*, larger duct; *b*, beginning of duct; *c, d*, acini with cells belonging to the corresponding duct-radicles in their centers; *c*, acinus, cut just beyond the lumen; *f*, interalveolar cell-group (?); *g*, fibrous connective tissue, forming the interstitial tissue of the organ.

interest to the physician as being points particularly liable to infection. The solitary follicles of the stomach and of the small and large intestine, and the collections of such follicles forming the patches of Peyer, are the sites which are most vulnerable to invasion by pathogenic bacteria in the digestive tract, though they are probably protected to a considerable extent by the germicidal powers of the acid gastric juice. This is not always capable of guarding them from infection by the typhoid and tubercle bacilli, and in the diseases of the intestinal canal occasioned by those bacteria the follicles and Peyer's patches are the seat of the earliest and most extensive ulcerations. The tonsils, which have the same general structure, are still more prone to infection of various kinds, for they are more directly exposed to the action of bacteria that may gain access to the mouth.

The reason for this vulnerability appears to lie in the close proximity of the lymphatics to the surface and their meagre protection by a thin layer of epithelium liable to abrasion or destruction. The solitary follicles of the intestine, for example, are covered with a single layer of columnar epithelium (Fig. 121).

The lymphadenoid tissue of the tonsil, it is true, is protected by a layer of stratified epithelium; but the surface of the tonsil is invaginated to form the crypts of that organ, and within those crypts it

FIG. 123.

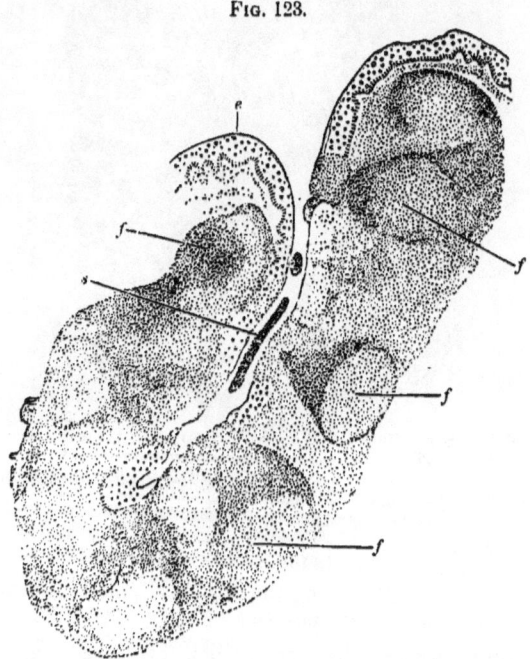

Section through one of the crypts of the tonsil. (Stöhr.) *e*, stratified epithelium of the general surface, continued into the crypt; *f*, follicles containing germinal foci. Between the follicles is a more diffusely arranged lymphadenoid tissue. *s*, material within the crypt, composed in part of lymphoid corpuscles that have wandered through the stratified epithelium.

is possible for bacteria to multiply and produce such an accumulation of poisonous products as to destroy the integrity of the epithelium and so permit an invasion of the lymphadenoid tissue beneath. We therefore find the tonsils specially prone to such inflammatory

THE DIGESTIVE ORGANS. 145

Fig. 124.

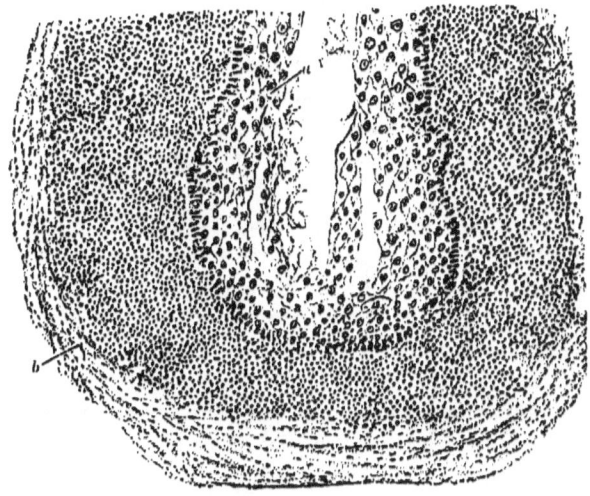

Section through the fundus of a crypt. (Benda and Guenther's *Atlas*.) *a*, stratified epithelium, desquamating at its surface; *b*, deep portion of the lymphadenoid tissue, in which proliferation of lymphoid cells takes place as well as in the follicles represented in Fig. 123.

processes as tonsillitis and diphtheritic inflammation (Figs. 123 and 124).

## CHAPTER XI.

### THE LIVER.

THAT portion of the liver which is exposed in the abdominal cavity is covered by a reflection of the peritoneum, closely attached to the organ, because its deeper side is continuous with the fibrous structures or interstitial tissue of the liver itself. This serous covering is so thin that the substance of the liver can be readily seen through it.

At the portal fissure, the serous coat having been reflected from it, the liver is covered with a loose areolar tissue in which the main trunks of all but one of the vessels connected with it are situated: namely, the portal vein, hepatic artery, gall-duct, and lymphatics. These vessels enter the liver together at this place, and are closely associated with each other in all their ramifications, being supported throughout by areolar tissue, which is continuous with that at the portal fissure and with the interstitial tissue of the liver.

These vessels, with their supporting fibrous investment, called Glisson's capsule, ramify in the liver in such a way as to resemble a tree with a multitude of branches and twigs, each composed of divisions of all the vessels named.

The hepatic vein enters the liver at a different place, and also suffers a tree-like subdivision; but its branches are surrounded by a very much smaller amount of fibrous tissue, which may be regarded as but a slightly reinforced portion of the interstitial tissue of the organ.

Sections of the liver (Fig. 125) will reveal portions of these two trees, cut in various directions with respect to their axes. It will be observed that the twigs and larger branches of the trees are nowhere in close relations to each other, showing that the hepatic vein, in all its ramifications, is separated from the other vessels by the parenchyma of the organ. If we select some part of a section which contains one of the smallest branches of the hepatic vein, and cut across its axis so that its lumen appears round, we shall notice that at about equal distances from it there are sections of two, three, or four twigs of the compound tree. In these the gall-duct can be identified by its distinct lining of columnar or cubical epithelium, and the hepatic artery distinguished from the portal vein by its relatively thick wall as compared with the size of its lumen. These vessels are collectively known as the *interlobular vessels*. Between and around them is the areolar fibrous tissue, which forms a part of Glisson's capsule, and which is abundantly supplied with

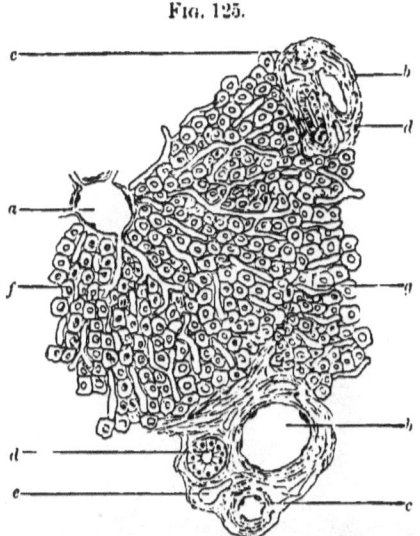

Fig. 125.

Diagrammatic sketch of a section of liver: *a*, central vein (radicle of the hepatic vein); *b, b*, branches of the portal vein; *c, c*, branches of the hepatic artery; *d, d*, small bile-ducts; *e*, lymphatic vessel; *b, c, d, e* are enclosed in areolar tissue, which is continuous with Glisson's capsule; *f*, liver-cells; *g*, line indicating the junction and blending of two neighboring lobules.

lymphatic spaces and vessels in the fibrous tissue. The lymphatics appear as clear spaces with smooth walls, some of them with distinct endothelial linings, but almost devoid of any other wall.

The parenchyma may be subdivided into portions which surround the smallest branches of the hepatic vein, and are bounded by imaginary lines connecting the groups of interlobular vessels. These subdivisions are called "lobules" of the liver. In the human liver they blend at their peripheries, between the masses of connective tissue enclosing the interlobular vessel; but in the liver of the pig these lobules are veritable subdivisions of the liver, and are separated by septa of fibrous tissue, the interlobular vessels lying in the lines formed by the junction of three such septa.

Connecting the branches of the portal vein with the hepatic vein is a plexus of capillaries, called the *intralobular vessels*, through which the blood passes from the portal vessels to the radicles of the hepatic vein and thence into the general circulation. These intralobular vessels also receive blood from the hepatic artery, the capillaries from which join them at a little distance from the periphery of the lobule. The radicles of the hepatic vein are called the *central veins*, from their situation in the axes of the lobules, which are conceived as having a somewhat cylindrical shape (Fig. 126).

Fig. 126.

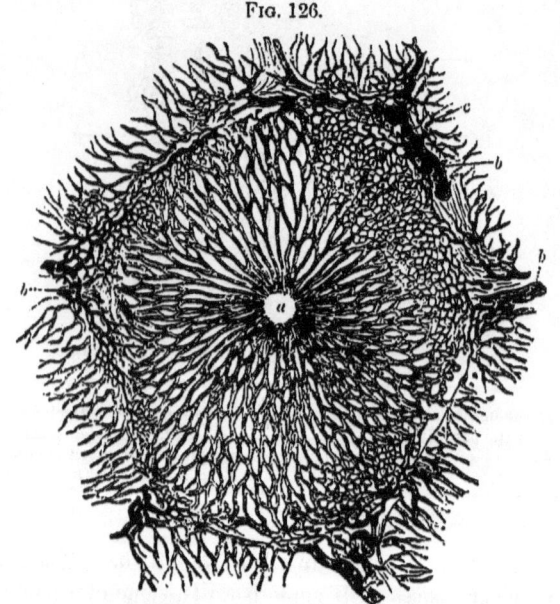

Vessels and bile-ducts of a lobule of a rabbit's liver in transverse section. (Cadiat.) *a*, central vein; *b, b*, interlobular veins (branches of the portal vein); *c*, interlobular bile-duct, receiving capillary bile-ducts from the lobule. Between *a* and *b* is the capillary plexus called the intralobular vessels. The biliary radicles are not represented throughout the figure, and the branches of the hepatic artery have been wholly omitted.

Between the interlobular capillaries are rows of epithelial cells, which constitute the functional part of the liver, its parenchyma. They appear to touch the walls of the capillaries, but are, in reality, separated from them by a narrow lymph-space (Fig. 127). In the human liver the epithelial cells of the parenchyma form a plexus lying in the meshes of the capillary network of the interlobular vessels.

It requires an effort of the imagination to conceive of a third plexus within the lobule, but such a plexus exists, being formed of the radicles of the gall-duct. These are minute channels situated between contiguous epithelial cells, each of which is grooved upon its surface to form half of the tiny canal. The cells themselves have fine channels running from the bile-capillaries into their cytoplasm and ending there in little rounded expansions. It is difficult to detect these bile-capillaries in ordinary sections of the liver, unless they have been previously injected through the main duct; but with a high power their cross-sections may sometimes be clearly seen, appearing as little round or oval spaces at the junction of two

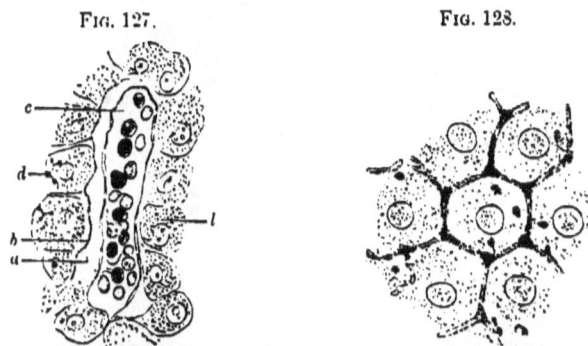

Fig. 127.—Perivascular lymphatic of the human liver. (Disse.) c, capillary in longitudinal section; a, lymphatic space between the capillary and row of epithelial cells; b, wall of the lymphatic space, slightly separated from the liver-cells and drawn a little emphatically; l, liver-cells; d, bile-capillaries in cross-section, with their intracellular ramifications.

Fig. 128.—Bile-capillaries between the liver-cells, with minute channels penetrating the cells and communicating with secretory vacuoles within the cytoplasm. Injected liver of the rabbit. (Pfeiffer.)

epithelial cells, midway between the nearest capillary bloodvessels. Throughout their whole course they appear to be separated from the nearest bloodvessels by a distance approximately equal to half the diameter of one of the epithelial cells. It is this fact that makes it so difficult to frame a mental picture of their distribution in the lobule (Fig. 128).

The nerves supplying the liver ramify in extremely delicate, non-medullated fibrils, which ramify throughout the substance of the liver and terminate in minute twigs among its epithelial cells.

The epithelial cells of the liver have a cubical shape, the grooved and other surfaces that come in contact with neighboring cells being flat, while the remaining surfaces may be somewhat rounded. The cytoplasm is granular, and, except after a considerable period of starvation, more or less abundantly infiltrated with irregular granules and masses of glycogen and globules of fat (Fig. 129). The

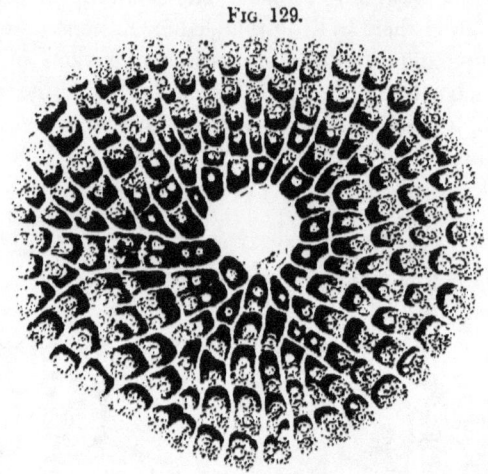

FIG. 129.

Portion of hepatic lobule of the rabbit; cells infiltrated with glycogen. (Barfurth.) The animal had been fed for twenty-four hours on wheat-bread, to promote the storage of glycogen within the liver-cells. The cells in close proximity to the central vein contain the largest amount of glycogen, which appears to fill the cytoplasm. Further from the central vein the cells contain less glycogen, which is most abundant in that portion of the cell turned toward the centre of the lobule. Fat-globules are most abundant in the cells at the periphery of the lobule. No fat-globules are represented in this figure.

glycogen dissolves out of the cells during the ordinary processes of fixation and hardening preparatory to the preparation of sections, leaving spaces in the cytoplasm, which cause it to have a coarsely reticulated appearance in cases where the glycogen was abundant. This reticulation would render it impossible to distinguish the minute intracellular bile-passages. Each cell has a round vesicular nucleus near its centre. In rare instances two nuclei may be found in a single cell.

It will, perhaps, make the structure of the liver a little more comprehensible if it is stated that the liver of some of the lower animals is a tubular gland, the tubes of which are lined with a layer of epithelium. In the human liver this tubular structure is disguised by the facts that the tubules anastomose with each other, and that their lumina are very minute and bounded by only two cells when seen in cross-section. So inconspicuous are these lumina that a casual glance at a section of a liver would not reveal the fact that it was a glandular organ.

The interstitial tissue of the liver consists of a few sparsely distributed fibres continuous with those of Glisson's capsule.

The intricate structure of the liver prepares us for the fact that its function is an extremely complex one. It is a secreting gland, elaborating the bile and discharging it into the duodenum. But the bile has more than one purpose. It aids in the digestion and absorption of food, and it also contains excrementitious matters destined to leave the body through the alimentary tract. Even the secretory function of the liver, therefore, serves a double purpose: the supply of substances useful to the organism and the elimination of products that would be detrimental if retained.

But the function of the liver is not confined to the elaboration of the bile. It also acts as a reservoir for the storage of nourishment, which can be drawn upon as needed by the organism. This is the meaning of the glycogen and fat which have infiltrated the cells.

The food-materials that are absorbed from the digestive tract pass into the system through two channels: the lymphatic and the portal circulations. The latter carries them to the liver, where some of the fat, probably after desaponification, is taken up by the epithelial cells, which also appropriate a portion of the sugar in the portal blood, transforming it into glycogen and holding it in that form until a relative deficiency of glucose in the blood reveals its need by the system.

The blood comes into such close relations with the epithelial cells of the liver that an interchange of soluble substances between them appears to be about as easy a matter as the interchange of gases between the blood and the air in the lungs; and, as in the latter case, this interchange is mutual: some matter passing from the blood to the liver-cells and some from the cells to the blood. In the lung there is a gaseous regeneration of the blood; in the liver, a renovation as to certain of its soluble constituents.

**The Gall-bladder.**—The bile is secreted continuously by the liver, for it is an excrement; but it is discharged intermittently into the alimentary tract, as required by the digestive processes. In the interval it is stored in the gall-bladder.

The gall-bladder is lined with columnar epithelium, capable of secreting mucus. Beneath this is a layer of fibrous tissue, which becomes areolar and supports the chief bloodvessels and lymphatics. Beneath this is the wall of the organ, composed of interlacing bands of fibrous and smooth muscular tissues. The surface is invested by a portion of the peritoneum. The excretory bile-duct has a similar structure.

## CHAPTER XII.

### THE URINARY ORGANS.

THE urine is secreted by the kidney, whence it passes successively through the renal pelvis, ureter, bladder, and urethra into the outer world.

1. **The kidney** is made up of homologous parts or lobes, which are readily distinguished in early life by the superficial furrows marking their lines of junction. In later years these depressions on the surface of the kidney disappear. Each of the lobes corresponds to one of the papillæ of the kidney and the pelvic calix that embraces it. In some of the lower animals—*e. g.*, the rabbit—the kidney has but one papilla, so that the whole renal pelvis in those animals corresponds to a single calix in man.

The kidney is a compound tubular gland of peculiar construction, the tubules taking origin from little spherical bodies, called Malpighian bodies, instead of from simple blind extremities, and, after running a definite and somewhat complicated course, uniting successively with several others to form the excretory ducts, called the "collecting tubules," which open into the calices near the tips of the papillæ.

If a section of the organ be made through its convexity down to the pelvis, the papillæ will be seen projecting into the calices of the pelvis, and it will be noticed that each papilla forms the apex of a pyramidal portion of tissue having a different tint and texture from the rest of the kidney. These pyramids form the "medulla" of the organ (Fig. 130).

The bloodvessels supplying nearly all its substance enter the kidney near the bases of the pyramids, having approached the organ through the fat that lies around the calices. Within the kidney they break up into branches that run along the base of each pyramid in that portion of the organ which is called the "boundary zone." Between that zone and the convex surface of the kidney the tissue is known as the "cortex."

The arrangement of the renal tubules, which make up the chief bulk of the kidney, can be most easily understood if they are traced back from their openings at the apex of the pyramid to their

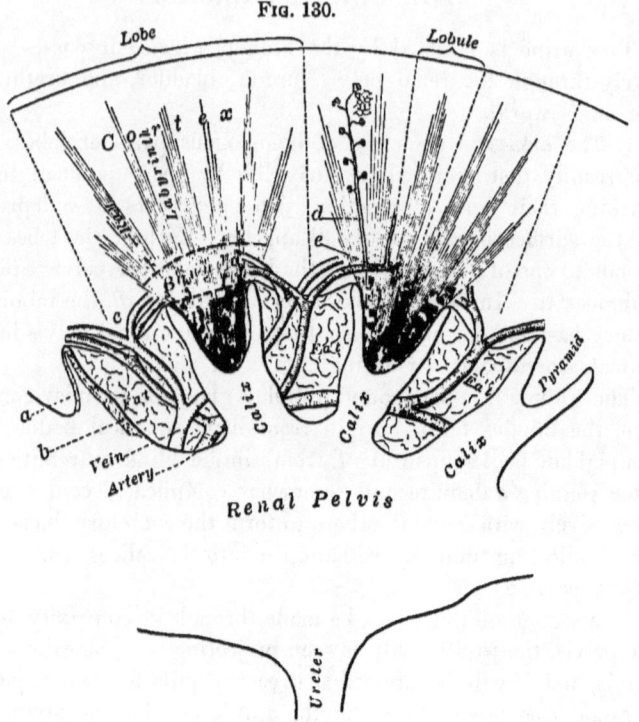

Fig. 130.

Diagrammatic sketch of a section of the kidney: *a*, columnar epithelium covering the external surface of the pyramid and continuous on the one hand with the columnar epithelium lining the collecting tubules within the pyramid, and on the other hand with the transitional epithelium lining the calices and renal pelvis. This transitional epithelium is indicated at *b*. It rests upon the fibrous tissue of the calices and pelvis, which becomes continuous with the fibrous capsule of the kidney at the junction of the calices with that organ. Outside of this capsule is the perinephric fat, indicated in the figure between the calices. The vessels approach the kidney through this fat, entering its substance near the bases of the pyramids and forming the vascular arcades (*c*, arterial arcade). From these arcades the interlobular vessels proceed, between the medullary rays and in the labyrinth, toward the convex surface of the kidney. *d*, interlobular artery, giving off branches, the afferent vessels, to the Malpighian bodies. The extensions of the cortical substance between the pyramids, *e*, are known as the columns of Bertini. During infancy the lobes of the kidney are marked by sulci upon the surface of the organ. With the growth of the organ these lobes blend with each other, and the sulci between them become indistinct or are wholly obliterated. The columns of Bertini are made up of the blended lateral portions of the cortex of two contiguous lobes.

origins in the Malpighian bodies. The different portions of the tubules present somewhat different characters, and have received special names.

The collecting tubes, which open into the calix at the apex of the pyramid, are straight, and lie nearly parallel to each other and to the axis of the pyramid, and, therefore, nearly perpendicular to the base of the pyramid. As they are followed from the apex, in a direction the reverse of that taken by the urine in flowing through them, they branch dichotomously, and the branches become progressively smaller. At the base of the pyramid these straight tubules are collected into bundles that radiate toward the convex surface of the kidney, and are called the " medullary rays." In these, and in the part of the pyramid that is near the boundary-zone, the collecting tubes are associated with other straight portions of the tubules, " Henle's tubes," which will be described presently. From the medullary rays the tubules pass into the region between those rays in the cortical portion of the kidney. This region of the cortex is known as the " labyrinth." Here the tubules lose their straight character and become much contorted, forming the " second convoluted tubules." They then re-enter the medullary rays, which they descend for a variable distance into the pyramid, constituting the "ascending branches of Henle's tubes," which make a sharp turn, " Henle's loop," and then retrace their course up the medullary rays into the cortical portion of the kidney, "descending branches of Henle's tube." They then pass again into the labyrinth and form the " first convoluted tubules," which finally merge into the structure of the Malpighian bodies, also situated in the labyrinth. In consequence of the passage of tubules from them into the surrounding labyrinth the medullary rays become smaller as they are followed from the base of the pyramid, and eventually disappear before the capsule of the kidney is reached. They are completely surrounded by the labyrinth.

If we now follow the course of the urine in its way from the Malpighian body to the outlet of the tubule, we shall find that it passes through the following divisions of the tubule: 1, the " first convoluted tubule;" 2, the " descending branch of Henle's tube;" 3, " Henle's loop;" 4, the " ascending branch of Henle's tube;" 5, the "second convoluted tubule;" 6, the "collecting tube." Of these, the two convoluted tubules are situated in the labyrinth; all the rest in the medullary rays and pyramid. All of the portions, with the exception of the convoluted tubules and the loop, are straight and lie parallel to each other (Fig. 131).

Before entering more particularly into the structure of the renal

FIG. 131.

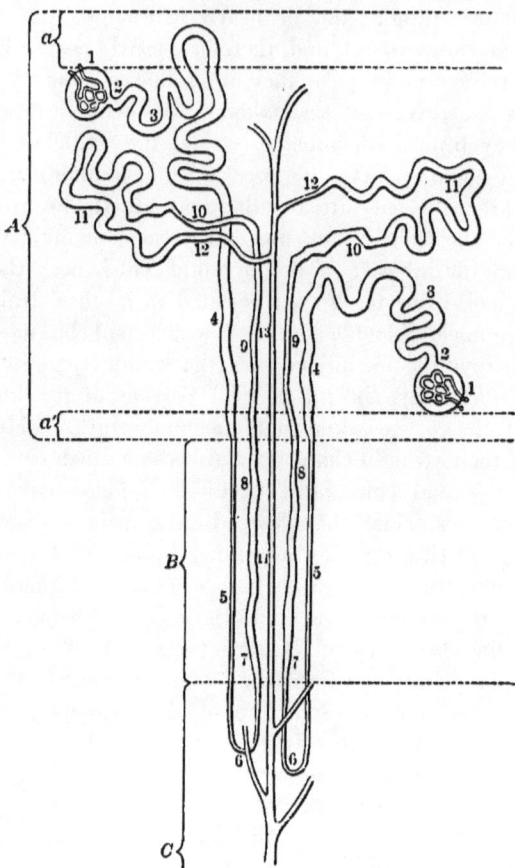

Diagram showing the course of the renal tubules within the kidney. (Klein.) *A*, cortex: *a*, subcapsular portion destitute of Malpighian bodies; *a'*, inner portion, also devoid of Malpighian bodies. *B*, boundary. *C*, portion of the medulla at the base of the pyramid. 1, Bowman's capsule surrounding the glomerulus; 2, neck of the capsule and beginning of the uriniferous tubule; 3, first convoluted tubule; 4, spiral portion of the first convoluted tubule in the medullary ray; 5, descending limb of Henle's tube; 6, Henle's loop; 7, 8, 9, ascending limb of Henle's tube; 10, irregular transition to the second convoluted tubule; 11, second convoluted tubule; 12, transition from second convoluted tubule to the collecting tubule; 13, 14, collecting tubule, joined below by others to form the excretory duct, which opens at the apex of the pyramid.

tubule, it will be best to complete this general sketch by considering the course of the bloodvessels.

As has already been said, the vessels enter the kidney between the calices and pyramids and are distributed in branches that lie parallel to the bases of the latter, and, therefore, to the convex surface of the organ, and are situated in the boundary-zone. The arterial branches in this location form the "arterial arcade." From this arcade perpendicular branches, the "interlobular arteries," pass toward the capsule, taking a straight course through the labyrinth between the medullary rays. In this course they give off branches, the "afferent vessels," which go to the Malpighian bodies.

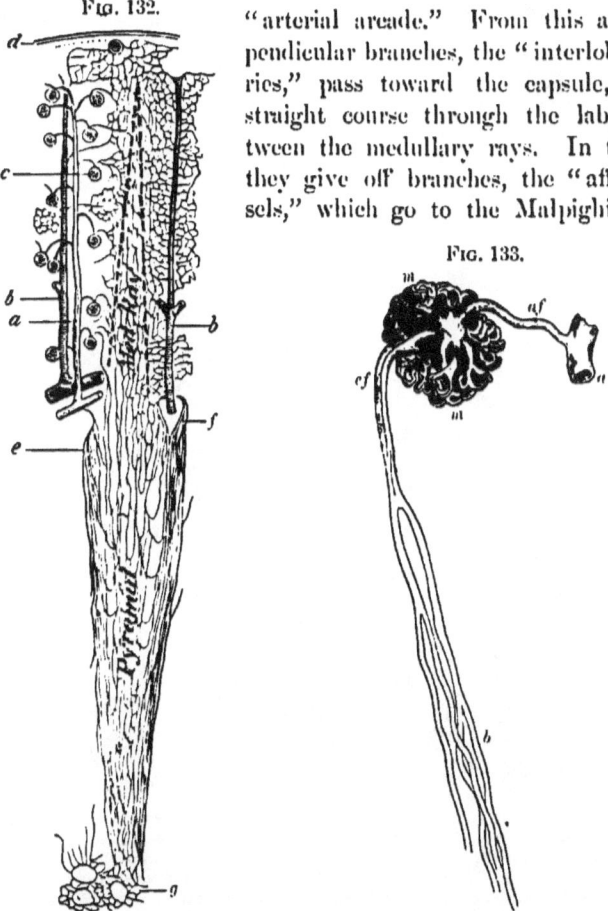

Fig. 132.

Fig. 133.

Fig. 132.—Diagram showing the course of the bloodvessels within the kidney. (Ludwig.) *a*, interlobular artery; *b*, interlobular vein; *c*, Malpighian body, with the afferent vessel entering it from the interlobular artery, and the efferent vessel leaving it to take part in the formation of the capillary plexus between the renal tubules; *d*, vena stellata; *e*, arteriæ rectæ; *f*, venæ rectæ; *g*, capillary plexus around the mouths of the excretory ducts.

Fig. 133.—Injected glomerulus from the horse. (Kölliker, after Bowman.) *a*, interlobular artery; *af*, afferent vessel; *m, m*, capillary loops forming the glomerulus; *ef*, efferent vessel; *b*, capillary network in the labyrinth and medullary rays.

The main artery becomes smaller in giving off these branches, and finally ends in terminal afferent vessels (Fig. 132).

Within the Malpighian body the afferent vessel divides abruptly into a number of capillary loops, which are compacted together to form a globular mass, called the "glomerulus" (Fig. 133). These loops rejoin to form the "efferent" vessel, which is somewhat smaller than the afferent vessel, and leaves the Malpighian body at a point close to that at which the afferent vessel enters it.

FIG. 134.

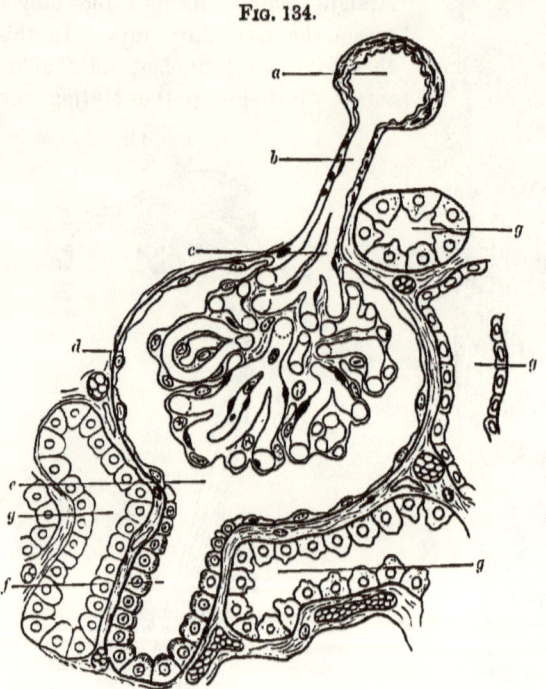

Sketch of a Malpighian body from kidney of a rabbit: *a*, interlobular artery; *b*, afferent vessel; *c*, capillary springing from afferent vessel; *d*, Bowman's capsule, with epithelial lining reflected upon the surface of the glomerulus; *e*, cavity of the capsule into which the watery constituents of the urine are first discharged; *f*, beginning of a uriniferous tubule; *g*, convoluted tubules of the labyrinth. Between these tubules and the capsule are capillary bloodvessels derived from the efferent vessel (which is not shown, but emerges from the capsule near the afferent vessel, on a different level from that represented). These and other structures are held in place by an areolar tissue, containing lymphatic spaces, some of which are represented.

Soon after leaving the Malpighian body the efferent vessel breaks up into a second set of capillaries, which lie among the convoluted tubules of the labyrinth and also penetrate into the medullary rays, to be distributed between the tubules composing them. This capillary network extends also into the pyramid, in which the capillaries run, for the most part, parallel to the renal tubules, with comparatively few transverse anastomosing branches. For this reason they have been called the "vasa recta." They also receive blood from little twigs given off from the arterial arcade.

The blood from the intertubular capillaries is collected in veins, which run a course parallel to that of the arteries and lie in close proximity to them. They have received names similar to those of the corresponding arteries: "interlobular veins," "venæ rectæ," and "venous arcade." Relatively large veins also leave the kidney from beneath the capsule on the convex surface of the organ. They are called the "stellate veins."

The Malpighian body is enclosed by a thin fibrous capsule (Bowman's capsule), which is perforated at two opposite points to permit the passage on the one hand of the afferent and efferent vessels, and on the other hand to allow of a communication between its cavity and the beginning of the uriniferous tubule. When distended with blood the glomerulus nearly fills this capsule, but when collapsed it is retracted toward the attachment formed by the vessels that pierce the capsule. It is covered by a single layer of epi-

Fig. 135.         Fig. 136.

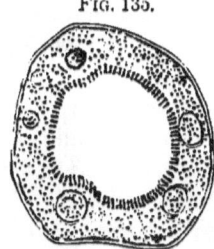
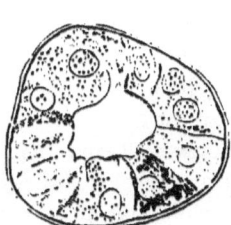

Cross-sections of convoluted tubules lined with cells in different states of activity. (Disse.)
Fig. 135.—From a criminal directly after execution. Cells in a state of rest. The cells are low and granular, and present a striation of their free ends resembling cilia.
Fig. 136.—From a cat. The cells are enlarged, because charged with material to be excreted, and the striated border is nearly obliterated. Similar appearances have been observed in the human kidney. In one of the lower cells in this figure a faint striation of the attached end is just discernible. This increases in distinctness as the cell becomes surcharged with excretory material, when the more central portion of the cytoplasm becomes hyaline and contains the nucleus.

thelial cells, which is reflected at that attachment and forms a lining for the inner surface of the capsule to the point where its cavity opens into the lumen of the renal tubule. Here the epithelial lining becomes continuous with that of the tubule (Fig. 134).

The different portions of the uriniferous tubule differ in their external diameters, the diameters of their lumina, and the character of their epithelial linings. The appearance of the epithelial cells differs, however, in accordance with their state of functional activity (Figs. 135 and 136).

The first convoluted tubule is relatively large, and is lined with large epithelial cells, which project into the tubule about one-third of its diameter. The cells have round nuclei situated near their centres, and are granular, with an appearance of radiate striation in their deeper halves when charged with secretion.

The descending branch of Henle's tube has a smaller diameter, but its lumen is wide in consequence of the thinness of the clear epithelial cells lining it. In the ascending branch the lumen is again smaller, although the diameter of the tube is larger, because the lining cells are thicker, somewhat resembling those of the first convoluted tubule. The transition from the character of the descending to that of the ascending branch does not always take place exactly at the loop.

The second convoluted tubule is a little smaller than the first, and is lined with cells that are not quite so granular and a little more highly refracting.

The collecting tubules are lined with columnar epithelium, the cells of which become longer as the diameter of the tube increases in its progress toward the apex of the pyramid.

The epithelial lining throughout the course of the renal tubule is said to rest upon a thin, homogeneous basement-membrane interposed between it and the interstitial fibrous tissue. The latter is present in small amount, and partakes of the character of an areolar tissue, holding the tubules and bloodvessels in place. It is rather abundantly supplied with lymphatics.

For the study of the uriniferous tubules sections made transverse to the course of the straight tubules will be found very useful. In the cortex the medullary rays, with their descending and ascending branches of Henle's tubes and their collecting tubules, will appear surrounded by the labyrinth, made up of the convoluted tubules, Malpighian bodies, and larger vessels, the latter in cross-section. Near the apex of the pyramid cross-sections of the larger collecting tubes and of the vasa recta will be seen; and near its base the smaller collecting tubes and the two limbs of Henle's tube, with, possibly, here and there a "loop" in nearly longitudinal section, will appear. Among all these sections of the tubules the interstitial tissue with its capillaries and lymphatics will complete the picture (Figs. 137 and 138).

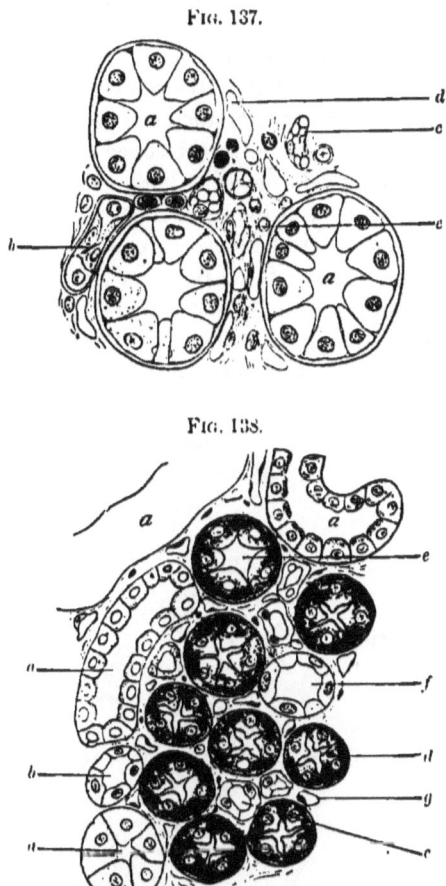

Fig. 137.

Fig. 138.

Sections from a rabbit's kidney, made perpendicular to the course of the straight tubules.

Fig. 137.—Through a portion of the pyramid: *a*, lower portions of the collecting tubules (excretory ducts); *b*, Henle's loop in tangential section; *c*, capillary bloodvessels; *d*, lymphatic; *e*, descending limb of Henle's tube.

Fig. 138.—Through part of a medullary ray and the adjoining labyrinth: *a, a, a, a*, convoluted tubules in the labyrinth; *b*, spiral tubule; *c*, descending limb of Henle's tube; *d*, ascending limb of Henle's tube; *e*, irregular tubule; *f*, collecting tubule; *g*, capillary bloodvessel.

The nerves of the kidney are small and apparently not abundant. Their larger branches follow the courses of the arteries.

162  NORMAL HISTOLOGY.

The external surface of the kidney is covered with a capsule of fibrous tissue, which on its deeper surface becomes continuous with the interstitial tissue, so that its vascular supply communicates with the capillaries in the superficial portions of the kidney.

The fibrous capsule of the kidney becomes continuous at the hilum of that organ with the fibrous coats of the calices and pelvis, and, through these, with those of the ureter and bladder.

The columnar epithelium lining the collecting tubes is continuous with a layer of similar cells covering the papillæ.

The watery constituent of the urine is secreted in the Malpighian body, where it passes from the blood through the capillary walls of the glomerulus into the cavity of Bowman's capsule. Under normal conditions it is free from albumin, and, therefore, is unlike the serum that passes through the walls of the capillaries in other parts of the body. It has been thought that this difference was attrib-

Fig. 139.

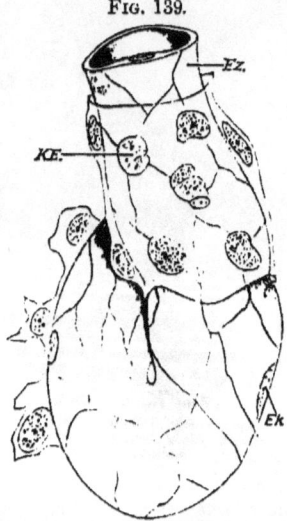

Capillary loop from the glomerulus of the frog. (Nussbaum.) *Ez*, endothelial wall of the capillary bloodvessel; *Ek*, nucleus of one of the endothelial cells (only three such nuclei are shown in the figure); *KE*, nucleus of one of the epithelial cells investing the capillary. The boundaries of these cells are not reproduced in the figure. At the left of the cut three epithelial cells have been partially reflected away from the capillary wall.

utable to the functional action of the endothelium in the glomerulus, though morphologically it is similar to that throughout the body. It is more probable that the epithelium covering the glomerulus has something to do with the prevention of a loss of albumin (Fig. 139). In disease of the kidney, alterations in the glomerulus and, perhaps, in other parts of the kidney permit albumin to pass into the secretion.

The epithelium lining the uriniferous tubules discharges its secretion into the lumen of the tubules, whence it is carried by the stream flowing from the Malpighian bodies. The epithelial cells lining the convoluted tubules and the ascending branches of Henle's tubes appear to be those most active in carrying on the eliminative function of the kidney.

2. **The pelvis** of the kidney and its calices are lined with transitional epithelium. It consists of only three or four layers of epithelial cells of different shapes. The most superficial layer is composed of rather large flattened cells, having ridges upon their lower surfaces, which fill the spaces between the tops of the next layer. This is made up of pear-shaped or caudate cells, the hemispherical tops of which fit into the cavities between the ridges on the layer above, while their slender processes penetrate between

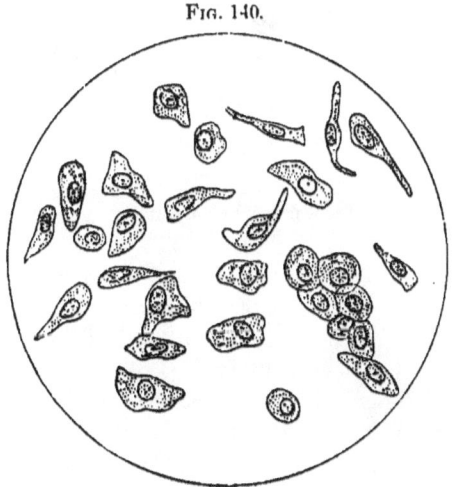

Fig. 140.

Epithelial cells from the pelvis of a human kidney. (Rieder.)

the oval or round cells that make up the deepest layers of the epithelial covering (Fig. 140).

Beneath the epithelium is a coat of fibrous tissue, denser near the epithelium and more areolar in its deeper portions. Here it is interlaced with smooth muscular fibres, outside of which is the external coat of fibrous tissue.

3. **The ureters** closely resemble in structure the pelvis of the kidney; but the muscular fibres have a somewhat more definite arrangement, being disposed in an inner imperfect coat of longitudinal and an external layer of circular fibres, outside of which a few supplementary longitudinal fibres are, here and there, added (Fig. 141).

4. **The bladder** also has a lining of transitional epithelium (Fig.

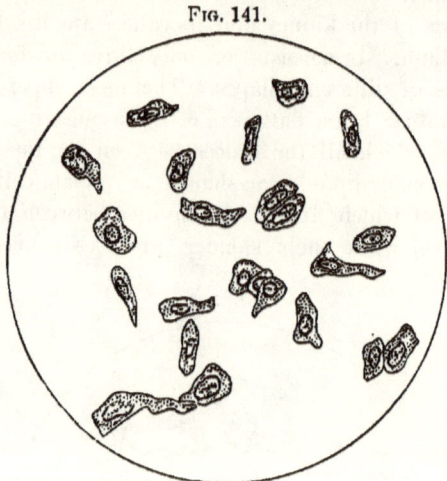

FIG. 141.

Epithelial cells from the human ureter. (Rieder.)

40), beneath which is a layer of fibrous tissue resembling that of the renal pelvis, but of greater thickness. The muscular coat, which comes next, is thick and composed of bundles of smooth muscular fibres, interlacing in various directions or disposed in more or less well-defined strata. External to the muscular coat is a fibrous coat, which is covered by a reflection of the peritoneum for a part of its extent, and in other situations passes into the surrounding areolar tissue.

The spear-shaped cells of the transitional epithelium of the bladder have thicker processes than those of the pelvis or ureter; but when detached and macerated in the urine it is often very difficult to determine from their appearance from what part of the urinary tract such cells were derived (Figs. 142 and 143).

THE URINARY ORGANS. 165

5. **The urethra** differs in structure in the two sexes. In the male the prostatic portion is lined with epithelium resembling that

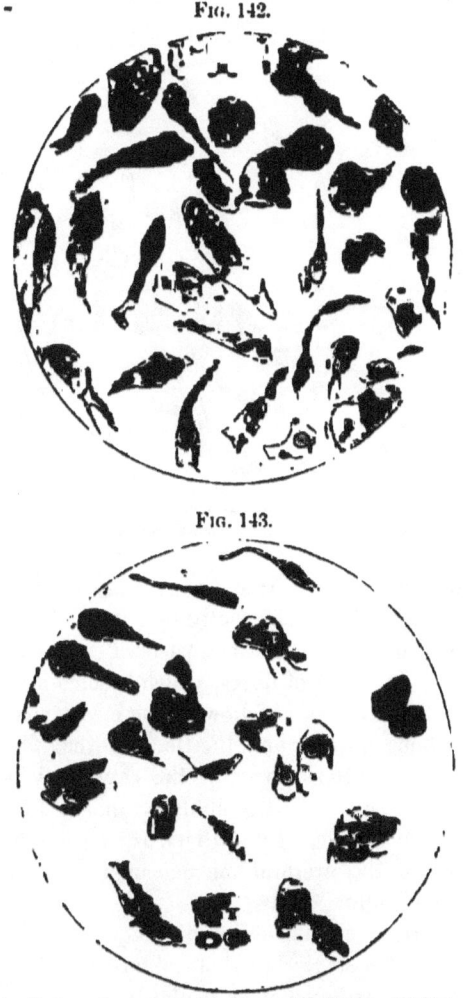

Fig. 142.

Fig. 143.

Epithelial cells from the mucous membrane of the human bladder. (Rieder.)
Fig. 142.—From the urinary sediment from a case of cystitis. The cells are somewhat swollen after maceration in the altered urine.
Fig. 143.—Removed from the internal surface of a normal bladder.

of the bladder. Further forward, it gradually passes into cylindrical epithelium, at first more than one layer thick; but in the cavernous portion of the urethra it consists of but a single layer. The stratified epithelium covering the glans extends for a short distance from the meatus into the urethra (Fig. 144). The epithe-

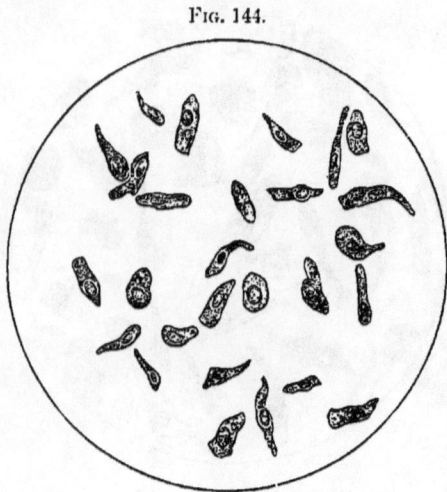

Fig. 144.

Epithelium from the human male urethra. (Rieder.)

lial lining rests upon fibrous tissue containing a number of elastic fibres, and this is bounded externally by a muscular coat. In the prostatic portion the muscular coat consists of an inner longitudinal and an outer circular layer of fibres, which become less well marked as the course of the urethra is followed, the circular coat disappearing in the bulbous portion and the longitudinal fibres becoming scattered toward the anterior part of the cavernous portion. The mucous membrane contains little tubular glands, "Littré's glands," some of which are simple, while others are compounded. In the collapsed condition the urethral mucous membrane is thrown into one or more longitudinal folds.

In the female the epithelial lining of the urethra is either stratified or composed of a single layer of columnar cells. The glands are more sparsely distributed than in the male, except for a group situated near the meatus. On the other hand, the muscular coat is thicker and consists throughout the course of the urethra of a well-defined internal longitudinal and external circular layer of fibres.

From the pelvis of the kidney to the stratified epithelium of the meatus the mucous membranes are capable of secreting mucus, which is much increased in amount under the influence of irritating substances, such as concentrated urine or the various causes of inflammation. The bloodvessels are most numerous and of largest size in the areolar tissue beneath the epithelium, and are accompanied by the lymphatics. The nerves are distributed chiefly to the muscular coats, but also extend into the fibrous tissue, up to and into the epithelium. The cells of the latter are connected by little protoplasmic bridges, as in the case of the epidermis, leaving minute channels between the cells for the passage of nutrient fluids.

## CHAPTER XIII.
### THE RESPIRATORY ORGANS.

THE respiratory tract consists of the larynx, trachea, bronchi, and lungs.

1. **The Larynx.**—The interior of the larynx is lined with ciliated columnar epithelium, which extends over the false vocal cords and about half-way up the epiglottis above, and is continuous below with a similar lining throughout the trachea and bronchi. This lining is interrupted over the true vocal cords by a covering of stratified epithelium, and at its upper limits passes into the stratified epithelium lining the buccal cavity and pharynx and covering the tongue. Opening upon this epithelial surface, except upon the true vocal cords and in the smallest bronchi, are mucous glands, varying in number in different situations. Some of the columnar cells upon the surface are also mucigenous, discharging their secretion upon the free surface of the mucous membrane.

The thyroid, cricoid, and most of the arytenoid cartilages are composed of the hyaline variety of that tissue: the epiglottis, cornicula laryngis, and the apices of the arytenoids, of elastic cartilage.

Beneath the epithelium lining the laryngeal ventricle is a considerable layer of lymphadenoid tissue. In other situations the epithelium rests upon fibrous tissue.

2. **The Trachea.**—The tracheal wall may be divided into four coats: *a*, the mucous membrane; *b*, the submucous coat; *c*, the cartilage; *d*, the fibrous coat (Fig. 145).

*a*. The mucous membrane is covered with ciliated columnar epithelium resting upon a nearly homogeneous basement-membrane, beneath which is a layer of fibrous tissue. This may be divided into two portions: an outer one, next to the basement-membrane, which is areolar in character, with a large admixture of elastic fibres and lymphadenoid tissue, and an abundant supply of bloodvessels; and an inner one, less highly vascularized, and composed chiefly of elastic fibres running a longitudinal course.

*b.* The submucous coat is of areolar fibrous tissue, supporting the mucous glands that open into the trachea, and the bloodvessels, lymphatics, and nerves, and also little masses of adipose tissue. In the neighborhood of the cartilages this fibrous tissue becomes condensed to form the perichondrium.

*c.* The cartilages are composed of the hyaline variety of that

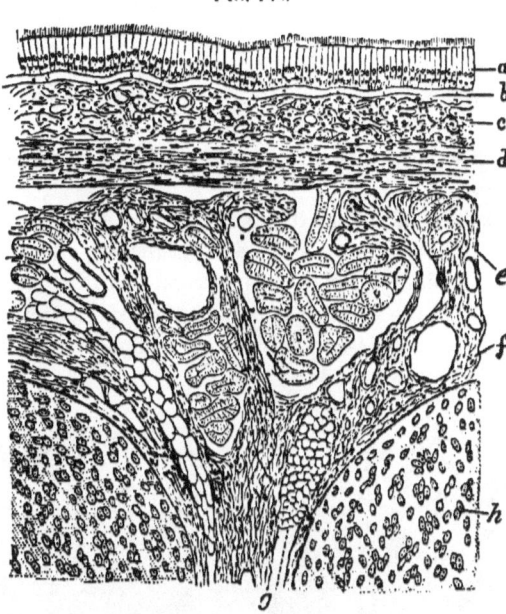

Fig. 145.

From a longitudinal section through the trachea of a child. (Klein.) *a*, the stratified columnar ciliated epithelium of the internal free surface; *b*, the basement-membrane; *c*, the mucosa (tunica propria); *d*, the network of longitudinal elastic fibres (the oval nuclei between them indicate connective-tissue corpuscles); *e*, the submucous tissue, containing mucous glands; *f*, large bloodvessels; *g*, fat-cells; *h*, hyaline cartilage of the tracheal rings. (Only a part of the tracheal wall is given in the figure.)

tissue, and are incomplete rings, interrupted behind, where the two ends are united by a band of smooth muscular tissue.

*d.* The fibrous coat is of areolar tissue beyond the bounds of the perichondrium, and serves to connect the trachea with its surroundings.

3. **The Bronchi.**—The main bronchi branching from the trachea have a structure similar to that organ, but the cartilaginous rings become more delicate as the tubes diminish in size.

The smaller bronchi differ in structure from the trachea in possessing a muscularis mucosæ, with its fibres disposed in a circular direction, and having irregular cartilaginous plates in their walls, instead of C-shaped, imperfect rings. The four coats may be enumerated as follows:

*a.* Mucous membrane, covered with ciliated columnar epithelium resting upon a basement-membrane, beneath which is a fibrous tissue containing numerous elastic fibres lying parallel to the axis of the bronchus. Under this are the circular fibres of the muscularis mucosæ.

*b.* Submucous coat, similar to that of the trachea and larger bronchi.

*c.* Cartilaginous coat, containing the plates of cartilage that support the walls.

*d.* Fibrous coat of areolar tissue, containing a little adipose tissue and passing into the areolar tissue of neighboring structures.

As the bronchi subdivide and become smaller the coats get thinner, and first the cartilaginous and then the muscular coat disappears. Those air-passages which are without cartilage, but have

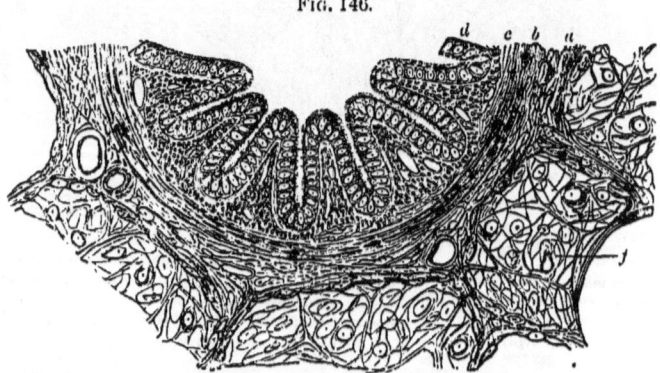

Fig. 146.

Portion of a cross-section of a bronchiole from the lung of a pig. (Schultze.) *a*, areolar external coat; *b*, muscularis mucosæ; *c*, subepithelial areolar tissue, containing numerous longitudinal elastic fibres, represented here in cross-section; *d*, ciliated epithelium, forming the most superficial layer of the mucous membrane; *f*, walls of the neighboring pulmonary alveoli. In these walls branching and anastomosing elastic fibres are shown; the capillary plexus has been omitted.

a muscularis mucosæ, are called "bronchioles" (Fig. 146). The still smaller branches, which have lost their muscular tissue, are known as the "alveolar passages." In the latter the columnar epithelium lining the bronchi gives place to a pavement-epithelium, composed of small flattened cells disposed in a single layer. The elastic tissue of the mucous membrane is continued through all the divisions of the air-passages, and becomes a constituent part of the alveolar walls of the lung itself.

The alveolar passages open into spaces, called the "infundibula," in the sides of which are the openings into the alveoli of the lung, the ultimate destination of the inspired air. Here and there

Fig. 147.

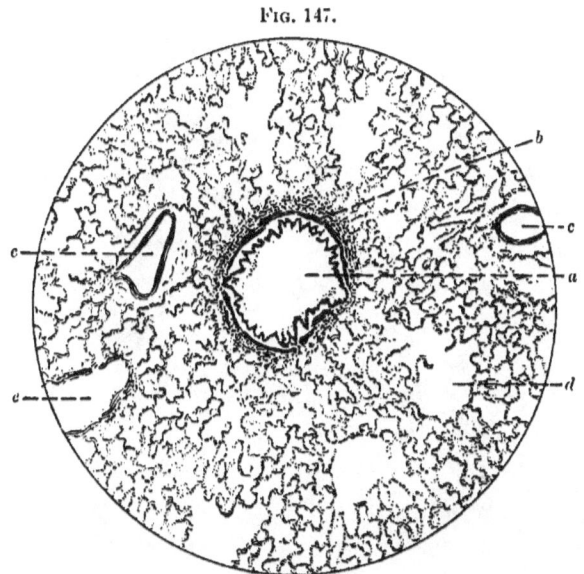

Section of lung of the dog, showing a transverse section of a bronchiole; a, bronchiole (a little mucus covers the epithelial lining); b, muscular layer of the mucous membrane; c, c, radicles of the pulmonary vein; d, alveolar passage, just at its division to form infundibula. An infundibulum extends from this passage toward the bronchiole. The wall of the alveolar passage at this point is similar in structure to that of the pulmonary alveoli. e, alveolar passage in oblique section. This passage is cut at a point further from its opening into the infundibula, and has a somewhat thicker wall than d. The rest of the section is made up of infundibula (the larger spaces) and pulmonary alveoli.

stray alveoli open directly into the alveolar passages (Figs. 147, 148, and 149).

4. **The pulmonary alveoli** and the smaller air-passages are so arranged that there are no vacant spaces; and neighboring alveoli, whether they belong to a group of infundibula springing from the same alveolar passages or to separate groups, are so closely situated that they have but one common wall dividing their cavities

172     NORMAL HISTOLOGY.

from each other. Notwithstanding this general compactness of arrangement, the lungs are divided by delicate septa of fibrous tissue into more or less well-defined lobules, corresponding to the smallest bronchi or the bronchioles.

The alveolar walls are made up of a delicate, loose areolar tissue, containing numerous elastic fibres and supporting the abundant capillary plexus in which the blood suffers the gaseous exchanges with the air that constitute the function of respiration (Fig. 150).

FIG. 148.

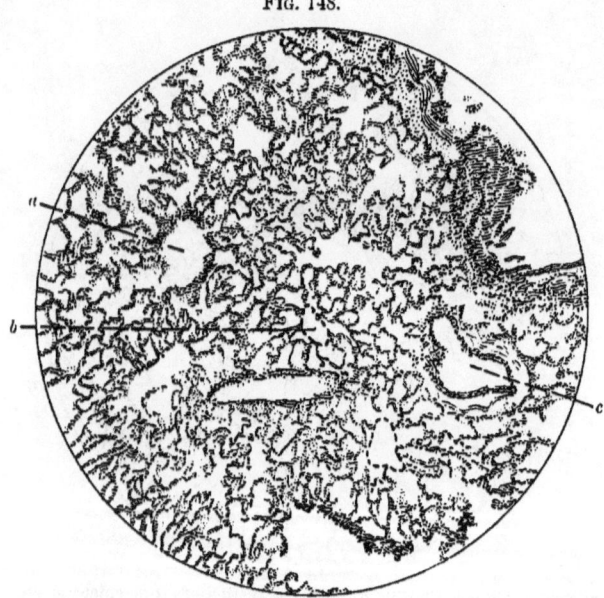

Section of lung of the dog: a, alveolar passage opening into an infundibulum and also into a solitary alveolus; b, cross-section of an infundibulum. The dotted line indicates the limits of the infundibular space. Opening into it are a number of alveoli. Were the dotted line removed, the infundibular cross-section and the alveoli around it would form a stellate space in the section. c, junction of two radicles of the pulmonary vein. At the top of the section, to the right, is an oblique section of a bronchiole.

Covering the two surfaces of the alveolar wall is a layer of very thin cellular plates (pavement-epithelium, see Fig. 30), among which are scattered a few cells resembling those lining the alveolar passages. This cellular investment is continuous with the lining of the infundibulum, which is of similar character, and thence with the epithelium covering the inner surface of the alveolar passage. It is to be regarded as a special modification of epithelium, fitting it for usefulness in this situation.

The lung receives blood from two sources: 1, venous blood, through the pulmonary artery, which is oxygenated in the walls of the alveoli; 2, arterial blood, through the bronchial arteries. This arterial blood serves for the nourishment of the tissues of the lung, and is distributed to the bronchi, interlobular connective tissue, lymph-glands, and walls of the vessels. Part of this blood returns through the pulmonary veins; the rest through the bronchial veins.

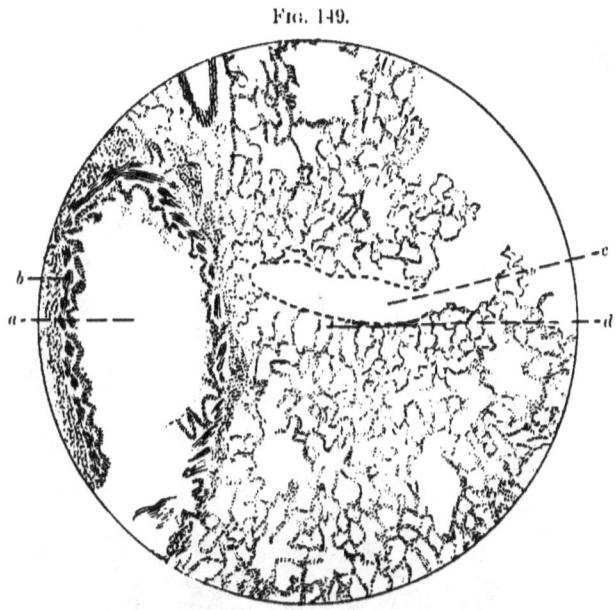

Fig. 149.

Section of lung of the dog: *a*, oblique section of a bronchiole; *b*, its muscular coat; *c*, longitudinal section of an infundibulum, communicating to the right with an alveolar passage (the wall of the latter is torn further to the right); *d*, one of the alveoli opening into *c*.

The lymphatics arise in the walls of the alveoli and bronchi and pass to the bronchial lymph-glands.

The nerves supplying the lung may be traced along the bronchi, where they occasionally connect with groups of ganglion-cells, and along the vessels. They are of both the medullated and the nonmedullated varieties.

The surface of the lung is covered with serous membrane, a portion of the pleura.

Little need be said about the functional activity of the lung. The cilia, belonging to the columnar epithelium lining nearly the

whole of the air-passages, possess a motion that urges particles
lodging in the mucus covering them toward the larynx, whence
they are either coughed out or are swallowed. Such solid particles
as pass beyond the regions guarded by ciliated epithelium are taken
up by leucocytes, which frequently migrate into the alveoli and the
air-passages, and are conveyed by them into the lymphatic vessels
or glands. Because of this the lymphatics and bronchial lymphatic
nodes are apt to be blackened by the deposition of carbon, except
in young individuals. The flow of air into the lung is the result of
atmospheric pressure, which tends to fill the thoracic cavity when the

Fig. 150.

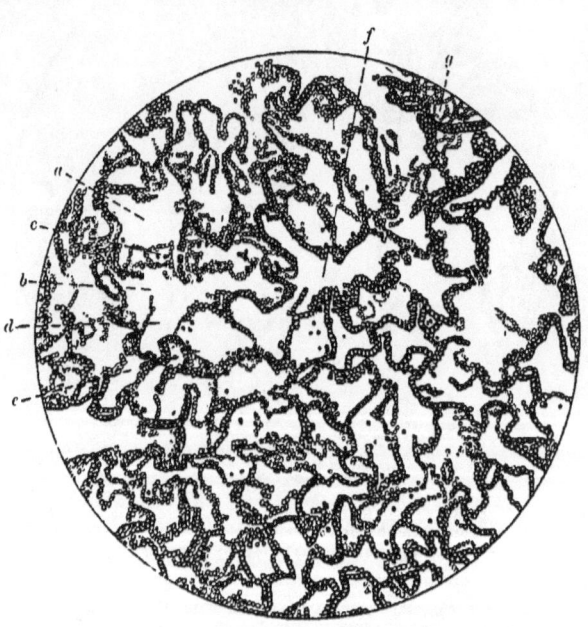

Section of the lung of a dog, killed by ether-narcosis. The lung was hyperæmic at the time
of death, and the capillaries retain their blood in the section. *a*, alveolus in cross-section, communicating with the infundibulum. *b*. A portion of the wall of the alveolus is
seen, in surface-view, at *c*. *d*, *e*, other alveoli opening into the same infundibulum; *f*,
cross-section of an infundibulum with alveoli opening into it; *g*, surface-aspect of an
alveolar wall, showing capillary plexus filled with red blood-corpuscles.

chest is expanded through the action of the muscles of respiration.
The air is expelled from the lungs when those muscles relax, partly
because of the pressure exerted by the thoracic walls, but chiefly
because of the contraction of the elastic fibres in the alveolar walls.

Because of their presence the lungs retract when the chest is opened.

When sections of the lung are examined under the microscope it is difficult, at first, to identify the different portions, which are cut in all directions. The smaller bronchi may be recognized by the presence of cartilage in their walls. The bronchioles possess no cartilage, but are surrounded by a band of smooth muscular tissue, the muscularis mucosae. This becomes thinner, then incomplete, and finally disappears as the infundibula are reached. The infundibulum, it will be remembered, is the space into which the alveoli open. When seen in section it will appear as a round, oval, or elongated space, according to the direction in which it has been cut, bounded by scallops, each of which is the cavity of an alveolus. In every section there will be many alveoli which have been so cut that their openings into the infundibulum will not be included in the section. These alveoli have a continuous wall surrounding their cavities. Still other alveoli will have been cut in such a way that a portion of their walls will lie in the plane of the section and parallel to it, so that the flat surface of the alveolar wall will be visible, surrounded by an oblique or cross-section, where the wall meets the surface of the section. Those alveolar walls which have been cut perpendicular to their surfaces will appear thinner than those which have been cut obliquely. With these considerations in his mind, the student can have little difficulty in identifying the different portions of the section (see Figs. 147–150).

## CHAPTER XIV.

### THE SPLEEN.

NEARLY the whole surface of the spleen is invested with a covering of peritoneum similar to that which partially covers the liver. Beneath this is the true capsule of the spleen, which completely surrounds it. This capsule is composed of dense fibrous tissue, containing a large number of elastic fibres and a few of smooth muscular tissue. From its inner surface bands of the same tissue, called the "trabeculæ," penetrate into the substance of the organ, where they branch, and the branches join each other to form a coarse meshwork occupied by the parenchyma of the organ, the "pulp."

The bloodvessels of the spleen enter at the hilum and pass into the large trabeculæ, which start from the capsule at that point and enclose the vessels until they divide into small branches. The vessels then leave the trabeculæ and penetrate the pulp, where they break up into capillaries, which do not anastomose with each other. There is some doubt as to the way in which these capillaries end. According to one view, they unite to form the venous radicles, so that the blood is confined within vessels throughout its course in the spleen. Another view, which is more probably correct, is that the walls of the capillaries become incomplete, clefts appearing between their endothelial cells, which finally change their form and become similar to those of the reticulum of the pulp. The veins, according to this view, arise in a manner similar to the endings of the arteries. The result of this structure would be that the blood is discharged, from the capillary terminations of the arteries, directly into the meshes of the pulp, after which it is taken up by the capillary origins of the veins (Figs. 151 and 152).

The pulp consists of a fine reticulum of delicate fibres and cells, with branching and communicating processes, in the meshes of which there are red blood-corpuscles, leucocytes in greater number than normally present in the blood, and free amœboid cells considerably larger than leucocytes, called the "splenic cells."

## THE SPLEEN.

The adventitia of the arteries contains considerable lymphadenoid tissue, which after the exit of the vessels from the trabeculæ is

Fig. 151.

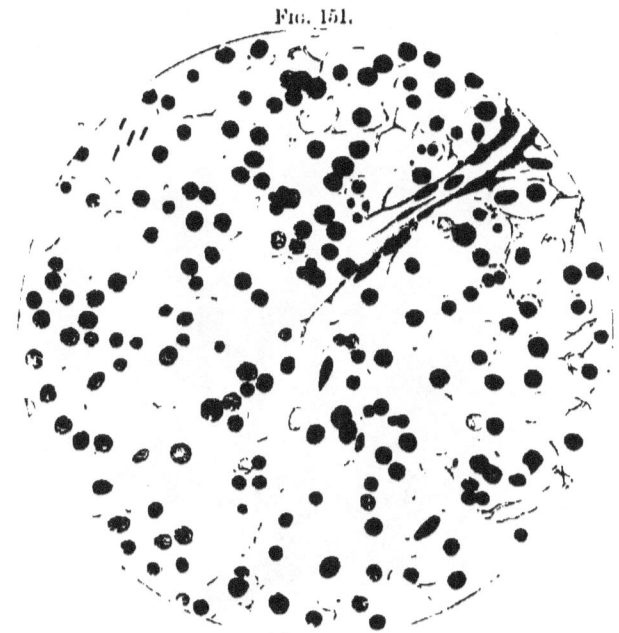

Section from the spleen of the cat. (Baunwarth.) Termination of an arterial capillary in the pulp.

expanded at intervals to form spherical bodies, about 1 mm. in diameter, called the "Malpighian bodies" or "corpuscles." These are

Fig. 152.

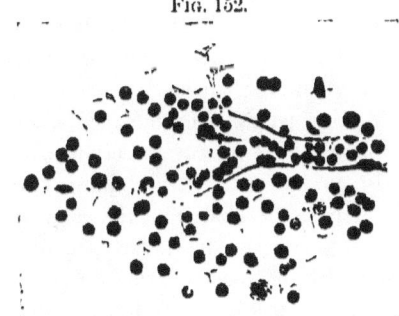

Section from the spleen of the cat. (Baunwarth.) Beginning of a capillary venous radicle.

like little lymph-follicles, through which the artery takes its course. The reticulum in these Malpighian corpuscles is scanty and inconspicuous near their centres, so that the lymphoid cells it contains appear densely crowded; but toward their peripheries the reticulum is more pronounced and the cells a trifle more separated. At the surface of the Malpighian body its reticulum becomes continuous with that of the pulp surrounding it (Fig. 153).

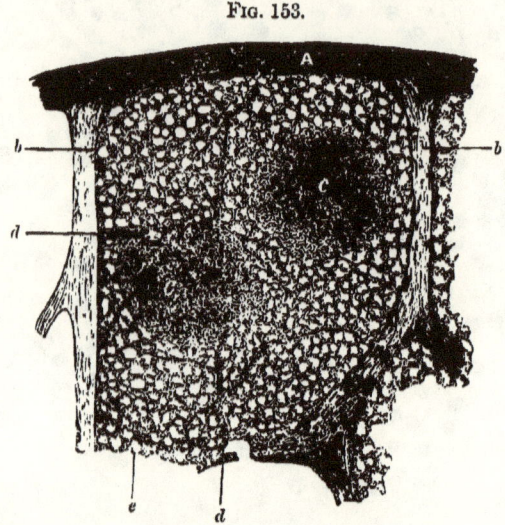

FIG. 153.

Section from human spleen. (Kölliker.) A, capsule; *b, b*, trabeculæ; *c, c*, Malpighian bodies (lymph-follicles), traversed by arterial twigs. In the follicle to the left, part of the arterial twig is seen in longitudinal section; in that to the right, it appears in cross-section to the right of the centre of the follicle. *d*, arterial branches; *e*, splenic pulp. The section is taken from an injected spleen.

The relations between the spleen and the blood flowing through it appear to be very similar to those between the lymphatic glands and the lymph passing through them. It seems to act as a species of filter, in which foreign particles or damaged red blood-corpuscles are arrested and destroyed. In many infectious diseases the splenic pulp is increased in amount and highly charged with granules of pigment that appear to be derived from the coloring-matter of the blood. This is notably the case in malaria, in which the red corpuscles are destroyed by the plasmodium occasioning the disease. When bacteria gain access to the blood they are apt to be especially abundant in the splenic pulp, and it is said that monkeys, which are normally immune against relapsing fever, may acquire the disease if the spleen be removed before inoculation with the spirillum which is the cause of that disease. These observations all tend to confirm the view that the function of the spleen is to assist in maintaining the functional integrity of the blood. The lymphadenoid tissue within the spleen also enriches the blood with an additional number of leucocytes.

# CHAPTER XV.

## THE DUCTLESS GLANDS.

THE organs included in this group possess, at some stage of their development or in the adult, a structure analogous to that of the secreting glands. Those which retain this structure after complete development differ from the other glandular organs in being devoid of ducts, through which the materials elaborated by their parenchyma could be discharged. Of these organs the thyroid is the most striking example. Other members of this group, notably the thymus, become greatly modified as development advances, and after a

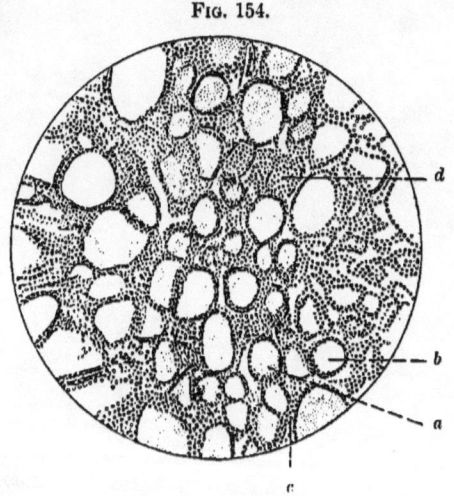

FIG. 154.

Section of human thyroid gland: *a*, alveolus filled with colloid; *b*, alveolus containing a serous fluid; *c*, interalveolar areolar tissue; *d*, tangential section of an alveolus, giving a superficial view of the epithelial cells.

while retain mere vestiges of their original epithelial character; the chief bulk of the organ being composed of lymphadenoid tissue.

The following organs and structures will be considered as belonging to the general group of ductless glands: the thyroid gland, the parathyroids, the adrenal bodies, the pituitary body, the thymus, and the carotid and coccygeal bodies.

1. **The Thyroid Gland** (Fig. 154).—This consists of a number of alveoli or closed vesicles, lined with cubical epithelial cells arranged in a single layer upon the delicate, vascularized areolar tissue which forms their walls and separates the neighboring alveoli from each other. This fibrous tissue is more abundant in places, where it serves to divide the gland into a number of imperfectly defined lobes. At the periphery of the organ its connective tissue becomes continuous with a thin but moderately dense fibrous capsule.

The individual alveoli differ both in respect to their size and their contents. Many are more or less completely filled with a nearly homogeneous, glairy substance, of a slight yellowish tint, called "colloid," while others appear to be occupied by a serous fluid.

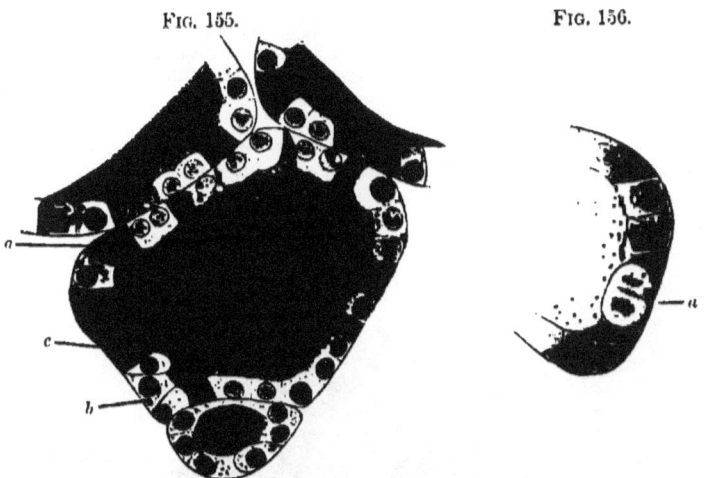

Fig. 155.  Fig. 156.

Sections of thyroid gland. (Schmid.)

Fig. 155.—From a dog: *a*, colloid or secreting cells; *b*, reserve cells (these differ only in their states of activity); *c*, cells containing less colloid than *a*.

Fig. 156.—From a cat: *a*, daughter-cells arising from the division of an epithelial cell.

The elaboration of this colloid material seems to be the function of the organ, though it may have other less obvious duties.

The cells lining the alveoli may be divided into two classes, which differ in appearance (Fig. 155): first, those engaged in the production of colloid, secreting cells; and, second, those in which no colloid is present, and which are regarded as reserve cells. The latter are capable of multiplication, thereby replacing such of the secreting cells as may be destroyed (Fig. 156). The colloid material is produced within the cytoplasm of the secreting cells,

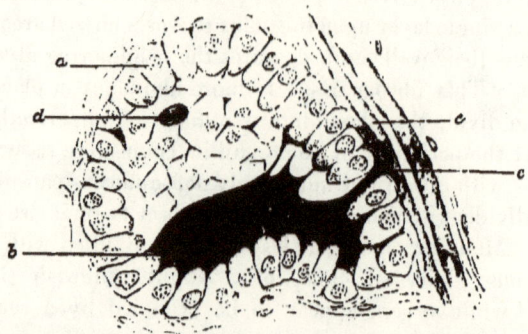

FIG. 157.

Section from thyroid of dog, illustrating the egress of colloid from the alveoli. (Bozzi.) *a*, epithelial cells lining the alveolus, seen in section. The internal ends of similar cells are seen in superficial aspect below. *b*, colloid within the alveolus; *c*, exit of colloid between two epithelial cells; *e*, lymphatic vessel; *d*, end of a colloid or secreting cell in the epithelial lining of the alveolus.

whence it is either expelled into the lumen of the alveolus, or the whole cell becomes detached from the alveolar wall and suffers col-

FIG. 158.

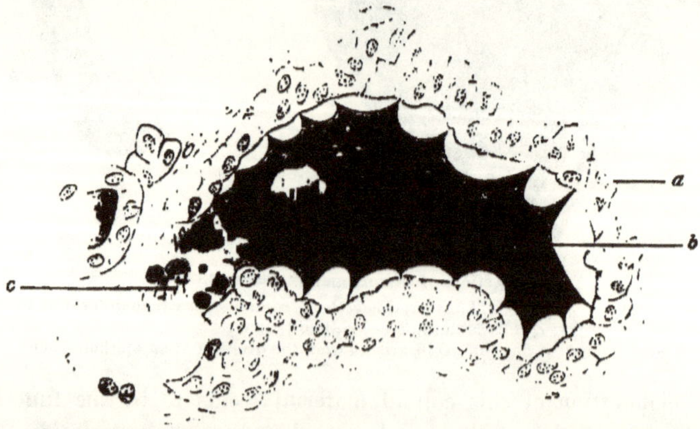

Section from thyroid of dog, illustrating the egress of colloid from the alveoli. (Bozzi.) *a* epithelial lining of the alveolus; *b*, colloid; *c*, escape of colloid through a defect in the wall occasioned by the colloid metamorphosis of some of the epithelial cells, the nuclei of which are discernible within the colloid near *c*.

loid degeneration, with destruction of the nucleus, within the alveolar cavity.

The colloid material subsequently finds its way into the general circulation, either by passing between the intact cells of the alveolus (Fig. 157), or after a passage has been prepared for it through alterations in certain of those cells (Fig. 158). The colloid is then taken up by the lymphatics, through which it reaches the general circulation. This is an example of internal secretion which presents much of interest. It is probable that a similar, but much less obvious, process takes place in some of the ordinary secreting glands of the body, certain elaborated materials being returned to the circulation by the cells of the gland, while others are utilized for their nourishment and for the elaboration of the more obvious secretion.

That the secretion of the thyroid gland is of importance to the general organism is shown by the effects of disease or removal of the gland upon the general nutrition. Total extirpation of the thyroid, together with the parathyroids, occasions the death of an animal within a few days, after symptoms of grave disturbances in the central nervous system, among which are tetanic convulsions. A partial removal of the gland, or its removal without that of the parathyroids, causes profound disturbances of nutrition, grouped under the title "cachexia strumipriva." The animal becomes weak, drowsy, and emaciated; the skin dry and scaly, with a loosening of the hairs. In young animals the growth is retarded, especially the development of the bones, through degenerative changes in the epiphysial cartilages. In these, the intercellular substance becomes swollen and disintegrated; the cells atrophied or destroyed. Marked changes, designated as myxœdema, also appear in the subcutaneous tissue, which is converted into a species of mucoid tissue, probably as the result of an altered metabolism within the pre-existent cells of the tissue. The functional activity of the kidney is modified; after a while, albuminuria results. Exactly similar disturbances have been observed in people suffering from disease of the thyroid gland.

The foregoing facts are cited here in order to emphasize by a striking example the statement previously made, that the organs of the body are mutually dependent upon each other.

Experimentation and clinical study have further shown that the symptoms of myxœdema may be moderated or perhaps entirely arrested by feeding with thyroid extracts, or still more markedly by injecting extracts from thyroid glands beneath the skin, where they would speedily pass into the lymphatics and thence into the general circulation.

Chemical examination has revealed the presence of a substance called "thyroiodin" in the alveoli of the thyroid gland. This is a proteid containing a large amount of iodine. Its production by the thyroid gland may be increased by feeding with substances containing considerable iodine or by administering iodide of potassium. Injections of thyroiodin serve to mitigate the effects of thyroidectomy, very much as do injections of thyroid extracts. It is by no means clear, however, that the thyroiodin is the only substance elaborated by the thyroid gland which may be of use to the tissues

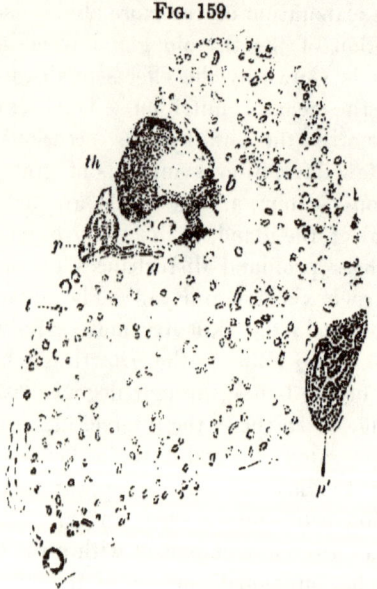

FIG. 159.

Section of the thyroid gland of a kitten two months old. (Kohn.) Showing the positions of the outer and inner parathyroid bodies and a thymus follicle: *t*, thyroid gland; *p*, inner parathyroid; *p'*, outer parathyroid; *th*, thymus follicle; *a*, portion of the section showing the intimate relations between the thyroid and the inner parathyroid; *b*, portion demonstrating a similar intimate relation between the thyroid and the tissues of the thymus follicle.

of other organs, or that the thyroid may not also remove injurious substances from the circulation and thus indirectly benefit other structures in the body.[1]

[1] Attention is also called to the possibility that an excessive or morbid thyroid secretion may cause symptoms of disease attributable to disturbances in the functions of other organs, and may also occasion disturbances in nutrition.

The bloodvessels of the thyroid are abundant, and form a rich plexus in the areolar tissue between the alveoli. The lymphatics are also abundant and large, forming a network of rather large vessels in the same situation. The nerves accompany the vessels, are destitute of ganglia, and have been traced to the bases of the epithelial cells, whence they may occasionally send minute terminal twigs with enlarged ends between the epithelial cells.

2. **The Parathyroids** (Figs. 159, 160, 161).—These are two bodies

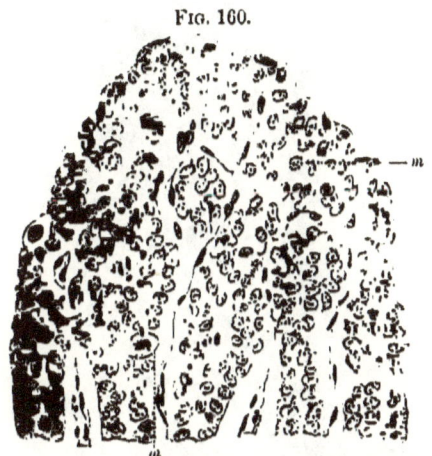

Fig. 160.

Section of a portion of the external parathyroid of a kitten two months old. (Kohn.) Showing the columns of epithelial cells separated by a delicate, vascular areolar tissue. The nuclei between the columns of epithelium belong chiefly to capillary bloodvessels. *m*, *m*, nuclei exhibiting karyokinetic figures.

of identical structure, which are developed in conjunction with the thyroid gland; but, while the latter progresses in its development until it attains the structure already described, the parathyroids retain a structure similar to that of the embryonic thyroid. They are composed of solid columns of epithelial cells, which anastomose with each other, but are elsewhere separated by a small amount of vascular areolar tissue. They are enclosed in a very thin capsule of areolar tissue, but are in very close relation to the neighboring tissues of the thyroid gland (Figs. 159 and 161), and frequently also with isolated follicles of thymus tissue.

Different observers vary in their opinions respecting the parathyroids. Some regard them as reserve thyroid tissue, remaining dormant while the thyroid is functionally competent, but developing into thyroid tissue when the gland furnishes an insufficient supply of secretion. Other observers deny this and regard the parathyroids as embryonic rudiments, nearly, if not quite, devoid of function. It is certain that in some cases of thyroidectomy the parathyroids become enlarged, and that the cachexia strumipriva is not certain to develop after the removal of the thyroid gland unless the parathy-

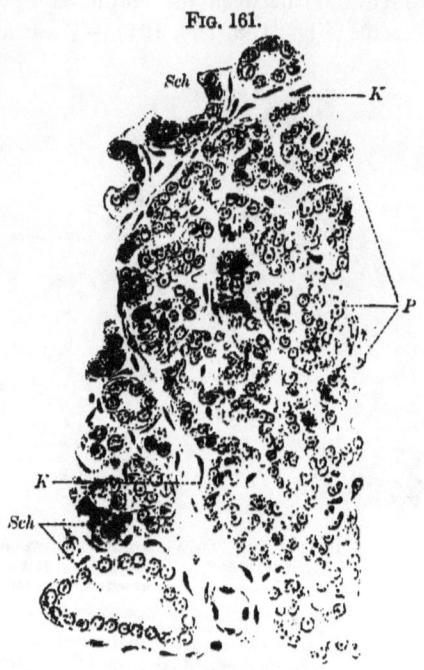

FIG. 161.

Section of the inner parathyroid of a kitten two months old. (Kohn.) Showing its close connection with the tissues of the thyroid gland: *Sch*, alveoli of the thyroid; *P*, epithelial columns of the parathyroid; *K*, capsule separating the two.

roids are also removed. Histological studies of the parathyroids in such cases have, however, failed to reveal a tendency on their part to develop into true thyroid tissue. Their relations to the thyroid, therefore, still remain undetermined.

In some animals—*e. g.*, the cat—there are four parathyroid bodies, two associated with each lobe of the thyroid.

3. **The Adrenal Bodies** (Fig. 162).—The adrenal bodies, or suprarenal capsules, possess a fibrous capsule, which is more areolar externally, where it frequently merges into the perinephric fat,

and denser internally, where it is reinforced in some animals by smooth muscular fibres. From this capsule septa of areolar tissue penetrate into the substance of the organ and constitute its interstitial tissue. The parenchyma of the organ consists of columns of epithelial cells, which are differently arranged and have a somewhat different appearance in different parts. As the result of these differences the organ has been divided into a cortical and a medullary portion.

In the cortical portion the cells are arranged in solid columns having their long axes perpendicular to the surface of the organ. Toward the capsule these columns lose their parallel arrangement and appear in vertical sections as islets of cells surrounded by areolar tissue, the "zona glomerulosa." In the deep portion of the cortex the cellular columns form a meshwork and completely lose their fascicular arrangement. This region is called the "zona reticularis." The epithelial cells in the cortical portion are polyhedral, and are frequently infiltrated with numerous globules of oil or fat, which give that part of the organ a yellow color.

Fig. 162.

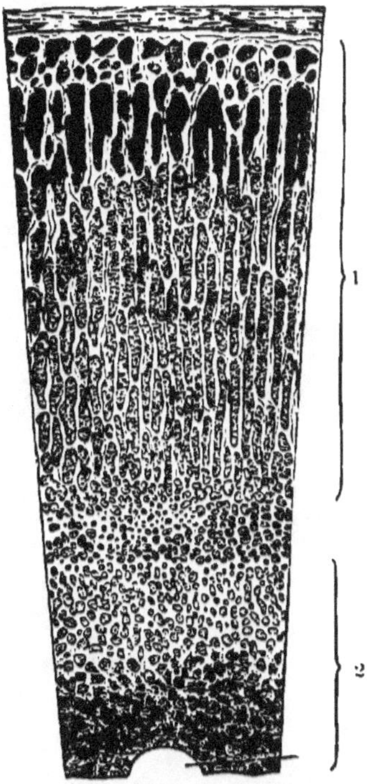

Vertical section of human adrenal body. (Eberth.) 1, cortex; 2, medulla; *a*, capsule; *b*, zona glomerulosa; *c*, zona fasciculata; *d*, zona reticularis; *e*, groups of medullary cells; *f*, partial section of a large vein.

In the medulla the interstitial tissue of the organ encloses groups of epithelial cells, which differ from those of the cortex in being free from fat. They are also larger than those cortical cells which contain no fat (Fig. 163).

The arteries of the adrenal bodies enter as numerous small twigs at the surface of the organ and divide into capillaries within its fibrous septa. These open into a venous plexus in the medulla, which communicates with a single vein leaving the organ.

The nervous supply of the adrenal bodies is very abundant. The

Fig. 163.

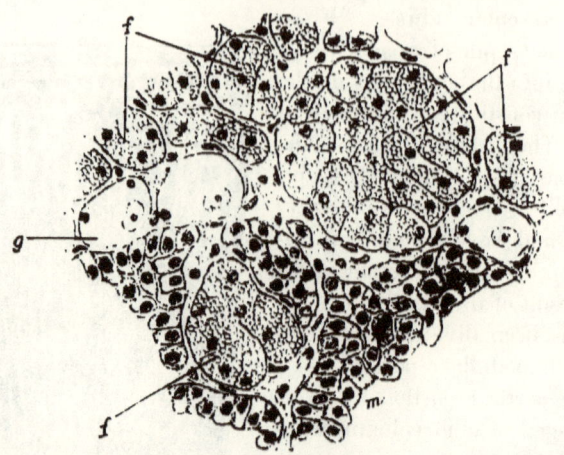

Section through the boundary between cortex and medulla in the adrenal body of the horse. (Dostoiewsky.) *f,f,f,* cells of the cortex, infiltrated with fat-globules; *g*, ganglion-cells; *m*, epithelial cells of the medulla.

nerve-fibres are chiefly of the medullated variety, and their bundles contain numerous ganglia before entering the organ. Here the fibres ramify abundantly in the cortex, whence they penetrate into

Fig. 164.

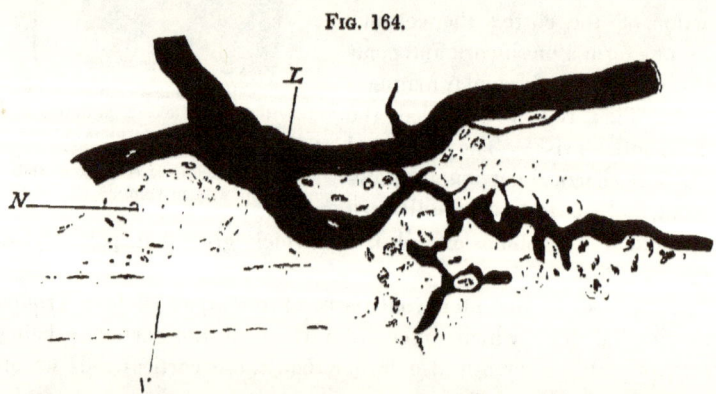

Injected lymphatics in an adrenal body of the ox. (Stilling.) *L*, injection-mass within the lymphatic vessels; *N*, cross-section of a nerve; *V*, longitudinal section of a vein. Lymphatic radicles are seen among the epithelial cells (cortical variety free from fat) to the right of the figure.

the medulla. At the junction of the medulla and cortex the nerve-fibres are connected with ganglion-cells. The nerve-terminations are distributed to the walls of the vessels and penetrate between the epithelial cells of the parenchyma.

As in the case of the thyroid gland, the relations of the epithelial cells of the adrenal bodies to the lymphatics appear of special interest. The lymphatic vessels are abundant and large, and accompany the bloodvessels lying in the areolar tissue of the septa. Here they come into close relations with the columns of epithelial cells, and, at least in the cortex, send minute terminal branches into those columns, where they end among the epithelial cells (Fig. 164). This arrangement of the lymphatics appears to point to the elaboration of an internal secretion as the function of the adrenal bodies. Small masses of lymphadenoid tissue are occasionally observed in the cortical portion of the adrenal body.

4. **The Pituitary Body.**—The pituitary body (hypophysis cerebri) is divisible into two portions, which differ both in their structure and in their embryonic origins. The posterior, or nervous, lobe is derived from a prolongation of the third cerebral ventricle. The anterior, or glandular, lobe develops from a tubular prolongation, lined with epithelial cells, from the buccal cavity of the embryo. This partially or completely invests the nervous portion of the body, but its chief bulk is situated in front. The connection with the buccal cavity is obliterated, and, in the further development of the detached portion, a number of anastomosing columns of epithelial cells are formed, which are separated from each other by septa of vascular areolar tissue. These septa become continuous at the periphery with a thin fibrous capsule furnished by the pia mater.

The cells of the epithelial strands in the glandular lobe appear to be of two sorts, which, like those in the thyroid gland, probably represent different stages of functional activity. The darker sort of cell yields microchemical reactions resembling those of colloid; and little masses of colloid, presumably derived from those cells, are of not infrequent occurrence within or at the margins of the epithelial columns (Figs. 165 and 166).

The glandular lobe is richly supplied with capillary bloodvessels in intimate relations with the epithelium, from which they often appear to be separated by only a thin basement-membrane, and the existence of this is doubtful in some situations (Fig. 167).

The above description shows that the structure of the hypophysis is similar to that of the other ductless glands already considered.

# 190 NORMAL HISTOLOGY.

Fig. 165.

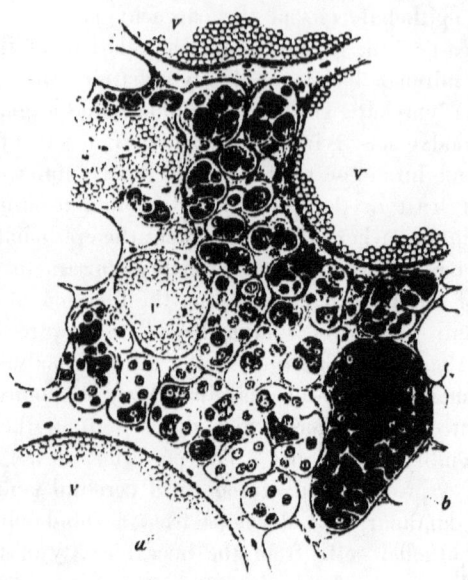

Section from the hypophysis of the ox. (Dostoiewsky.) *v*, veins; *a*, alveoli or cell-columns, with pale, relatively clear cytoplasm; *b*, alveoli or columns of darker granular cells. Other cell-groups contain both varieties of cell.

Fig. 166.

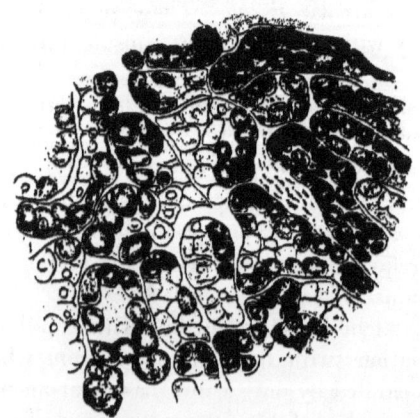

Section from the glandular lobe of the hypophysis; horse. (Lothringer.) Showing the darker cells at the periphery of the epithelial strands, and the clearer cells, for the most part, in their centres.

Its function is still very obscure; but it appears, in cases of experimental thyroidectomy and in disease of the thyroid in the human subject, to enlarge when the function of the thyroid gland is abolished and to assume vicariously the duties of that organ. In how far this points to a normal similarity in function of the two organs must, at present, be left undetermined. In cases of enlargement of the pituitary body profound changes in nutrition, characterized chiefly by overgrowth, frequently take place in the bones of the skeleton (acromegaly).

The nervous supply of the anterior lobe consists of non-medul-

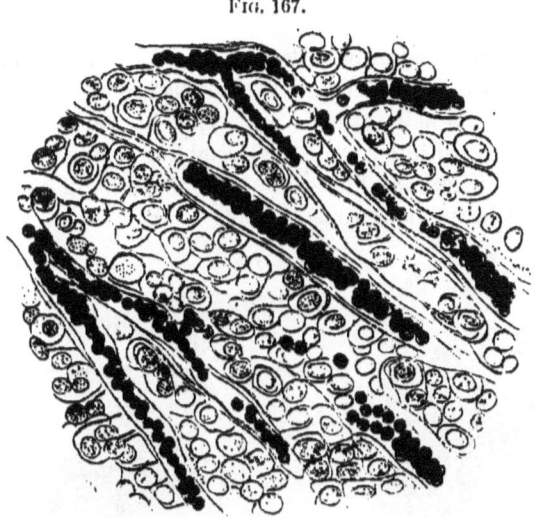

FIG. 167.

Section from the glandular lobe of the hypophysis; child six months old. (Lothringer.) The close relations between the epithelial cells and the capillary bloodvessels, and the differences in the cells, are indicated in this figure. The red blood-corpuscles within the capillaries have been stained dark.

lated fibres, destitute of ganglion-cells, which ramify about the vessels and send some of their terminal twigs between the epithelial cells.

The posterior lobe consists of tissues resembling those of the central nervous system: ganglion-cells, non-medullated fibrils, and neuroglia-cells. Within its substance there are also peculiar oval bodies surrounded by nervous terminations, to which sensory functions have been attributed, and small follicles, lined with cubical epithelium.

5. **The Thymus.**—This organ reaches its fullest development at about the second year of life, after which retrograde changes, ending in the substitution of fibrous and adipose tissues for its proper structure, take place. Its development begins as an ingrowth of epithelium from the branchial clefts. This epithelium forms a

FIG. 168.

Two concentric corpuscles of Hassall, from the fœtal thymus. (Klein.)

branching, solid column of cells surrounded by embryonic connective tissue, which develops into lymphadenoid tissue. In the meantime the epithelial strands are broken up and the whole organ becomes converted into a structure resembling a collection of lymph-follicles, but with this difference: that remnants of the epithelial strands remain in the centres of many of the follicles, where their

FIG. 169.

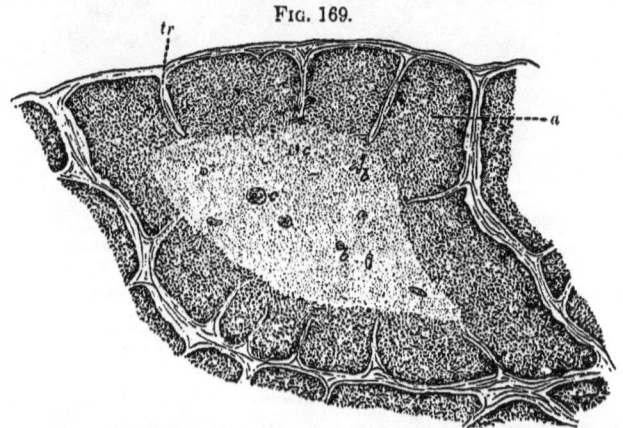

Lobule from the thymus of a child. (Schäffer.) *tr*, trabecula; *a*, nodule of denser lymphadenoid tissue at periphery ("cortex"); *b, b*, sections of vessels within the less dense lymphadenoid tissue in the centre ("medulla"); *c, c*, concentric corpuscles of Hassall.

cells become flattened and imbricated. These epithelial masses are known as the concentric corpuscles of Hassall (Fig. 168).

The thymus is enclosed in a fibrous capsule, which penetrates its substance, dividing it into lobes and lobules. Each of these lobules closely resembles a lymph-follicle, but it is doubtful whether lymph-sinuses, corresponding to those in the lymphatic nodes, are present in the thymus (Fig. 169).

The function of the thymus is still a matter of doubt. It has been regarded as one of the sites in which red blood-corpuscles are formed, and also as a temporary lymphadenoid organ playing the part of the lymph-nodes until these have become fully developed in other parts of the body.

The thymus is connected with the thyroid by a strand of thymus-tissue, and isolated thymus-lobules are found embedded in the edges of the thyroid, near the parathyroid body (see Fig. 159).

The bloodvessels ramify in the septa of the organ and send branches into the lymphoid follicles. The lymphatic vessels accompany the bloodvessels and surround the lobules, but do not appear

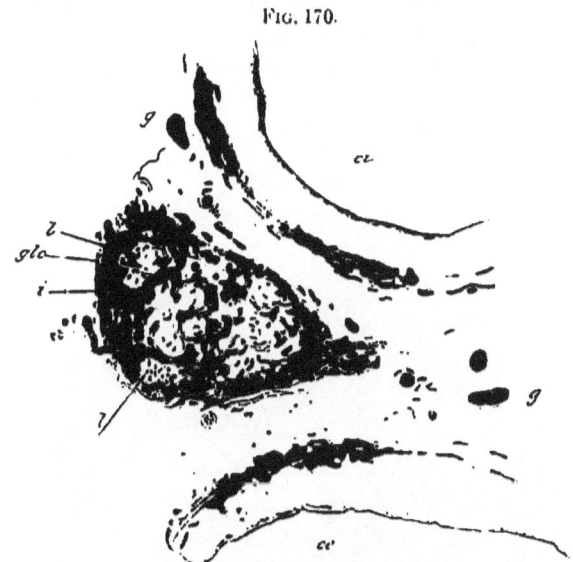

FIG. 170.

Section of the carotid gland and carotid arteries near their origin. (Marchand.) *ci*, internal carotid; *ce*, external carotid; *gle*, carotid gland; *l, l*, groups of epithelial cells; *i*, fibrous tissue between the epithelial groups; *g*, bloodvessel. Numerous vessels are also seen within the gland.

to penetrate into the lymphadenoid tissue. The nerves are small and not numerous. They accompany the bloodvessels, but nervous terminations have not been traced as distributed to the lymphadenoid tissue.

The involution of the gland appears to be accomplished through

194      NORMAL HISTOLOGY.

Fig. 171.

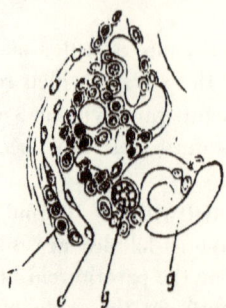

Portion of the same gland as Fig. 170, more highly magnified: *p*, epithelial cells; *g*, capillary bloodvessels; *e*, endothelium forming the capillary wall.

a proliferation of the fibrous tissue around the lobules, which encroaches upon the lymphadenoid tissue and gradually replaces it. This fibrous tissue subsequently becomes, in great measure, converted into adipose tissue. It appears as though the endothelium of the bloodvessels also proliferated, giving rise to masses of imbri-

Fig. 172.

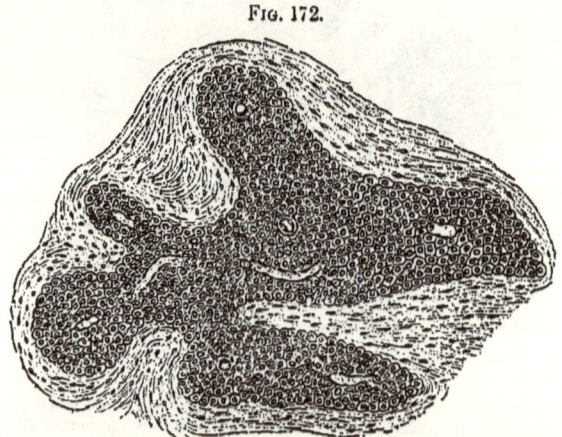

Section of the coccygeal gland. (Sertoli.) The group of cells, apparently of epithelial nature, is traversed by small bloodvessels and enclosed by fibrous tissue.

cated cells within the follicles and leading to an obliteration of the vascular lumen.

6. **The Carotid Glands.**—These consist of groups or islets of epithelial cells, surrounded by fibrous tissue from which numerous capil-

lary bloodvessels are distributed in close relation with the epithelial cells (Figs. 170 and 171). Their function is unknown.

7. **The Coccygeal Gland.**—This body is made up of groups and strands of cells, probably of epithelial nature, closely applied to the walls of capillary bloodvessels and surrounded by fibrous tissue. Its function and mode of origin are both unknown (Fig. 172).

# CHAPTER XVI.

## THE SKIN.

THE skin consists of a deeper, fibrous portion, the *corium*, or true skin, and a superficial, epithelial layer, the *epidermis*. As a part of the latter, and developing from it, the skin contains two sorts of glands, the sebaceous and the sweat-glands, and two kinds of appendages, the hairs and nails.

The corium is composed of vascularized fibrous tissue, which is

FIG. 173.

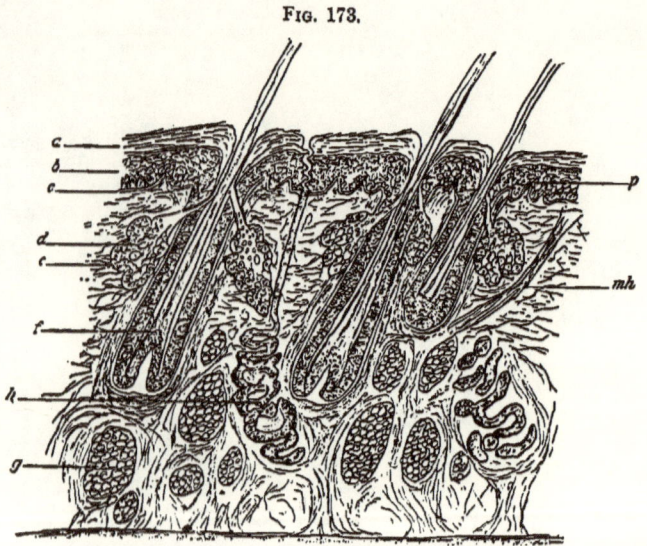

Section of skin perpendicular to the surface. (Arloing.) *a*, horny layer of the epidermis; *b*, rete mucosum; *c*, surface of the corium; *d*, sebaceous gland; *e*, areolar tissue of the corium; *f*, hair-shaft within the hair-follicle; *g*, lobule of adipose tissue in the subcutaneous tissue; *h*, sweat-gland; *mh*, arrector pili; *p*, papilla of the corium extending into the rete mucosum. The lower limit of the corium is not marked by a plane parallel to that of the surface of the skin. The corium may be said to end where the fat of the subcutaneous tissue begins.

made up of bundles loosely arranged in its deeper portions, where it becomes continuous with the subcutaneous areolar tissue, and contains a variable amount of fat, but more compactly disposed in the superficial portions, where it comes in contact with the epidermis, into which it projects in the form of papillæ. Some of these papillæ contain loops of capillary bloodvessels, while others are occupied in their centres by peculiar nerve-endings, called "tactile corpuscles." In some situations, notably upon the palms and soles, the papillæ of the corium are arranged in rows. In most parts of the skin they are irregularly scattered over the surface of the corium (Fig. 173).

The **epidermis** (Fig. 174) is a layer of stratified epithelium in

FIG. 174.

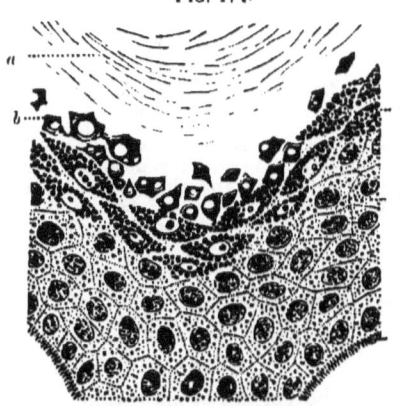

Vertical section of the epidermis of the finger. (Ranvier.) *a*, stratum corneum, or horny layer; *b*, stratum lucidum; *c*, stratum granulosum; *d*, rete mucosum; *e*, "prickles" on the cells bordering on the corium, which is not represented.

which the cells multiply, where they are situated near the corium, and gradually suffer a conversion into horny scales as they are pushed toward the surface, where they are eventually desquamated. The changes the cells undergo in their journey from the deeper layers of the epidermis to its surface cause variations in their appearances which have occasioned a division of the epidermis into a number of more or less well-defined strata. The deepest stratum, where the cells multiply and grow, is called the "rete mucosum." It is composed of cells which gradually enlarge, becoming rich in cytoplasm, and are connected with each other by minute cytoplasmic "prickles," between which there is a space affording a channel for the circulation of nutrient fluids (Fig. 39). Above the rete mucosum the cells appear more granular, owing to the formation of a substance, called "eleidin," within the cytoplasm (Fig. 175). These cells form the "stratum granulosum." The eleidin appears to be produced at the expense of the cytoplasm, the process being a form of degeneration, so that after a while the whole cell is converted into a homogeneous material in which the nucleus persists in a form deprived of chromatin, and therefore insusceptible of staining. The presence of these cells gives rise to the formation of the "stratum lucidum" immediately above the stratum granulosum. Within this stratum the eleidin appears to pass into a closely related substance of a horny nature, keratin, and the cells become con-

FIG. 175.

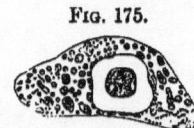

Cell from the stratum granulosum of the epidermis of the scalp. (Rabl.) The cytoplasm of the cell has been in great measure converted into granules of eleidin; the chromatin of the nucleus has retracted into a compact mass in the centre of the nuclear region, and is destined to disappear. This cell is from a section made parallel to the surface of the epidermis, which accounts for its shape and apparent size.

verted into firmly compacted scales, which make up the most superficial or horny layer of the epidermis.

The **sweat-glands** are simple tubular glands, the deep ends of which are irregularly coiled to form a globular mass situated in the deeper portion of the corium or at various depths in the subcutaneous tissue. From these coils the excretory duct passes through the corium to the epidermis, where it opens into a spiral channel between the epidermal cells, ending in an orifice at the surface of the skin.

The epithelial lining of the sweat-gland is a continuation of the stratum mucosum, from which it is derived, and consists of two or more layers of cubical cells in the duct and of a single layer of more columnar cells in the deeper, secreting portion of the gland. In the duct these cells rest upon a homogeneous basement-membrane, but in the secreting portion there is a more or less complete layer of elongated cells, similar in appearance to those of smooth muscular tissue, which lie between the epithelial cells and the basement-membrane (Fig. 176). It is doubtful whether these are really muscle-cells. The loops of the glandular coil are surrounded by fibrous tissue, which contains the bloodvessels supplied to the gland and serves to support it in its globular form.

The sebaceous glands can best be described in connection with the hairs and their follicles.

The bulbous attachment, or "root," of **the hair**, and the adjacent portion of its shaft, are contained in an invagination of the corium and epidermis, called the "hair-follicle" (Fig. 173, *f*). This is surrounded by fibrous tissue, forming its external coat, which may be imperfectly distinguished into an outer layer, containing relatively abundant longitudinal fibres, and an inner layer, in which encircling

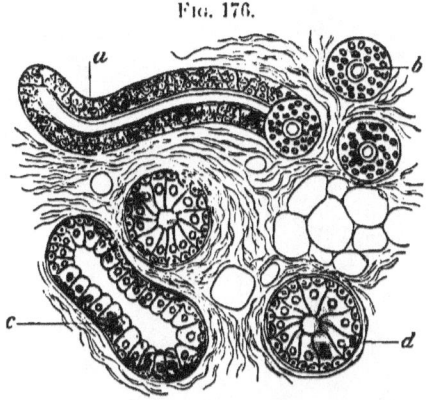

Fig. 176.

Section through the coiled end of a sweat-gland. (Klein.) *a, b*, duct in longitudinal and cross-section; *c, d*, sections of the secretory portion of the tubule. Above *d* is a little adipose tissue. The rest of the section is composed of vascularized areolar tissue.

fibres predominate. At the bottom of the follicle this fibrous tissue becomes continuous with that of a vascularized papilla, similar to those existing on the surface of the corium, which projects into the root of the hair.

The fibrous sac constituting the outer part of the hair-follicle is lined with a continuation of the epidermis, leaving a cylindrical cavity occupied by the hair. This layer of epithelium is reflected upon the surface of the papilla, where it forms the root of the hair, and then passes into the shaft, which is made up of cells, derived from those of the root, that have suffered keratoid degeneration.

The epithelium lining the follicle, as well as that which composes the hair, is not of uniform character throughout, and has been divided into a number of layers, to which different observers have given special names. The group of cells surrounding the papilla are the seat of the multiplication which results in the growth of the hair. Upon the surface of the shaft these cells become transformed into thin scales, each of which overlaps that above it. This very thin

FIG. 177.

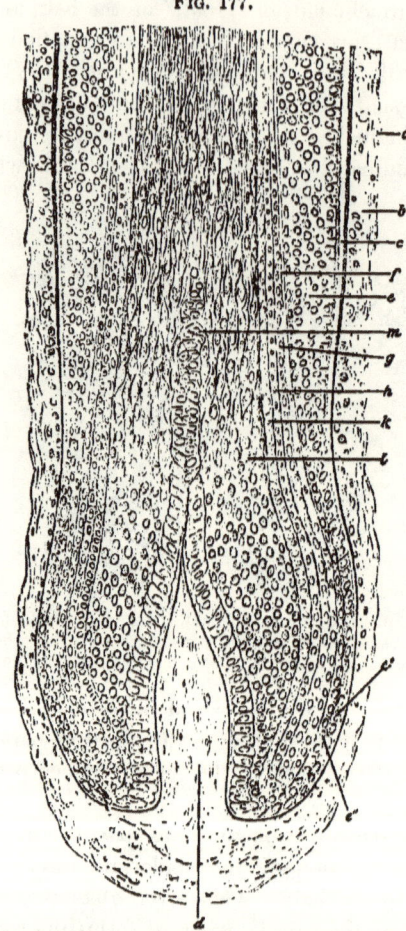

Hair-follicle from the human scalp. (Mertsching.) Longitudinal axial section through the fundus: *a, b,* longitudinal and encircling layers of the fibrous coat; *c,* hyaline layer, formed of an outer faintly fibrillated and an inner more homogeneous lamina; *d,* papilla; *e,* outer root-sheath, continuous with rete mucosum of epidermis; *e′,* its outer layer, continuous with deepest cells of rete and with columnar cells covering the papilla; *e″,* its inner layer, continuous with the cortical cells of hair; *f,* Henle's sheath; *g,* Huxley's layer; *h,* cuticle of root-sheath; *k,* cuticle of hair; *l,* cortical cells of the hair; *m,* medulla.

layer is called the "cuticle" of the hair. Beneath the cuticle the cells are crowded together into fusiform or fibrous elements, which make up the chief mass of the hair-shaft. In the centre of this mass there is sometimes a line of more loosely aggregated cells, forming the "medulla" of the hair. When this is present the surrounding part of the shaft, between it and the cuticle, is known as the "cortex" (Figs. 177 and 178).

The **sebaceous glands** (Fig. 173, *d*) are sacculations in the corium near the hair-follicles, which are filled with epithelial cells. The cells at the periphery divide, and, as they increase in size, push

Fig. 178.

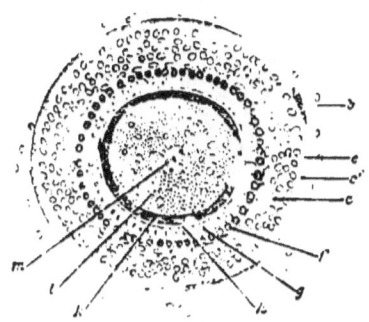

Hair-follicle from the human scalp. (Mertsching.) Cross-section from middle third of the follicle: *b*, longitudinal and encircling layers of the fibrous coat; *c*, hyaline layer, formed of an outer faintly fibrillated and an inner more homogeneous lamina, *c'*; *e*, outer root-sheath, continuous with rete mucosum of epidermis; *f*, Henle's sheath; *g*, Huxley's layer; *h*, cuticle of root-sheath; *k*, cuticle of hair; *l*, cortical cells of the hair; *m*, medulla.

each other toward the centres of the sacs. Here they undergo a fatty degeneration, ending in destruction of the cells and the formation of an oily secretion, the sebum, which is discharged into the hair-follicle a short distance below its opening on the surface of the skin. The sebum is a lubricant for both the hair and the epidermis (Fig. 179).

The color of the epidermis and of the hair is due to a pigmentation of the cells in the deeper layers of the rete mucosum and those composing the hair. The whiteness of the hair which comes with years is due to little spaces which appear in unusual numbers between the cells of the cortex, and are filled with air, reflecting the light and masking the pigmentation of the cells.

The **nails** are especially thick and condensed masses of epithelial cells which have undergone keratoid degeneration and are closely compacted. They are produced at the root of the nail, and as they

### Fig. 179.

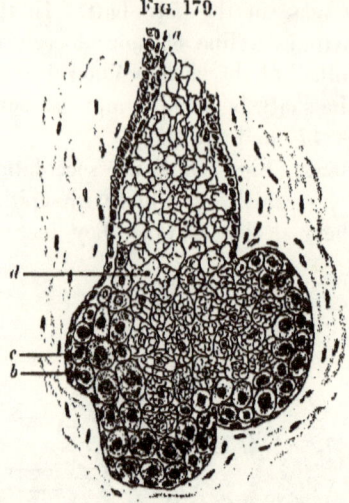

Sebaceous gland from the external auditory canal. (Benda and Guenther's *Atlas*.) *a*, epithelium continuous with that lining the hair-follicle; *b*, layer of proliferating epithelium lining the sac of the gland; *c*, enlarged cell beginning to undergo fatty metamorphosis of the cytoplasm; *d*, mass of sebum derived from a single epithelial cell.

accumulate push the body of the nail forward. They, therefore,

### Fig. 180.

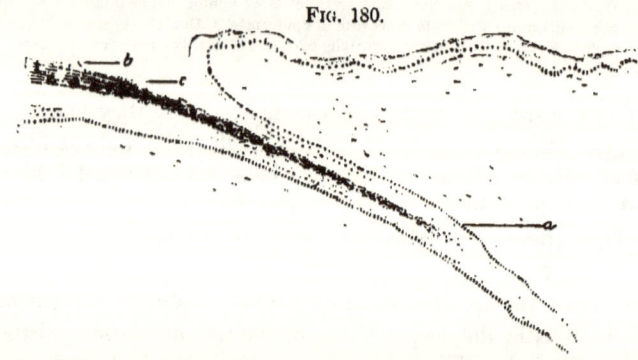

Section through the root of the nail of a sixth-months fœtus. (Ernst.) *a*, matrix of the nail formed by an invagination of the rete mucosum. Near the point indicated by the letter the epithelial cells have begun to change into keratoid material. *b*, loosened scales of the surface of the nail; *c*, remains of the fœtal cuticle which have not become keratoid. The letter *a* and line proceeding from it both lie in the corium.

correspond to the horny layer of the epidermis, which has become modified to form these special structures (Fig. 180).

The skin contains little muscular bands, the arrectores pili (Fig. 173, *mh*), composed of smooth muscular fibres, which are attached to the fibrous coat of the hair-follicles near their deep extremities and to the superficial layer of the corium on the side of the follicle toward which the hair leans. The action of these muscles is to cause the hair to assume a more vertical position, and to raise it and the follicle, producing the effect known as "goose flesh." By their contraction they may also aid in the discharge of sebum, since their fibres often partially invest the sebaceous glands.

The functions of the skin have reference to its being the organ coming in contact with the external world. The epidermis protects the underlying tissues from mechanical and chemical injury and from desiccation. The keratin in its horny layer forms an impervious and tough investment of the body, which is highly resistant toward chemical action and mechanical abrasion, and is constantly renewed from the layers that lie beneath it. It is kept in a pliable condition by the sebum discharged upon its surface and by the moisture proceeding from the sweat-glands, the "insensible perspiration." The skin also plays a prominent *rôle* in the regulation of the bodily temperature. When its vessels are contracted the amount of heat given off from the surface of the body is reduced; when they are dilated, it is increased. A further loss of heat is occasioned by an increased secretion of sweat, which bathes the surface of the skin and abstracts from the body the heat required to convert it into vapor. Under the influence of sudden and marked cold the vessels of the skin become much contracted and the arrectores pili shorten, occasioning the production of a roughness of the skin, goose-flesh, and probably also a discharge of sebum, which reduce the evaporation from the skin. At the same time a reflex rhythmical contraction and relaxation of the voluntary muscles is brought about—shivering, which increases the liberation of stored energy within the body, and causes it to appear as heat. In conjunction with these functions the skin is also an organ of tactile and thermal sensation, functions which are not merely beneficial in themselves, but are useful auxiliaries in the furthering of the other functions exercised by the skin. It is a common experience that the sensation of cold stimulates the desire for muscular exercise, of which the liberation of heat is a result. The sensation of pain often gives timely warning of exposure to an injury sufficiently great to overcome the usual protective powers of the epidermis. Thus we see that when the automatic action of the skin is inadequate for the performance of its functions it calls forth

FIG. 181.

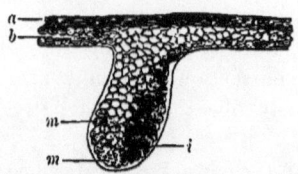

Hair-rudiment from an embryo of six weeks. (Kölliker.) *a*, horny layer of epidermis; *b*, Malpighian layer, rete mucosum; *i*, limiting membrane; *m, m*, cells extending from the rete mucosum to fill the future hair-follicles. The elongated cells near the base of the sac are those from which hair is developed. The secreting glands of the body arise from some epithelial layer in a similar manner.

an auxiliary activity of other organs, through the medium of the nervous system.

The **hair-follicles** are developed from the rete mucosum of the epidermis, and first appear as little masses of cells growing into the

FIG. 182.

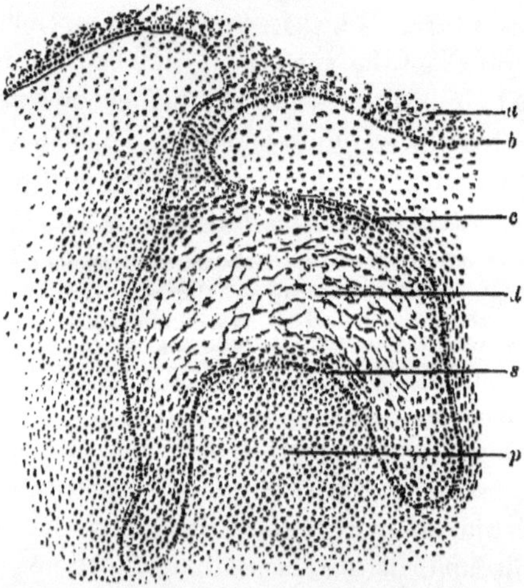

Section of developing tooth. From embryo of sheep. (Böhm and Davidoff.) *a*, epithelium of the gum; *b*, its deepest layer; *c*, superficial cells of the enamel-pulp; *d*, enamel-pulp formed of modified epithelial cells; *s*, cells of the enamel-pulp destined to produce the enamel ("adamantoblasts"); *p*, dental papilla.

underlying connective tissues (Fig. 181). The sebaceous glands arise as offshoots from these cellular masses.

**The Teeth.**—The development of the teeth presents close analogies to that of the hairs. They also first appear as little masses of cells, growing into the connective tissues of the alveolar processes from the stratified epithelium covering them. Into the bases of these masses connective-tissue papillæ are developed, which eventually become differentiated into the pulp of the tooth-cavities. The epithelial cells which immediately surround these papillæ become elongated to a columnar form and then become converted

FIG. 183.

Section of developing tooth. From embryo of rabbit. (Freund.) *ep*, epithelium of gum; *sh*, epithelial cells forming outer layer of the enamel-pulp of the temporary tooth; *L*, similar layer belonging to the rudiment of the permanent tooth; *Sr*, enamel-pulp; *p*, dental pulp of the tooth-cavity; *d*, dentin; *v*, bloodvessels; *B*, rudiment of second or permanent tooth; *a*, embryonic connective tissue of the alveolar process.

into or elaborate the tissue of the enamel. The superficial cells of the papillæ likewise elongate and produce the dentin. The cement which constitutes the outer layer of the root of the tooth is bone, and is developed from the fœtal connective tissue in that region (Figs. 182 and 183).

Only a brief description of the structures entering into the formation of the fully developed tooth can be given here. For a more detailed account of them the student is referred to special works on the subject.

The centre of the tooth is hollow, and the cavity opens by a small orifice at the tip of the root. This cavity is filled with a highly vascular delicate areolar tissue, richly supplied with nerves. Where this pulp is in contact with the tooth its outer layer is made up of modified connective-tissue cells, odontoblasts, which are capable of elaborating dentin. The body of the tooth is composed of dentin. This contains minute canals, analogous to the canaliculi in bone, but much longer. They extend from the pulp-cavity nearly, if not quite, to the outer boundary of the dentin, and, toward their terminations, give off branches. These canals are occupied by long fibrous processes of the odontoblasts already mentioned.

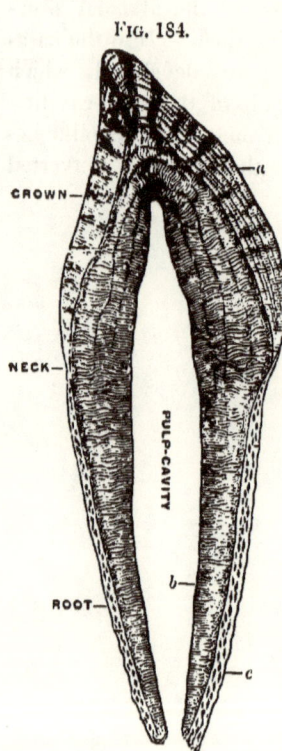

Fig. 184.

Axial section of a human tooth having but one root: *a*, enamel; *b*, dentin; *c*, cement.

The crown of the tooth, down to its neck, is covered with enamel. This is a tissue derived from epithelium, and is composed of long, prismatic elements extending from the surface of the tooth to the dentin. These prisms have a polygonal cross-section and are held together by a hard cement-substance. They are not perfectly rectilinear, but pursue a wavy course, being disposed in laminæ or bundles, in which the prisms have not quite the same direction.

The root of the tooth, below the point where the enamel ends, is covered with cement, which has the structure of ordinary bone, but is usually devoid of Haversian canals (Fig. 184).

# CHAPTER XVII.

## THE REPRODUCTIVE ORGANS.

### I. IN THE FEMALE.

The female reproductive organs are: (1) the ovary, in which the egg is produced; (2) the Fallopian tube, through which it is conveyed to (3) the uterus, where it develops into the fœtus, and from which the child at maturity passes through (4) the vagina and (5) external genitals into the external world.

1. **The Ovary** (Fig. 185).—The free surface of the ovary is covered with a single layer of columnar epithelium, called the "germinal epithelium." Beneath this the substance of the organ is composed of a vascularized fibrous tissue, the "stroma," which is slightly different in the details of its structure in different parts of the organ. Immediately beneath the germinal epithelium it is slightly richer in intercellular substance than in the subjacent parts, so that the organ appears to have a proper fibrous coat. This coat is not distinct, however, and gradually passes into a highly cellular form of fibrous tissue, in which the spindle-shaped cells are separated by only a small amount of a delicate fibrous intercellular substance. Toward the hilum of the ovary this connective tissue passes into a more distinctly fibrous tissue, containing a larger amount of intercellular substance and cells that are less prominent. In this portion of the stroma the larger vessels supplying the organ are situated, and from it they send smaller branches throughout the stroma of the organ. Within the more cellular regions of the stroma are the structures known as the Graafian follicles, each of which contains an ovum. In order to understand the structure of these Graafian follicles it will be well to trace the history of their development.

The Graafian follicles and ova are derived during fœtal life from the germinal epithelium covering the ovary. From this layer of cells little columns of epithelium make their way into the stroma, where they become broken up into small isolated groups, in each of which one of the cells develops into an ovum, while the rest contribute to the formation of the Graafian follicle. This mode of origin may serve to explain the fact that the younger Graafian follicles are most abundant in the peripheral portion of the stroma. At first the Graafian follicle consists of a large central cell, the ovum,

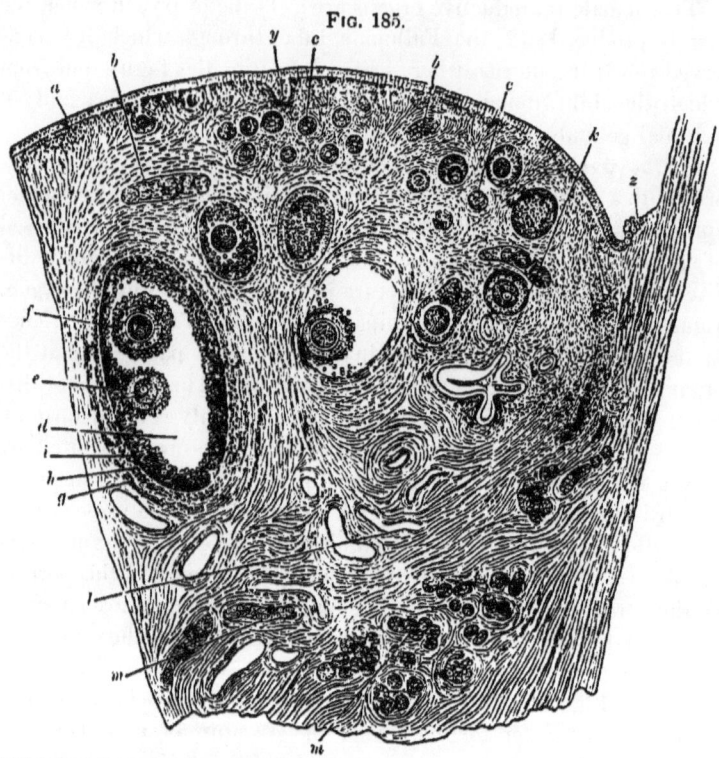

Fig. 185.

Section from the ovary of an adult bitch. (Waldeyer.) *a*, germinal epithelium; *b, b*, columns of germinal epithelium within the stroma; *c, c*, small follicles; *d*, much more advanced follicle; *e*, discus proligerus and ovum; *f*, second ovum in same follicle (a rare occurrence); *g*, fibrous coat of the follicle; *h*, basement-membrane; *i*, membrana granulosa of epithelium; *d*, liquor folliculi; *k*, old follicle from which the ovum has been discharged; *l*, bloodvessels; *m, m*, sections of the parovarium; *y*, ingrowth from the germinal epithelium; *z*, transition from the germinal epithelium to the peritoneal endothelium.

surrounded by an envelope of somewhat flattened epithelial cells, which are in direct contact externally with the unmodified, highly cellular tissue of the stroma (Fig. 186).

As the Graafian follicle develops, its position in the ovary becomes more central, and the cells around the ovum lose their flattened shape and divide, forming a double layer of cubical or columnar cells. These two layers then become separated by a clear fluid, the liquor folliculi, so that the outer layer forms the wall of a sac, while the inner layer remains as a close investment of the ovum. The cells of these two layers multiply; those surrounding the ovum forming the "discus proligerus," and those lining the sac the "tunica granulosa"; but they blend with each other at one point on the wall of the follicle, so that the ovum retains a fixed position.

Meanwhile the tissue of the stroma undergoes modifications which

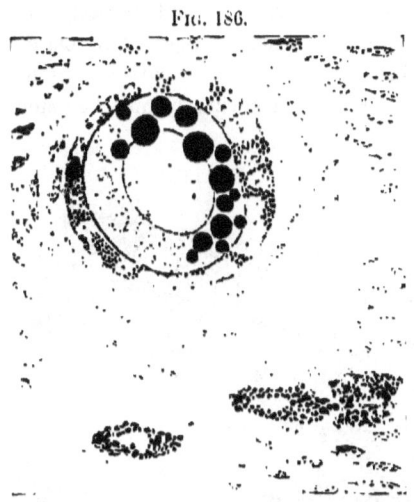

Fig. 186.

Graafian follicle and stroma in ovary of adult sow. (Plato.) The ovum occupies the centre of the follicle, appearing as a very large cell with a large vesicular nucleus ("germinal vesicle"), within which is a large nucleolus ("germinal spot"), exceeding in size the whole nucleus of the surrounding epithelial cells of the follicle. The cells of the stroma are arranged about the follicle as though to form the fibrous coat of the latter. In the lower portion of the figure are three large cytoplasmic cells, containing globules of fat and granules of pigment. These cells are analogous to those found in the interstitial tissue of the testis. The epithelium of the Graafian follicle, and the ovum, also contain globules of fat of various sizes, stained black by the osmic acid used in the preparation of the specimen.

contribute a clear basement-membrane and a fibrous envelope, the "membrana propria," to the structure of the follicle.

The follicle now enlarges, as the result of an increase in the amount of the liquor folliculi, eventually approaches the surface of the ovary at some point, and then ruptures, discharging the ovum.

After the rupture of the Graafian follicle and the escape of its contents a slight hemorrhage usually takes place into its cavity, which then appears filled with remains of the liquor folliculi mixed with coagulated blood. Into this, granulations[1] now

[1] See Chapter XXIV.

develop from the fibrous wall, replacing the clot and eventually producing a scar. This process is much more rapid in case the ovum is not impregnated (corpus hæmorrhagicum) than when impregnation has taken place. In the latter case the productive inflammation is more marked, and is accompanied by a fatty degeneration of the older granulations which gives them a yellowish tinge (corpus luteum). In the centre of this yellowish zone is the remainder of the clot, and about its periphery an envelope of fibrous tissue, which is usually irregular in contour. The corpus luteum finally becomes a mass of cicatricial tissue of greater size than that resulting from a corpus hæmorrhagicum (corpus album) (Figs. 187 and 188).

FIG. 187.

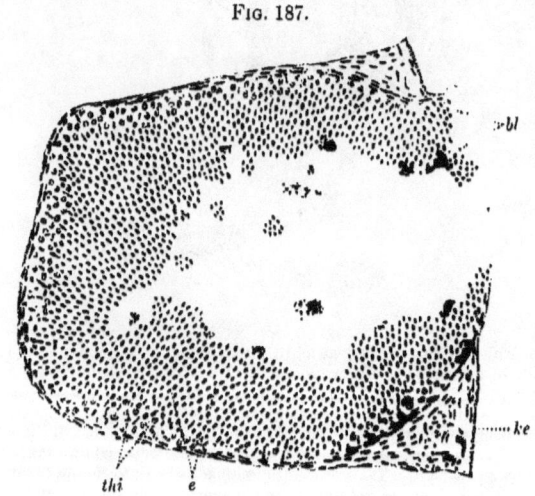

Section from rabbit's ovary, illustrating the formation of the corpus luteum. (Sobotta.) Recently ruptured Graafian follicle. *ke*, germinal epithelium; beneath it, the ovarian stroma. Bounding the follicle externally is the fibrous capsule of the follicle. Within this, *thi*, is a layer of proliferating fibrous tissue, composed of polyhedral cells with round nuclei. Among these are elongated nuclei belonging to endothelial cells springing from the capillaries, and destined to form the walls of future bloodvessels; *e*, epithelium of the membrana granulosa. Within this are the viscid remains of the liquor folliculi, containing a few red blood-corpuscles and some epithelial cells detached from the membrana granulosa, *M*, red blood-corpuscles. This section was prepared from an ovary about twenty-four hours after coitus, and the development of the layer *thi* probably took place within that time.

2. **The Fallopian Tube.**—The free surface of the Fallopian tube is covered by a serous membrane, continuous with the rest of the peritoneum. This rests upon fibrous tissue, in which the longitudinal bundles of smooth muscular tissue constituting the external

Fig. 188.

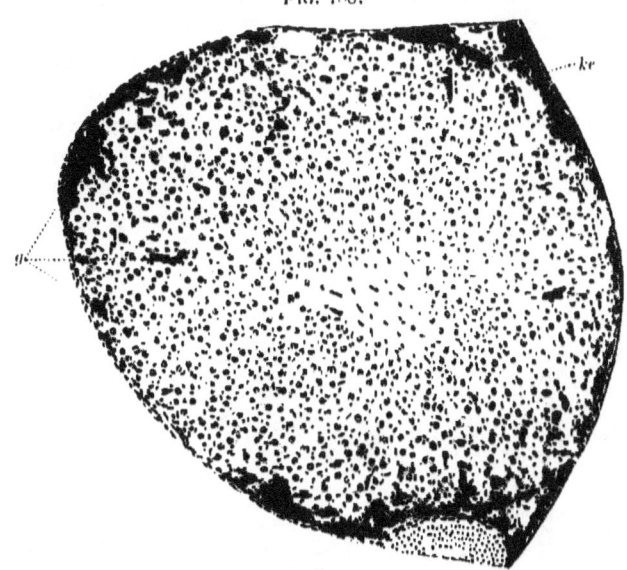

Section of young corpus luteum, four days after coitus. The proliferating connective tissue has nearly filled the cavity of the follicle, only a small mass of fibrin remaining in its centre. The young connective tissue is highly vascularized, the blood in some of the capillaries being represented, g. ke, germinal epithelium. Below is the margin of a Graafian follicle, with its membrana granulosa.

muscular coat are situated. This is followed by an internal muscular coat of encircling bundles of smooth muscular tissue, inside of which is the submucous coat of areolar tissue, containing a few scattered ganglion-cells.

The mucous membrane consists of a highly cellular connective tissue covered with ciliated columnar epithelium. During life these cilia propel toward the uterine cavity substances coming into contact with them. Toward and at the fimbriated extremity of the tube the mucous membrane is thrown into deep longitudinal folds, upon which are numerous secondary and tertiary folds, but further toward the uterus these folds give place to branching villous projections into the lumen (Fig. 189). Toward the uterine end of the tube these complicated folds and villi disappear and the lumen of the tube becomes round or stellate.

3. **The Uterus.**—The external surface of the uterus, throughout most of its extent, is covered by a reflection of the peritoneum. Beneath this are three distinct coats of smooth muscular tissue, the

outer two in close contact with each other; the two inner separated by a thin layer of areolar fibrous tissue, supporting large bloodvessels. This separation of the innermost layer from the middle layer leads to the inference that the former is analogous to the muscularis mucosæ found in other hollow viscera, although in the uterus it forms the chief mass of the muscular tissue of the organ. The outer layer is made up of bundles of fibres that have a general longitudinal position; the two inner layers have a general circular

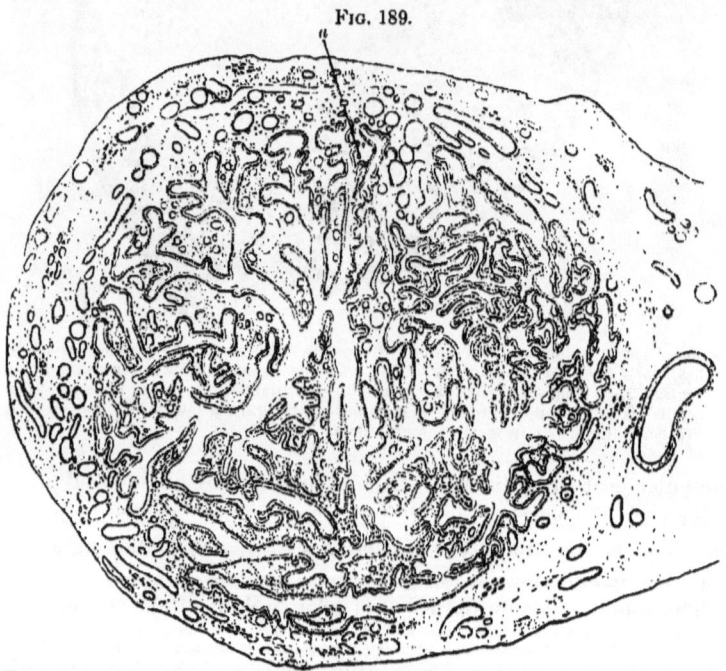

FIG. 189.

Transverse section of the Fallopian tube near its free end. (Orthmann.) Numerous branching villous projections of the wall, covered by ciliated columnar epithelium, extend into the lumen. The open spaces in these villous projections are sections of the bloodvessels.

disposition of their bundles, though the latter interlace with each other in various directions within the muscularis mucosæ, leaving masses of areolar tissue containing the larger bloodvessels between them.

Covering the surface of the muscularis mucosæ is a highly cellular connective tissue, not unlike granulation-tissue in appearance, except that it is less richly supplied with bloodvessels. It is composed of round and fusiform cells, lying in a small amount of intercellular substance, in which fibres can be distinguished only with difficulty. The surface of the mucous membrane is covered with a layer of ciliated columnar epithelium, which is continued into long tubular glands penetrating the superficial portions of the muscularis mucosae, where they frequently branch before terminating in blind extremities. It should be borne in mind that at the extremities of these glands the whole tubule is often filled with epithelial cells, so that no lumen is visible. In their course into the mucous membrane these glands are usually straight at first, but in their deeper portions become tortuous (Figs. 190 and 191).

FIG. 190.

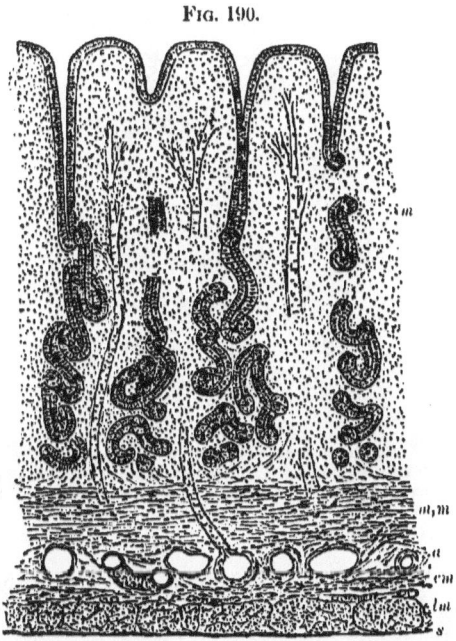

Section through the uterine wall of a rabbit, near one of the cornua. (Schäffer.) *m*, glandular portion of the mucous membrane; *m, m*, muscularis mucosae; *a*, submucosa of areolar tissue, containing the large bloodvessels which send branches into the stroma of the mucous membrane; *cm*, circular layer of the muscular coat; *lm*, longitudinal, thicker layer of the muscular coat; *s*, serous coat, derived from a reflection of the peritoneum.

During the childbearing period of life the portion of the mucous membrane resting upon the muscularis mucosae is the seat of active changes which pass through a cycle corresponding to each menstrual period, but interrupted by a special series of changes during pregnancy. These changes are of importance in their bearing upon the pathology of the organ, and must be briefly described.

At the menstrual period the superficial portion of the mucous membrane, down to its muscular coat, suffers a degeneration, which results in its disintegration and discharge, along with some blood derived from the exposed and damaged vessels of small size within its tissues. After this degeneration the membrane is restored by a proliferation of the elements contained between the bundles of the muscularis mucosæ, the glands being reformed from the remnants of their deep extremities. The mucous membrane slowly continues

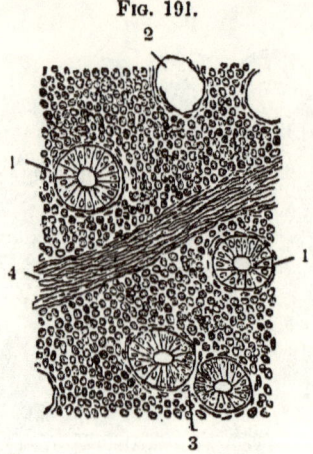

FIG. 191.

Section of the human uterine mucous membrane parallel to its surface. (Henle.) 1, 2, 3, uterine glands in cross-section. In 2, the basement-membrane alone is represented, the epithelium having fallen out of the section. 4, bloodvessel in longitudinal section. Between these structures is the highly cellular stroma of the mucous membrane, only the nuclei of its cells being represented.

to increase in thickness and the glands in tortuousness until the next menstruation, when the same process is repeated. It will be noticed that the connective tissue of the mucous membrane, in the absence of pregnancy, is subject to periodical degeneration and regeneration, which probably prevent its development into a mature fibrous tissue with an abundance of fibrillated intercellular substance.

If an ovum, discharged from the ovary, becomes fertilized, the menstrual cycle of changes in the superficial portion of the mucous membrane of the uterus is interrupted. That portion of the mucous membrane then undergoes extensive modifications in structure during the early months of the ensuing pregnancy. The intercellular tissue between the uterine glands becomes more hyperplastic than during the intervals separating the menstrual periods, and at the same time the cells composing it become hypertrophied, until they closely resemble large epithelial cells. These cells have been called "decidual cells." The ovum, when it reaches the cavity of the uterus, becomes embedded in this tissue, which grows around and encloses it, after which it is differentiated into three portions. The part beneath the ovum is called the decidua serotina; that which invests the ovum, the decidua reflexa; and that lining the rest of the uterine cavity, the decidua vera. While the decidual tissue is developing and its cells enlarging the uterine glands suffer changes. Their mouths become widened, and their lower portions down to the muscularis mucosae dilated, after which the epithelial lining atrophies and seems to disappear, so that the lumina of the glands appear as spaces in the decidual tissue. As the ovum enlarges, the decidua reflexa comes in contact with the decidua vera, and the two layers exert a mutual pressure upon each other, which flattens the spaces they contain and may obliterate many of them. The decidual tissue now consists of a number of flattened spaces which are separated from each other by thin walls of fibrous tissue produced by the further development of the decidual tissue. The decidua reflexa and the decidua vera blend with each other to form a part of the membranes that are expelled from the uterus, along with the placenta, after the birth of the child, the rest of the membranes and most of the placenta being derived from the fœtus. After the birth of the child and the expulsion of the membranes the mucous membrane is regenerated from the tissues remaining in the superficial layers of the muscularis mucosae.

The mucous membrane of the cervical portion of the uterus does not participate in these changes incident to menstruation and pregnancy, and the connective tissue underlying its epithelial lining is more fibrous in character than that in the corresponding part of the uterine body. About the middle of the cervical canal the ciliated epithelium, which is continuous with that of the body, passes into a stratified epithelium, which extends over the cervix uteri, the portio vaginalis, and the inner surface of the vagina to join that of the epidermis upon the labia minora. The fibrous tissue beneath this stratified epithelium possesses papillae similar to those upon the skin, and contains mucigenous glands, which secrete a tenacious mucus serving to close the cervical canal during pregnancy. The orifices of these glands sometimes become occluded, causing a cystic dilatation of the acini, due to accumulated secretion, "ovula Nabothi."

The muscular and other tissues of the uterine wall undergo hypertrophy during pregnancy, the individual muscular fibres becoming as much as thirty times their original bulk in the non-pregnant uterus. The bloodvessels also enlarge and acquire thicker walls. These retain much of this increase of size, even after the involution of the uterus following parturition, but the muscular fibres suffer a partial fatty degeneration, which restores them to nearly their original condition.

4. **The Vagina** (Fig. 192).—The subepithelial fibrous coat of the vagina is covered with small papillæ, which project into the epithe-

FIG. 192.

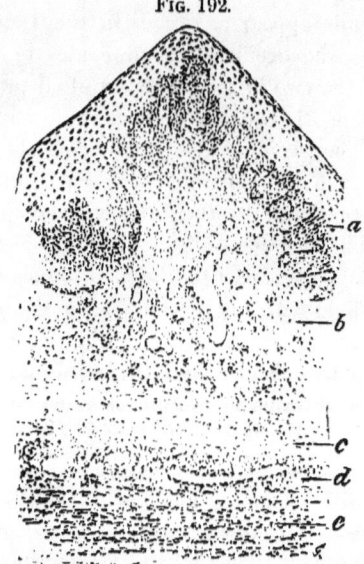

Portion of a longitudinal section of the vaginal wall. (Benda and Guenther's *Atlas*.) *a*, stratified epithelium; *b*, subepithelial areolar tissue; *c*, muscularis mucosæ; *d*, areolar submucosa containing vascular trunks; *e*, muscular coat. Outside of the latter is the ill-defined fibrous coat, not represented in the figure.

lium. Outside of this coat is one of smooth muscular tissue, which is not clearly divisible into layers, but in which the inner fibres are chiefly circular, forming an imperfectly defined muscularis mucosæ, while the outer have a longitudinal direction, and may be regarded as the true muscular coat of the vagina. Outside of the muscular coat is a layer of areolar tissue connecting the vagina with the neighboring parts, except at its posterior and upper part, where it is covered with a serous membrane, forming part of the peritoneum.

5. **The External Genitals.**—The hymen is a fold of the mucous membrane, and consists of fibrous tissue with a covering of stratified epithelium. The same general structure obtains also in the labia minora, prepuce, and labia majora; but the labia minora and prepuce are destitute of fat, while the labia majora contain considerable adipose tissue. All three organs are supplied with sebaceous glands, which are numerous beneath the prepuce and are associated with hairs only on the labia majora. The latter also contain fibres of smooth muscular tissue, corresponding to the analogous dartos of the scrotum. The bulbi vestibuli, crura of the clitoris, and the body and glans of that organ are composed of erectile tissue. The glands of Bartholin are compound racemose glands, in which the alveoli are lined with a columnar epithelium resembling in structure that of the mucous glands in other parts of the body. The epithelium lining their ducts is of the cubical variety.

The parovarium is a remnant of the Wolffian body of the fœtus, consisting of a series of blind tubules lined with epithelium (Fig. 185). It is situated between the Fallopian tube and the ovary. The remains of the Wolffian duct and of the duct of Müller, having a similar structure to the tubules of the parovarium, are sometimes persistent, the one connected with the parovarium, the other with the extremity of the Fallopian tube. These structures are of interest because tumors occasionally arise from them.

**The Maturation of the Ovum.**—Before the ovarian ovum is ready for fertilization it must undergo two divisions, during which the amount of chromatin left in the mature egg is reduced one-half. The first division results in the formation of two cells, which differ enormously in the amount of cytoplasm they possess, but which have equal shares of the chromatin in the original nucleus. The smaller of these two cells is known as the "first polar body." After its separation from the larger cell both cells divide again, without an intermediate growth of the chromatin. In this second division of the larger cell the two resulting cells are again very unequal in size, the smaller being the "second polar body." The first polar body having also divided, there result from these successive divisions one mature egg and three polar bodies, each with only half as many chromosomes in its nucleus as are commonly found in the general or "somatic" cells of the body (Fig. 193). The polar bodies perish, as does also the ovum, unless fertilized by the introduction of a spermatozoon. The latter, as we shall see, also contains half the number of chromosomes contained in the somatic cells; so that

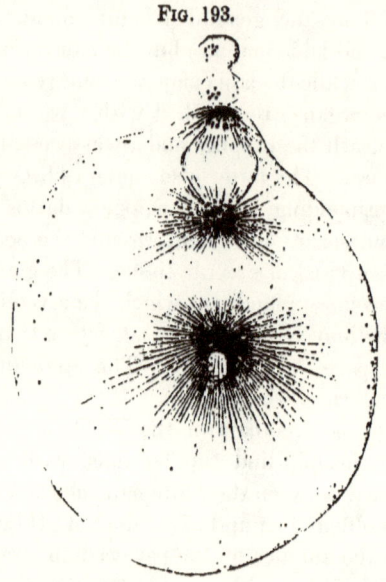

Fig. 193.

Maturing ovum of physa (fresh-water snail). (Kostanecki and Wierzejski.) Above are the two small cells resulting from the division of the first polar body. Below is the ovum, the nucleus of which is dividing to form the second polar body. Near the centre of the ovum is the nucleus of the spermatozoon, just above which is its (divided) centrosome with surrounding radiations in the cytoplasm. When the second polar body has been formed the chromosomes remaining in the ovum will be ready to participate with those of the spermatozoon in the further development of the then fertilized egg.

after its entrance into the mature ovum the latter acquires its full complement of chromosomes and is ready for development.

**The Mammary Gland.**—Each mamma consists of a group of about twenty similar compound racemose glands, opening by distinct orifices at the tip of the nipple, and separated and enclosed by fibrous tissue, in which there is a variable amount of fat. At the edges of the mamma this fibrous stroma becomes continuous with the tissues of the superficial fascia in which the breast is situated.

Each of the glands entering into the composition of the breast possesses a single main duct, the "galactiferous duct," which is lined with columnar epithelium, except near its orifice, where the stratified epithelium of the epidermis extends for a short distance into its lumen. A little below the base of the nipple the duct presents a fusiform dilatation, called the "ampulla," which serves as a reservoir for the comparatively small amount of milk secreted in the intervals between nursings.

The main duct branches in its course from the nipple into the deeper portions of the gland, and these branches give off twigs, which terminate in the alveoli of the gland. The columnar epithelium lining the main duct gradually passes into a cubical variety in the branches, and this becomes continuous with the epithelial lining of the alveoli. The terminal branches of the ducts are short, so that the alveoli opening into them lie close together and are collectively known as a "lobule" of the gland. These lobules are, in turn, grouped into lobes, each of which corresponds to one of the main ducts of the breast.

The individual alveoli and the lobules are surrounded by fibrous tissue, which may be subdivided into an intralobular and an interlobular portion, the latter more abundant than the former. This fibrous tissue supports the vessels and nerves supplied to the gland.

The character of the epithelium lining the alveoli varies with the functional activity of the gland.

Before puberty the secreting acini are only slightly, if at all, developed, the mamma consisting of a little fibrous tissue and the ducts of the gland, which possess slightly enlarged extremities.

When the gland has become fully developed, at or about puberty, the epithelial cells lining the acini are small and granular and nearly fill the diminutive lumina. The fibrous stroma is, at this period, abundant and makes up the chief bulk of the breast.

When the gland assumes functional activity the cells enlarge and multiply (Fig. 194), and the lumina of the acini become distinct and filled with a serous fluid. Into this fluid a few fat-globules are discharged from the epithelial lining, forming an imperfect milk, very poor in cream and differing in the proportions of the dissolved constituents from the milk that is produced after the function of the gland is fully established. This secretion is called "colostrum." Besides the scant supply of fat-globules which it contains, it is further characterized by the presence of so-called colostrum-corpuscles. These are leucocytes which have wandered into the acini of the gland from the bloodvessels in the interstitial tissue, and have taken some of the fat-globules of the secretion into their cytoplasm. This process results in an enlargement of the leucocyte, and, in extreme cases, to an obscuring of the nucleus and cytoplasm by fat-globules, so that the whole appears as though composed of an agglutination of numerous drops of fat (Fig. 195).

As the functional activity of the gland matures the epithelial

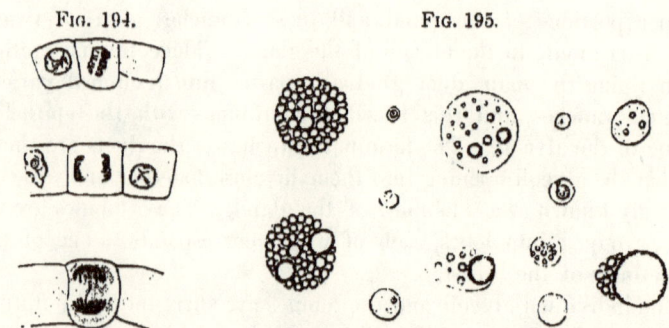

Fig. 194.—Dividing epithelial cells from the mammary gland of the guinea-pig. (Michaelis.) The figure represents the proliferation of the cells by the indirect mode before lactation has been established—*i. e.*, during the maturation of the gland.
Fig. 195.—Colostrum-corpuscles and leucocytes from the colostrum of a guinea-pig. (Michaelis.)

cells lining its acini produce drops of fat in the cytoplasm bordering on the lumen, and these are subsequently discharged into the lumen, forming the fat or cream of the milk. The casein of the milk appears to be produced in the following manner: it has been observed that during lactation the nuclei of some of the cells present changes in form that lead to the inference that they undergo division by the direct mode—*i. e.*, without passing through the phases of karyokinesis. It thus happens that some of the epithelial cells contain two nuclei. These cells, after a while, project into the lumen of the acinus, the two nuclei lying in a line perpendicular to its wall. It is supposed that the nuclei nearest the lumen become detached, together with some of the cytoplasm, and that the chemical constituents of the nucleus and cytoplasm enter into the formation of the casein. Such free nuclei have been observed in the lumina of the acini, and it is known that the chromatin which they contain disintegrates and eventually disappears (chromolysis), so that it is not found in the secreted milk. It is probable that the other constituents of the nucleus likewise undergo chemical changes (karyolysis) (Fig. 196).

When lactation is suspended the breast at first secretes a fluid in every way resembling colostrum, and eventually returns to the dormant state, in which the cells are again small and granular and the stroma is relatively abundant.

As the glandular portion of the breast enlarges during lactation, the whole breast becomes increased in size, but this increase is not proportional to the development of the alveoli, for the stroma is reduced in amount, so that the lobules of the gland are closer to each other. After the period of lactation is passed the alveoli return almost to their original size, but the stroma is not repro-

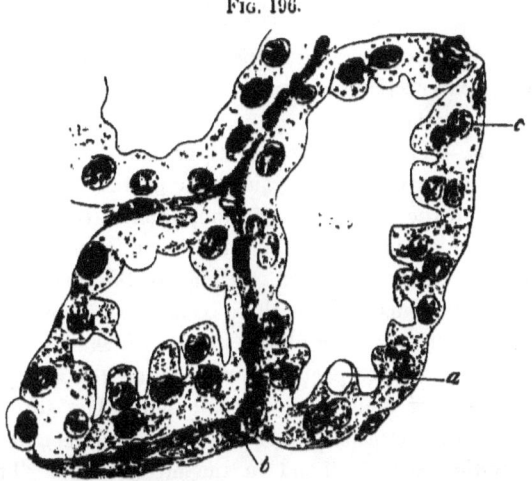

Fig. 196.

Section from the mammary gland of a guinea-pig during lactation. (Michaelis.) The figure represents sections of two acini and the margin of a third, separated by vascularized areolar tissue. *a*, fat-globule, separated from the lumen by a mere film of cytoplasm; *b*, projecting cell with two nuclei; *c*, two nuclei which appear to have been produced by constriction of a single pre-existent nucleus.

duced in fibrous form, but its place is taken by adipose tissue, the amount of which depends upon the individual, being great in those that are fat, and slight in those that are lean. In the latter, therefore, the breast becomes soft and pendulous after lactation has ceased.

It is important to bear the above changes in the normal gland in mind when examining the mamma for evidences of a tumor. When, for example, the stroma is abundant and the glandular structures undeveloped, as is the case before puberty, sections of the gland may be mistaken for those of a mammary fibroma.

The nipple is composed of fibrous tissue, with a considerable admixture of elastic fibres, in which there are scattered bundles of smooth muscular tissue lying parallel to the axis of the nipple. A circular bundle of the same tissue is found at the base of the nipple, and by its compression on the bloodvessels may be the cause of the erection of the nipple. The skin at the base of the nipple and in the areola surrounding it contains large sebaceous glands.

The mammary gland in the male is functionless, and, while it contains the same structures as in the female, it remains in a comparatively undeveloped condition.

## II. IN THE MALE.

The male organs of generation include the penis, prostate, vesiculæ seminales, vasa deferentia, epididymis, and testes, together with certain accessory glands.

1. **The Penis.**—This is formed by three parallel structures: the corpora cavernosa, lying side by side and partially blending in the median line, and the corpus spongiosum, situated beneath their line of junction and containing the urethra. At its anterior end the corpus spongiosum expands about the ends of the corpora cavernosa to form the glans penis. These three bodies, except over the glans, are firmly held together by fibrous tissue, which is condensed at their surfaces to form compact sheaths or external coats enveloping the erectile tissue of which each is composed. The sheaths of the corpora cavernosa are incomplete where they are in contact, permitting the erectile tissue to blend in the median line. This intercommunication is freer toward the anterior end of the penis than near its root, where the corpora cavernosa are more distinctly separated, preparatory to their divergence to form the crura.

The sheaths of the corpora cavernosa are composed of fibrous tissue containing an abundance of elastic fibres. From its inner surface each sheath gives off a number of fibrous bands, called "trabeculæ," which divide and anastomose with each other, forming the chief constituent of the erectile tissue. Within these trabeculæ are numerous bundles of smooth muscular tissue.

The erectile tissue is made up of these trabeculæ, which give it a spongy character and are covered with endothelial cells, converting the spaces between them into cavernous venous channels. These become engorged with blood during erection. The vessels supplying this blood are situated in the trabeculæ, and give off capillary branches, which open into the intertrabecular spaces, discharging blood into those enormously dilated venous radicles. Here and there arterial twigs, surrounded by an investment of fibrous tissue, project from the trabeculæ into the venous spaces. These, because of their twisted forms, have received the name helicine arteries (Figs. 197 and 198).

The structure of the corpus spongiosum is

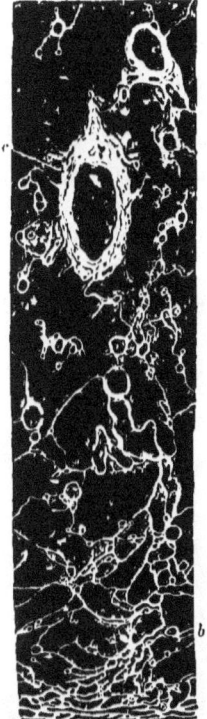

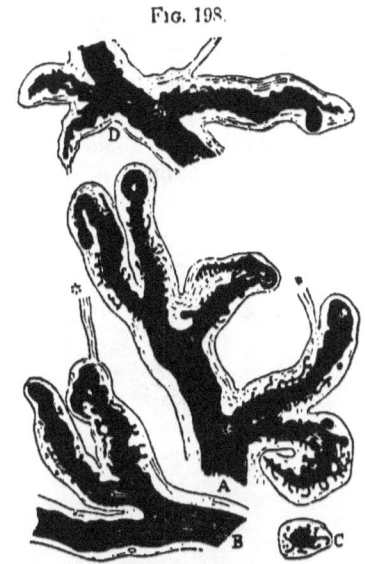

Fig. 198.

Fig. 197.—Section of injected corpus cavernosum. (Henle.) a, fibrous capsule; b, trabeculæ; c, section of the arteria profunda penis. All the spaces are filled with the material used for injection.
Fig. 198.—Helicine arteries. A, B, C, from the corpus cavernosum; D, from the corpus spongiosum; * *, fibrous bands forming a part of the trabecular network.

similiar to that of the corpora cavernosa, but the trabeculæ are more delicate and the spaces between them of more uniform size. Its sheath is studded with papillæ where it covers the glans, at the edge of which they are unusually large. They are covered with a layer of stratified epithelium, which conceals them over the surface of the glans, where they are comparatively small, but merely invests the larger ones at the corona. This layer of epithelium is continuous with that of the skin covering the rest of the penis, which is elsewhere loosely connected with the underlying structures by areolar tissue devoid of fat. The skin is without hairs on the anterior two-thirds of the penis, but contains sebaceous glands, which are especially numerous in the fold of the prepuce, where it is attached near the corona of the glans, glands of Tyson.

2. **The Prostate.**—This body is regarded as the analogue of the uterus, its utricle corresponding to the cavity of that organ. It has a fibrous investment, which merges into the areolar tissue connecting the prostate with the surrounding structures and, in its deeper portions, contains smooth muscular tissue, which accompanies it in forming the stroma of the organ. Within this stroma are the prostatic glands, composed of acini, lined with epithelium of the columnar variety, and opening into a series of ducts having their orifices in the floor of the urethra. The glandular alveoli frequently contain little concretions of a substance closely resembling amyloid, corpora amylacea, which often display a marked concentric lamination (Fig. 199).

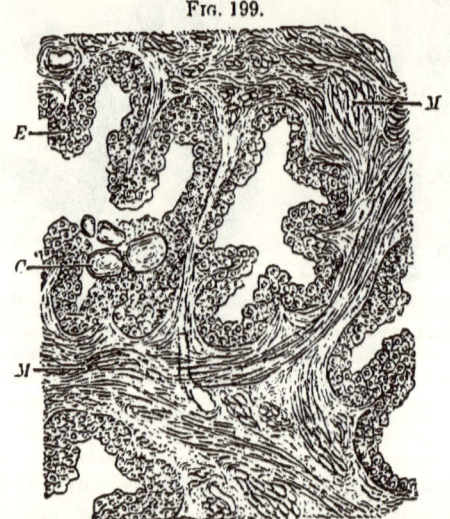

Fig. 199.

Section of the prostate. (Heitzmann.) Sections of one acinus and portions of three others are included in the figure. These are surrounded by fibrous tissue traversed by bundles of smooth muscular fibres. E, epithelial lining of the acini; M, M, smooth muscular tissue; C, concretions of amyloid material, showing concentric lamination.

The two ejaculatory ducts pass through the prostate to open into the urethra in its course within that organ. A little behind their orifices is the verumontanum, containing erectile tissue, which is supposed, during erection, to serve as a dam, preventing the entrance of semen into the bladder.

The ejaculatory ducts divide behind the prostate, one branch forming the duct of the seminal vesicle, while the other becomes continuous with the vas deferens.

3. **The Seminal Vesicles.**—These are tubular sacs ending in blind extremities, with occasional saccular branches given off from their sides. They are lined with a mucous membrane covered with columnar epithelium, resting upon areolar fibrous tissue. Outside of this is a muscular coat containing internal circular and external longitudinal fibres, and surrounded by an ill-defined fibrous coat that passes into the general areolar tissue of the region. The seminal vesicles sometimes contain semen, for which they may serve as a temporary reservoir, but they also secrete a fluid that is mixed with the semen at the time of ejaculation.

4. **The Vasa Deferentia.**—The vas deferens of each side resembles the seminal vesicle in structure. It is lined with columnar epithelium, beneath which is a layer of areolar fibrous tissue, resting upon the muscular coat. This is surrounded by fibrous tissue, becoming areolar as it blends with that of the neighboring parts. The muscular coat is thicker than that of the seminal vesicle, and is divisible into an inner layer of circular and an outer layer of longitudinal fibres. The mucous membrane, like that of the seminal vesicle, is thrown into folds, which are longitudinal throughout most of the course of the vas deferens, but are irregular in the sacculated distal portions of the tube, giving the surface a reticulated or alveolar appearance.

5. **The Epididymis.**—The vas deferens of each side becomes continuous with the canal of the epididymis, which is an enormously long tube, twenty feet, so convoluted and packed together as to occupy but little space. It is lined throughout with columnar epithelium, continuous with that of the vas deferens; but, except for a short distance from the junction with the vas, the cells possess cilia of considerable length, which induce currents toward the vas deferens. The muscular coat of the latter is continued in the epididymis, but is very thin. Opening into the canal of the epididymis are the vasa efferentia of the testis.

6. **The Testis.**—The testis is a compound tubular gland, of which the secretion contains the spermatozoa. The latter are derived from certain of the cells lining the tubules, and contain within their

structure a definite amount of chromatin and a centrosome. During the fertilization of the ovum this chromatin unites with a similar amount present in the egg-cell, and thus forms a complete cell, the nucleus of which contains equal amounts of chromatin from the male and female parents of the future offspring. We have seen (Chapter I.) that the nuclei of the cells throughout the body break up, during karyokinesis, into a definite and constant number of fragments, called "chromosomes," which split during metakinesis; one-half of each chromosome going to each of the daughter-nuclei. These chromosome-halves form a reticulum within the daughter-nuclei, and while in that form the chromatin appears to increase in amount, so that by the time the cell divides again the full supply of chromatin is present in its nucleus. During the two cell-divisions which immediately precede and result in the formation of the spermatozoa and the matured egg this growth of the chromatin does not take place, and, as we shall presently see, each spermatozoon or matured ovum contains but half of the chromosomes that are normally present in the somatic or general cells of the body. This "reduction of the chromatin" has been a matter of much study within the last few years, because of its probable bearing upon the problems of heredity. The fact of its occurrence is strongly confirmatory of the idea that the chromatin is the carrier of hereditary characteristics, the fertilized ovum receiving equal shares from both parents.

The tubular glands of the testis are enclosed in a strong fibrous capsule, made up of interlacing bands of fibrous tissue. This becomes continuous, behind, with a mass of areolar tissue containing the vascular supply of the organ and the epididymis, with the vasa efferentia opening into it. The fibrous capsule is called the "tunica albuginea." It is covered, except posteriorly, by the visceral portion of a serous membrane, the "tunica vaginalis." From the inner surface of the capsule numerous bands and strands of fibrous tissue, trabeculæ, traverse the glandular part of the organ, imperfectly dividing it into lobes, each of which contains several of the glandular or seminiferous tubes.

Upon the surfaces of the trabeculæ and upon the inner surface of the capsule the dense fibrous tissue of those structures passes into a delicate areolar tissue, which gives support to the numerous small bloodvessels and abundant lymphatics distributed within the organ. This vascular areolar tissue also penetrates between the seminiferous tubules, giving them support. In this region the interstitial tissue just mentioned contains large cytoplasmic cells of connective-tissue origin, which frequently contain globules of fat or granules of pigment, and in many instances, in man, have been observed to contain crystalloids of proteid nature. It has been surmised that these cells may serve for the storage of nutriment required by the active proliferation of the cells that produce the spermatozoa within the seminiferous tubes (Fig. 200).

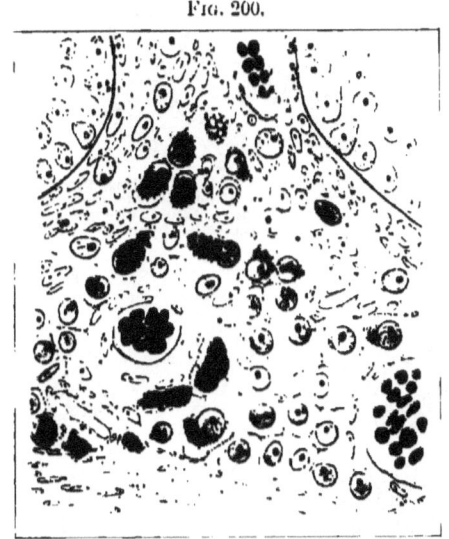

Fig. 200.

Interstitial tissue in the testis of the cat. (Plato.) Three bloodvessels are shown in either complete or partial section. Portions of two seminiferous tubules are represented at the upper corners. Between these structures is the interstitial tissue, containing large cytoplasmic cells. This tissue is rather more abundant in this instance than in the human subject.

Each seminiferous tube is provided with a basement-membrane, upon the inner surface of which are epithelial cells. These are divisible into three groups: first, a parietal layer of cells, the "spermatogonia," lying next to the basement-membrane; second, a layer of cells, often two or three deep, called the "spermatocytes," lying upon and derived from the spermatogonia; third, the "spermatids," lying most centrally. The spermatids are derived from the spermatocytes, and are the elements from which the spermatozoa develop, one spermatozoon being formed from each spermatid.

The cells of the parietal layer, that containing the spermatogonia, are not all alike. At intervals certain cells, called "sustentacular" cells, or the "cells of Sertoli," are differentiated from the others (Figs. 201-213). These sustentacular cells rest with a broad base, the

Fig. 201.

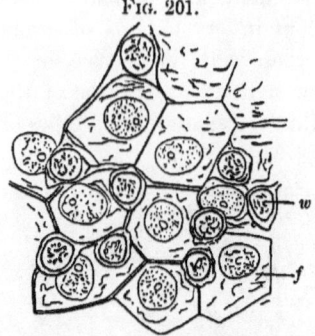

Superficial aspect of the parietal cells of the seminiferous tube; rat. (Ebner.) *f*, basal plates of the sustentacular cells (cells of Sertoli), each containing a large vesicular nucleus, poor in chromatin, and a distinct nucleolus of considerable size; *w*, spermatogonia resting upon the basal plates of the cells of Sertoli. Only a few of the spermatogonia are represented.

Fig. 202.              Fig. 203.

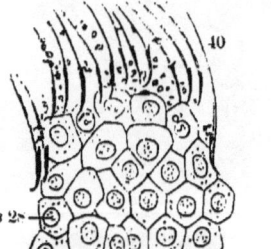 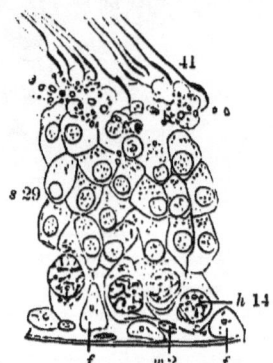

Sections from the testis of the rat, illustrating spermatogenesis. (Ebner.)

Figs. 202-213.—*w*, spermatogonia; *f*, sustentacular cells, or cells of Sertoli; *h*, spermatocytes; *s*, spermatids; *sp*, spermatids becoming transformed into spermatozoa; *w*1 to *w*10 traces the history of the spermatogonia from the resting condition to that in which they have grown to become primary spermatocytes. During this process they move from the parietal layer into that covering it. *h*11, a recently formed spermatocyte; *h*12 to *h*20, growth of the spermatocyte; *h*21, beginning of the division to form secondary spermatocytes; *h*22, its end; *h*23, secondary spermatocyte, with chromatin in open spirem; *h*24, division of the secondary spermatocyte to form two spermatids; *s*25, recently formed spermatid; *s*26 to *s*29, growth of the spermatid. (By this time the preceding crop of spermatozoa is fully developed and has been discharged into the lumen of the seminiferous tube.) *s*30 and *s*31, beginning transformation of the spermatids into spermatozoa. Their cytoplasm blends with that of the sustentacular cell. *sp*32 to *sp*39, stages in the differentiation of the spermatozoa; *40*, completed spermatozoon ready to pass into the lumen of the tube. *wf* (Fig. 212) and *wff* (Fig. 213) illustrate the division of the spermatogonia before they begin to develop into spermatocytes. It is supposed that the sustentacular cells aid in the nourishment of the spermatids during their transformation into spermatozoa, and that after the discharge of the latter the cytoplasmic process is retracted toward the basement-membrane, bringing with it the globules of fat and cytoplasmic fragments of the spermatids represented by dark spots and small round bodies in nearly all the figures. This retraction is taking place at *f*, Fig. 201. The cells of Sertoli do not appear to multiply; at least no karyokinetic figures have been observed in their nuclei.

## THE REPRODUCTIVE ORGANS.

Fig. 204.

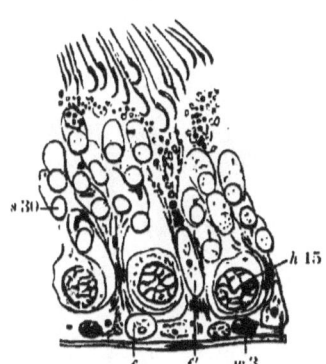

Fig. 205.

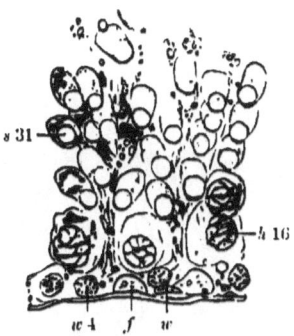

Fig. 206.

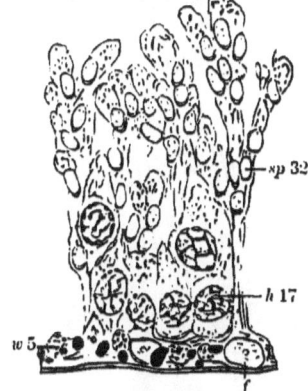

Fig. 207.

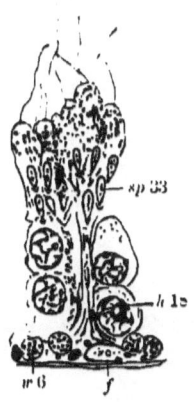

Fig. 208.

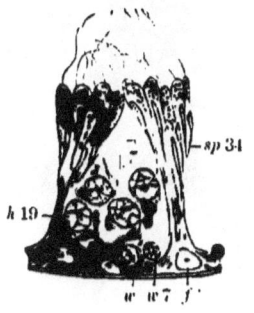

Fig. 209.

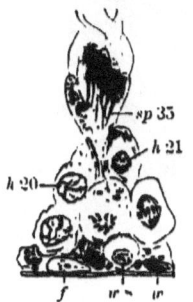

230  *NORMAL HISTOLOGY.*

Fig. 210.

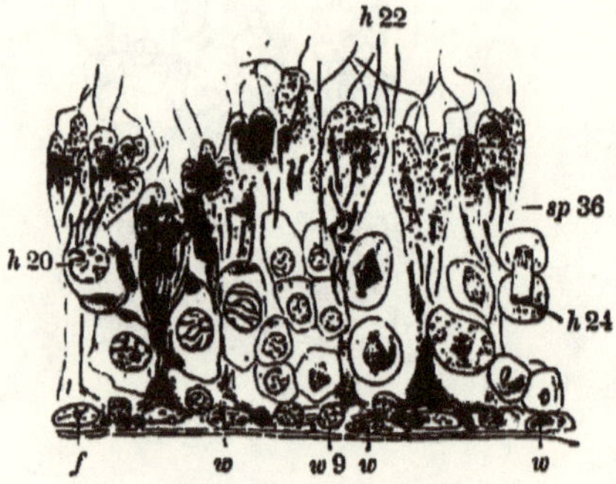

Fig. 211.

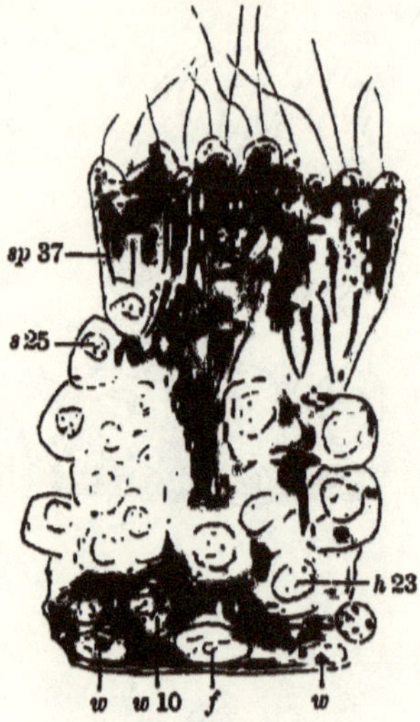

Fig. 212.

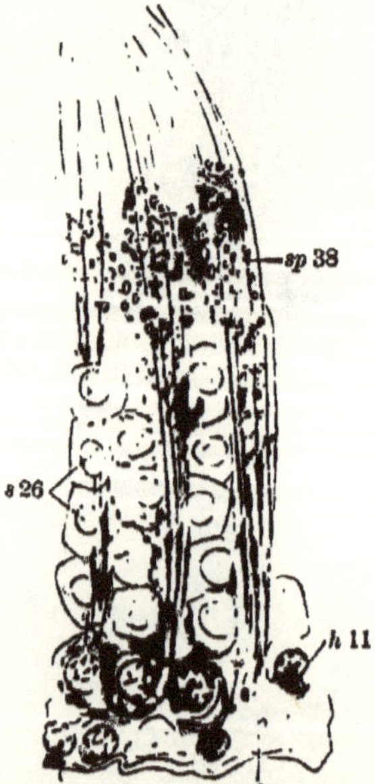

Fig. 213.

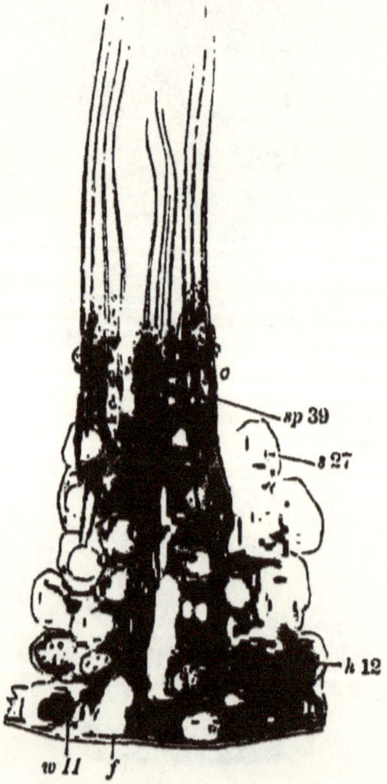

"basal plate," directly upon the basement-membrane, where the edges of the basal plates are in contact, forming a sort of bed with depressions in its upper surface, in which the spermatogonia find lodgement. The cells of Sertoli possess a thick cytoplasmic process, which extends toward the lumen of the tubule, and to which those spermatids which are developing into spermatozoa become attached. For this reason they are called sustentacular cells. Their nuclei differ from those of the neighboring spermatogonia in being less rich in chromatin and in possessing a single and prominent nucleolus.

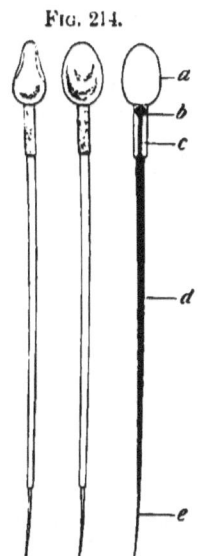

Fig. 214.

The appearances of the various cells enumerated depend upon the stage in their activity which happens to be under observation. The general course of development, ending in the formation of the spermatozoa, is as follows: the spermatogonia, between the cells of Sertoli, multiply until quite a collection of such cells is produced. Each division is followed by a period of rest, during which the chromatin increases in amount. When the final stage of rest is at an end and the cells have attained their maturity, they constitute what are called the primary spermatocytes. These now divide, each forming two secondary spermatocytes, which in turn divide, without an intermediate distinct resting-stage, to form two spermatids. Each primary spermatocyte, therefore, gives rise to four spermatids. It is during the division of the secondary spermatocytes that the reduction in chromatin, which was mentioned above, takes place (Figs. 202–213). Each spermatid receives, in addition to its portion of chromatin, a single centrosome.

Human spermatozoon. (Böhm and Davidoff, after Retzius and Jensen.) The left figure represents the side view and the middle figure surface-view of a spermatozoon. *a*, head (nucleus); *b*, end-knob (centrosome?); *c*, middle piece; *d*, tail of flagella; *e*, end-piece. The thickness of *d* may be owing to the presence of a sheath surrounding the actual flagella, which projects from the sheath at *e*.

The spermatozoon, then, is derived from a corpuscle, the spermatid, which contains all the essential organs of a cell, differing from the general cells of the body, the somatic cells, only in possessing half the usual number of chromosomes in its nucleus. It is unnecessary to pursue the chain of events through which the spermatid gives rise to the spermatozoon. It may suffice to state that the body of the latter consists of the chromatin of the nucleus; that the long cilium constituting the tail of the spermatozoon is developed from the cytoplasm; and that the centrosome of the spermatid is probably contained in the middle piece of the spermatozoon (Fig. 214). Even these conclusions are inferences from studies of spermatogenesis in the lower animals, and not from direct studies of that process in man. The latter undoubtedly conforms very closely to the former in all essential details.

To return to the histology of the testis: the epithelial cells of the seminiferous tubules rest upon a basement-membrane, which is divis-

Fig. 215.

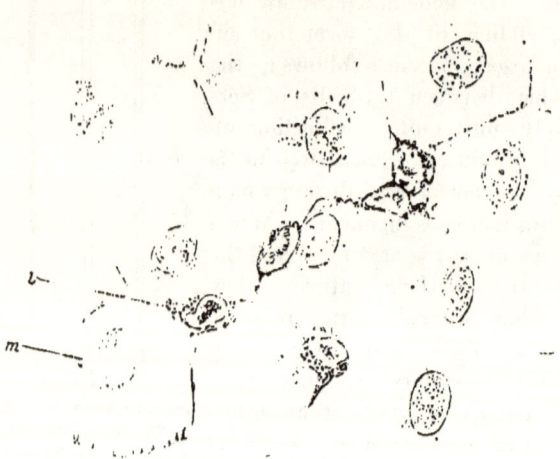

Basement-membrane from seminiferous tube of the rat. (Ebner.) *m*, endothelial cells composing the external layer; *l*, cells, presumably leucocytes, intercalated between the endothelial cells. The faint striations upon the endothelial cells represent wrinkles in the homogeneous membrane forming the inner surface of the basement-membrane; the wrinkling is probably due to a slight shrinkage of the endothelium.

ible into two layers: first, an internal, extremely delicate, homogeneous membrane, upon which the epithelial cells rest; and, second, a layer of endothelial cells (Fig. 215). The latter may bound, at least in places, the lymphatic spaces, which are abundant in the interstitial tissue of the testis.

Toward the back of the testis the seminiferous tubules unite with each other and open into a number of straight ducts of smaller diameter, called the "vasa recta." These are lined with a cubical epithelium resting upon an extension of the basement-membrane of the seminiferous tubes, and, in turn, open into a reticulum of tubules of larger diameter, situated in the mass of areolar tissue at the posterior aspect of the testis. This reticulum is called the "rete vasculosum," and the tubules composing it are lined with a low epithelium, apparently resting upon the surrounding fibrous tissue, without an intermediate basement-membrane. These tubes permit an accumulation of semen before it enters the vasa efferentia.

The vasa efferentia have a peculiar epithelial lining, which may be regarded as transitional between the cubical epithelium of the vasa recta and rete and the ciliated columnar variety lining the epididymis. It consists of alternating groups of cubical and ciliated columnar epithelial cells (Fig. 216).

FIG. 216.

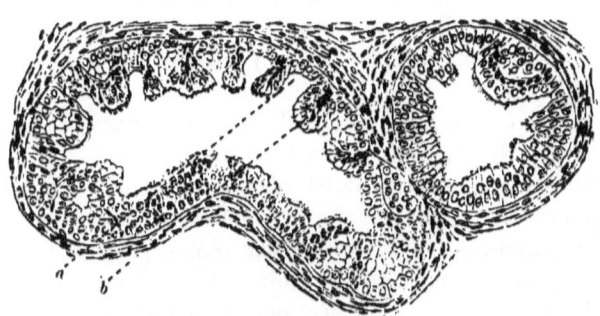

Section of vasa efferentia from human testis. (Böhm and Davidoff.) a, cubical or secretory epithelium; b, columnar ciliated epithelium, with deeper pyramidal cells beneath those that bear the cilia. This form of ciliated epithelium corresponds to that found in the epididymis where the cubical epithelium is absent.

The vasa efferentia, as already stated, open into the canal of the epididymis, through which their contents reach the vas deferens. The walls of the efferent tubes possess a layer of encircling smooth muscular fibres, which are reinforced in the epididymis by an additional external layer of longitudinal fibres.

The nerves supplied to the testis are destitute of ganglia, and are distributed to the vessels and surfaces of the seminiferous tubules. No terminations have been traced to the epithelial lining of those tubules.

## CHAPTER XVIII.

### THE CENTRAL NERVOUS SYSTEM.

The functional part, or parenchyma, of the central nervous system is composed of ganglion-cells with their processes. Some of these processes are of cytoplasmic nature, and, as explained in the chapter on the elementary tissues, are called the protoplasmic processes. From each ganglion-cell at least one process is given off which differs from the protoplasmic processes, and is called the "axis-cylinder process." This in most cases becomes the axis-cylinder of a nerve-fibre, and may be invested with a medullary sheath and neurilemma at some point near or at some distance from its exit from the cell.

It will be convenient, for the brief description of the central nervous system to which this chapter must be restricted, to adopt a special terminology for the different portions of the ganglion-cell and its processes, as follows: the term *ganglion-cell* will be restricted to the nucleus and the cytoplasm surrounding it; the protoplasmic processes will be called the *dendrites*, and their terminations the *teledendrites*. The axis-cylinder process will be termed the *neurite*; the delicate branches it may give off in its course, the *collaterals*; and the terminal filaments of the main trunk, collectively the *teleneurites*. The cell, with its processes and their terminations, will collectively constitute a *neuron*.

A complete neuron, then, consists of (1) certain teledendrites, which unite to form one or more dendrites connecting them with the ganglion-cell; (2) the cell itself; and (3) one or more neurites, which may give off collaterals and finally terminate in teleneurites (Fig. 217).

At the present time these neurons are believed to be without actual connection with each other, but to convey nervous stimuli by contact. The course of the nervous impulses is from the teledendrites to the nerve-cell, and thence, by way of the neurite, to the teleneurites, whence it is communicated, without a direct structural union, to the next tissue-element in the chain of nervous transmission. Those neurites which carry stimuli from the nerve-centres to the periphery, centrifugal impulses, form the axis-cylinders of some of the nerves. The axis-cylinders of those nerves which convey impulses from the periphery toward the nervous centres,

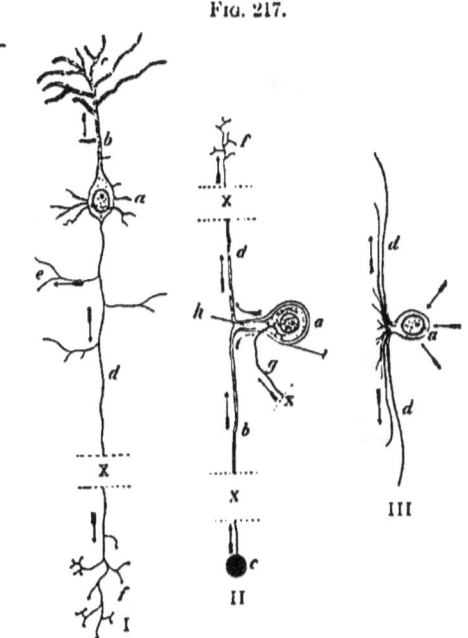

Fig. 217.

Sketch illustrating the composition of neurons. I, a neuron transmitting centrifugal impulses. II, a neuron receiving and transmitting centripetal impulses. III, a neuron, the function of which is supposed to be the distribution of impulses within the nerve-centre in which it is situated. *a*, ganglion-cell; *b*, dendrite; *c*, teledendrites; *d*, neurite; *e*, collaterals; *f*, teleneurites. In II the body *c* represents some sensory organ imparting nervous impulses to the teledendrites of a sensory nerve. The nervous filament *g* is a neurite, presumably derived from the sympathetic nervous system, leading to teleneurites applied to a ganglion-cell, *a*, of a posterior spinal ganglion. The portion *h* of the "nerve" springing from that cell is regarded as a portion of the cell itself. In the embryonic condition the dendrite and neurite both spring directly and separately from the body of the cell, the portion *h* being a subsequent development. *i*, endothelial envelope surrounding the ganglion-cell. III represents a ganglion-cell, apparently devoid of distinct dendrites, but having numerous processes that at first appear protoplasmic, but soon assume the characters of neurites. These cells are found in the retina and olfactory bulb, and have been termed spongioblasts, cellulas amacrinas, and parareticular cells. It is thought that nervous stimuli are received directly by the cytoplasm of the cell, without the intermediation of dendrites. *x* represents the omission of a portion of a fibre. The arrows indicate the directions taken by nervous impulses.

centripetal stimuli, may be the dendrites connected with ganglion-cells in or near those centres; *e. g.*, in the posterior root-ganglia of the spinal nerves, or they may be the neurites springing from peripheral ganglion-cells, as is exemplified in many, if not all, of the organs of special sense.

## I. THE SPINAL CORD.

The axis of the spinal cord is composed of a column of gray matter containing numerous ganglion-cells and nervous filaments held in position by a cement-substance, neuroglia-cells, the fibrous prolongation of the ependyma cells lining the central canal, and a little fibrous tissue accompanying the vessels derived from the pia mater.

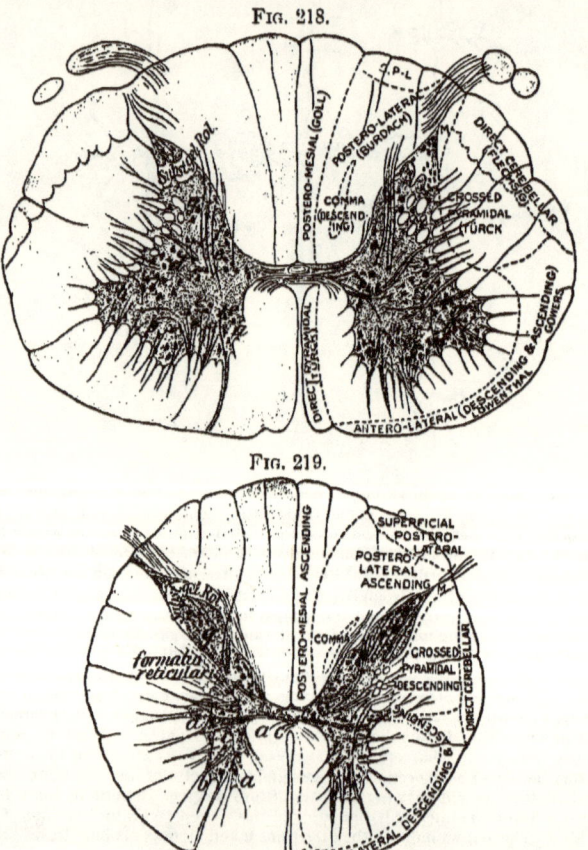

FIGS. 218 and 219.—Transverse sections of human spinal cord. (Schäfer.) Fig. 218, from the lower cervical region; Fig. 219, from the middle dorsal region. *a, b, c,* groups of ganglion-cells in the anterior horn; *d,* cells of the lateral horn; *e,* middle group of cells; *f,* cells of Clarke's column; *g,* cells of posterior horn; *c, c,* central canal; *a, c,* anterior commissure of white matter.

Fig. 220

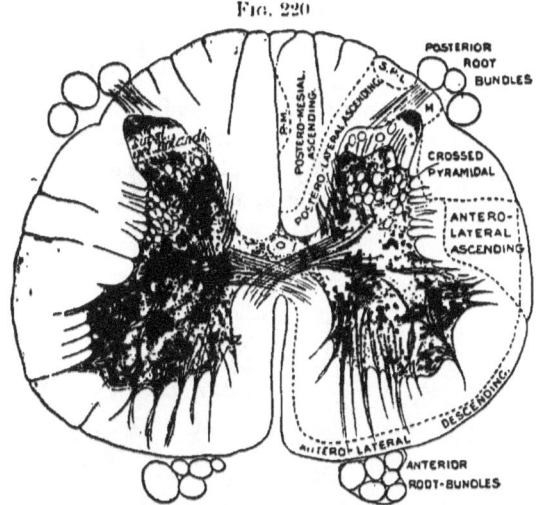

Transverse section of human spinal cord, from the middle lumbar region. (Schäfer.) *a b, c*, groups of ganglion-cells in the anterior horn; *d*, cells of the lateral horn; *e*, middle group of cells; *f*, cells of Clarke's column; *g*, cells of posterior horn; *c. c*, central canal; *a, c*, anterior commissure of white matter.

In cross-section this column of gray matter presents a transverse central portion, the gray commissure, near the middle of which is the central canal. At each side this gray commissure blends with masses of gray matter, occupying nearly the centre of each lateral half of the cord and having a general crescentic form. The ends of these crescentic masses form the anterior and posterior cornua of the gray matter, from which the anterior and posterior roots of the spinal nerves proceed. The anterior cornua are larger than the posterior and contain larger ganglion-cells.

Surrounding the column of gray matter everywhere, except at the bottom of the posterior median fissure of the cord, and the interruptions formed by the nerve-roots in their exit from the gray matter, is a layer of white matter, formed of medullated nerve-fibres running parallel with the axis of the cord and held together by neuroglia and delicate vascularized fibrous bands proceeding from the deep surface of the pia mater.

The white matter of the cord has been divided into a number of columns, for the most part indistinguishable through structural differences, but each containing fibres that play similar functional *rôles*. These columns, with their names, are indicated in Figs. 218, 219, and 220. The columns of Goll and Burdach, forming the posterior column of the white matter, between the posterior cornua and the posterior median fissure, conduct, for the most part, centripetal impulses. Impulses having the same upward direction are also conveyed by the direct cerebellar tract and the tract of Gowers in the lateral column of the white matter. Centrifugal impulses, motor stimuli, are conveyed by the fibres in the direct pyramidal tract of the anterior column and by those of the crossed pyramidal

FIG. 221.

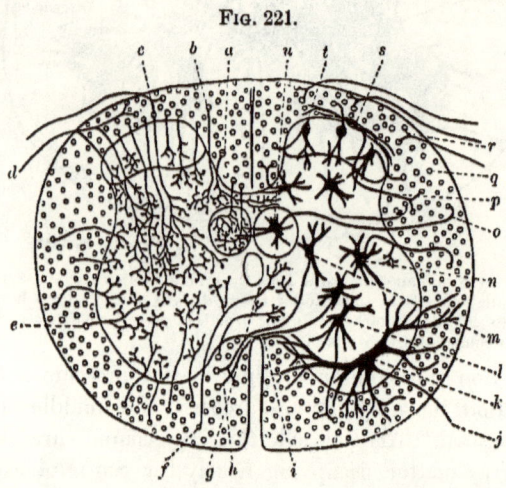

Diagram of spinal cord, illustrating the associations of its various nervous elements. (R. y Cajal.) *a*, collateral from Goll's tract, entering into the formation of the posterior commissure; *b*, collateral to the posterior horn; *c*, collateral to the formatio reticularis and the anterior horn; *d*, posterior nerve neurite, with its collaterals; *e*, collaterals from the lateral column; *f*, collaterals to the anterior commissure; *g*, central canal; *h*, neurite in the crossed pyramidal tract from the commissure-cell of the opposite side; *i*, its course in the commissure; *j*, neurite from a large motor cell in the anterior horn *k*; *l*, cell of the anterior horn, giving off a neurite dividing into an ascending and a descending branch (compare Fig. 224, *D*); *m*, commissure-cell; *n*, cell giving off a collateral within the gray matter; *o*, neurite of the cell *u*, in Clarke's column; *p*, neurite from the marginal cell *s*, of the substance of Rolando; *q*, cross-section of an axis-cylinder (neurite) in the white substance of the cord; *r*, division of a posterior nerve-fibre (neurite) into ascending and descending branches; *t*, small cell in the substance of Rolando. Aside from the cells indicated in the figure, the gray matter contains some that give off neurites which divide into two or three branches while in the gray matter, the branches going to different columns of white matter. There are also cells with very short neurites, which terminate in teleneurites within the gray matter, and probably distribute nervous impulses for short longitudinal distances.

tract in the lateral column. The tracts hitherto considered contain fibres that are continued into the higher nerve-centres of the brain and cerebellum, to or from which they convey nervous impulses. But the spinal cord is not merely a collection of such transmitting fibres. It is also a nerve-centre of complex constitution, in which neurons terminate in teleneurites or arise in teledendrites.

Some of the neurons within the cord are confined to its substance, and constitute nervous connections between the different parts at various levels. These may be termed longitudinal commissural neurons, or association-fibres. Portions of such neurons are represented in the diagram of a cross-section of the cord (Fig. 221), which also contains representations of some of the neurites in the posterior spinal nerve-roots, with their collaterals ending in teleneurites within the gray matter (*d*). On the right side of the figure, the nerve-cells, with their dendrites and the beginning of the neurites, are shown. On the left side the neurites connected with cells at another level are shown, re-entering the gray matter, where they terminate in teleneurites. In studying this figure it must be borne in mind that the teledendrites of the neurons on the right are in close relations with the teleneurites of other neurons, and that the teleneurites represented on the left are in close relations with the teledendrites of other neurons. These association-neurons are, therefore, merely links in chains of communicating neurons. They are again represented in Fig. 224, *D* and *E*.

Aside from these association-neurites, the gray matter of the cord receives innumerable collaterals from the neurites forming the axis-cylinders of the nerves in the various columns of the white matter. These collaterals terminate in teleneurites, which are in close relations with the teledendrites of the neurons arising in the cord. The distribution of these collaterals is represented in Fig. 222. The collaterals from the anterior column enter the anterior horn of the gray matter, where they are chiefly distributed about the large ganglion-cells in the antero-lateral portion of its substance (Fig. 218, *b*; Fig. 221, *j*), but may also extend to other parts of the gray matter. The collaterals from the fibres in the lateral columns of the white matter are most numerous near the posterior horn, which they enter, many of them passing through the gray matter behind the central canal and forming a part of the posterior or gray commissure of the cord (Fig. 222, I). The collaterals from the posterior column are divisible into four groups: first, those which are given off in the lateral portion of that column (Fig. 222, G), and are distributed in the outer portion of the posterior horn and in the substance of Rolando (Fig. 222, I); second, those which end in Clarke's column (Fig. 222, J); third, those which arise chiefly in the column of Goll, pass through the substance of Rolando, and then form an expanding bundle distributed in the anterior horn of the gray matter, where they are in association with the dendrites of the motor cells in that region (these fibres form the reflex bundle of Kölliker, Fig. 222, H); fourth, collaterals springing from fibres in the posterior column, passing

FIG. 222.

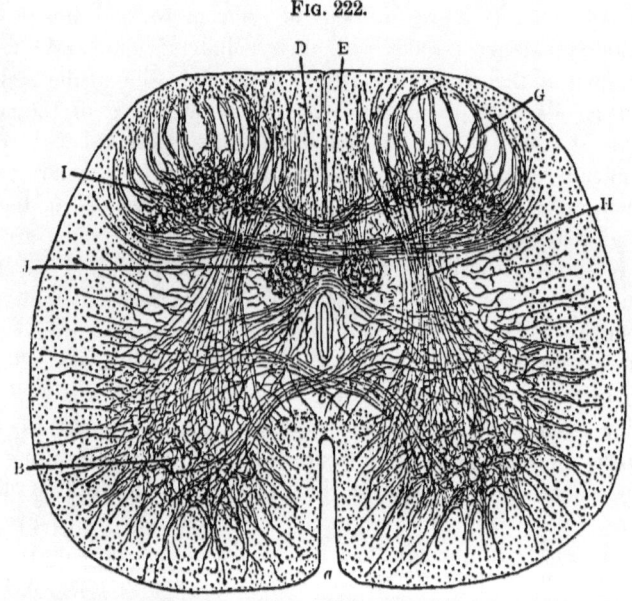

Cross-section of the spinal cord of a newborn child, showing the distribution within the gray matter of the collaterals from the neurites of the white matter. (R. y Cajal.) a, anterior fissure; B, pericellular branches of the collaterals from the anterior column; C, collaterals of the anterior commissure; D, posterior bundle of collaterals in the posterior commissure; E, middle bundle of the posterior commissure; f, anterior bundle; G, collaterals from the posterior column; H, senso-motory collaterals from the posterior column; I, pericellular terminations of collaterals in the posterior horn; J, collateral terminations in the column of Clarke.

through the posterior commissure of gray matter and ending in the substance of Rolando of the opposite side (Fig. 222, D).

The reflex collaterals arising in the posterior column are shown in Fig. 223, where their telencurites are in close relations with the teledendrites of the motor cells c.

The centripetal or sensory neurites of the posterior spinal nerve-roots spring from the ganglion-cells of the spinal ganglia. When they have entered the white matter of the spinal cord they divide into two branches (Fig. 221, c). One of these ascends in the white substance and the other descends. Both branches give off numerous collaterals, which penetrate the gray matter, ending in teleneurites associated with the teledendrites of the cells in both the anterior and the posterior horns, and the column of Clarke. The main branches of the sensory neurite also enter the gray matter, after

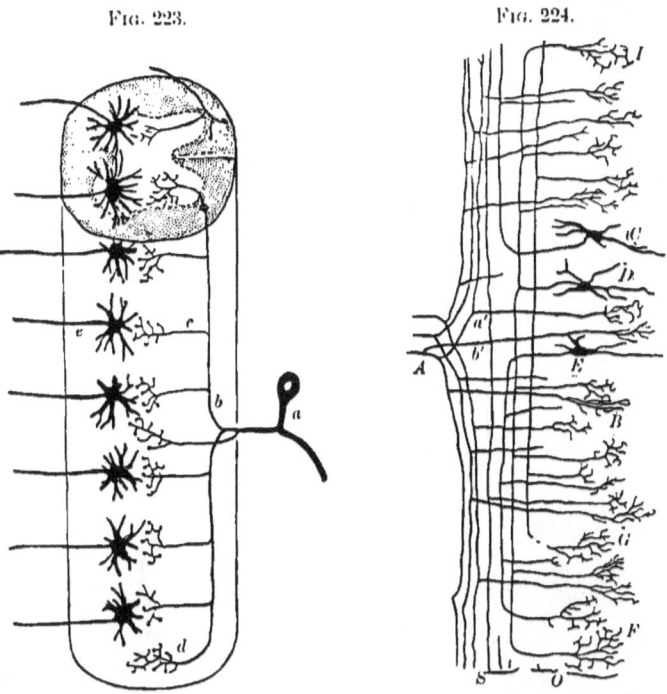

Fig. 223. Fig. 224.

Fig. 223.—Diagram of the senso-motory reflex collaterals in the cord. (R. y Cajal.) *a*, ganglion-cell of the posterior nerve-root; *b*, division of its neurite into ascending and descending branches; *c*, collaterals to anterior horn; *d*, terminal teleneurites in the posterior horn; *e*, motor cell of the anterior horn, with its processes.

Fig. 224.—Longitudinal section of a part of the spinal cord, including a posterior nerve-root. Semidiagrammatic. (R. y Cajal.) *A*, posterior nerve-root; *S*, white substance of the cord; *O*, gray matter; *B*, collateral teleneurites in the gray matter; *C*, cell with a single ascending neurite; *D*, cell with bifurcating neurite, terminating at *F* and *I*; *E*, cell with a single descending neurite; *F*, *G*, terminal teleneurites; *a'*, collateral from a branch of the posterior root-neurite; *b'*, collateral from the main neurite before its bifurcation.

following the posterior column for a short distance, and end in teleneurites among the cells of the posterior horn and the substance of Rolando. The collaterals which pass to the anterior horns (Fig. 222, H, and Fig. 223, c) have to do with the origin of reflex centrifugal impulses emanating from the motor cells in that region (Fig. 223, e, and Fig. 221, j). The further transmission of these centripetal stimuli toward the higher nerve-centres of the brain probably takes place: first, through the cells in the posterior horns, the neurites from which pass into the lateral columns and there ascend the cord; second, through the cells of Clarke's column, which also send neurites into the lateral column, where they enter the direct cerebellar tract (Fig. 221, o; see also Fig. 224). In addition to these centripetal or sensory neurites, the posterior nerve-roots contain a few centrifugal neurites.

FIG. 225.

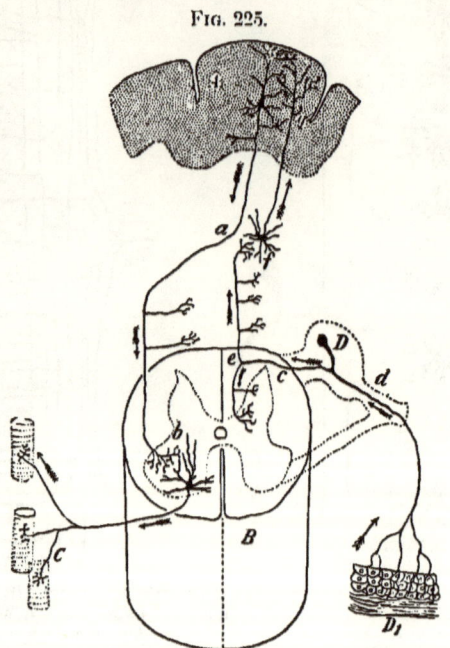

Diagram of a sensory and a motor tract. (R. y Cajal.) A, psycho-motor region in cerebral cortex; B, spinal cord; C, voluntary muscle; D, spinal ganglion; D', skin; a, axis-cylinder of a neuron extending from the cerebral cortex to the anterior horn of the spinal cord, where the terminal teleneurites are in relations with the teledendrites of the motor cell at b. The sensory stimulus arising in the skin, D', is transmitted by the neuron dDee to f, where it is communicated to the neuron fg. The point f may be in the cord or in the medulla oblongata.

In order to understand the origin of the anterior spinal nerve-roots we must first consider the course of the centrifugal neurites in the pyramidal tracts (Figs. 218, 219, 220). These enter the gray matter and end in teleneurites, which are associated with the teledendrites of the cells in the anterior horn, especially those which give off neurites to the anterior roots of the spinal nerves (Fig. 221, *j*).

The foregoing details may be summarized by means of the accompanying diagram (Fig. 225), in which the course of a nervous stimulus is traced from the organ of sense in, *e. g.*, the skin, to the cortex of the cerebrum, where it is translated into a nervous impulse, the course of which is traced to the motor plates of the voluntary muscles. The reflex mechanism which might at the same time be set into operation is not represented in the diagram, but will be sufficiently obvious from an inspection of Fig. 223. It will be noticed in Fig. 225 that both the sensory stimulus and the motor impulse are obliged to pass through at least two neurons before they reach the ends of their journeys. But the nervous currents are by no means entirely confined to the course marked by the arrows. Impulses may be transmitted in an incalculable number of delicate tracts through the collaterals given off from the neurites within the central nervous system, some of which are indicated in the diagram, and all of which end in teleneurites associated with the teledendrites of, perhaps, several neurons. One of these collateral tracts has already been considered, namely the senso-motory reflexes illustrated in Fig. 223.

## II. THE CEREBELLUM.

The cerebellum is subdivided into a number of laminæ by deep primary and shallow secondary fissures. The gray matter of the organ occupies the surfaces of these laminæ, while their central portions are composed of white matter. The gray matter may be divided into two layers: an external or superficial "molecular layer" and an inner "granular layer" (Figs. 226 and 227).

The molecular layer contains two forms of nerve-cells: first, the large cells of Purkinje; second, small stellate cells.

The cells of Purkinje have large, oval, or pear-shaped bodies lying at the deep margin of the molecular layer. Their dendrites form an intricate arborescent system of branches extending peripherally to the surface of the gray matter, and give off innumerable small teledendrites throughout their course. All these branches lie in one plane, perpendicular to the long axis of the lamina in which they are situated, and the teledendrites come into relations with certain longitudinal neurites springing from the cells of the granular layer, to be presently described. The neurites of the cells of Purkinje extend through the granular layer into the white matter and soon acquire medullary sheaths (Fig. 226, o); but before they leave the granular layer they give off collaterals, which re-ascend into the molecular layer, where their telencurites are in relations with the

FIG. 226.

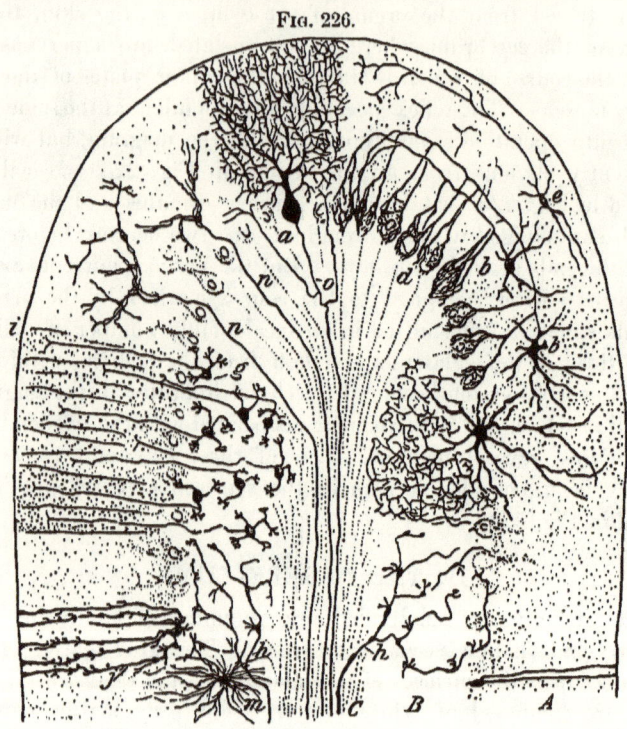

Section of a cerebellar lamina perpendicular to its axis. (R. y Cajal.) *A*, molecular layer of the gray matter; *B*, granular layer; *C*, white substance; *a*, cell of Purkinje; *o*, its neurite, giving off two recurrent collaterals; *b, b*, stellate cells of the molecular layer; *d*, basket-like distribution of the telencurites of one of their collaterals around the body of a cell of Purkinje; *e*, superficial stellate cell, which does not appear to come into relations with the bodies of the cells of Purkinje, but must lie close to their dendrites; *f*, large stellate cell of the granular layer; *g*, small stellate cell of the granular layer; *h*, centripetal neurite of a "moss" fibre; *n*, centripetal neurite distributed in the molecular layer; *j, m*, neuroglia-cells. The arborescent dendrites of only one of the cells of Purkinje are represented in the figure. Were those of the neighboring cells also represented, the molecular layer of the gray matter would display an enormously complex interdigitation of such filaments.

teledendrites of neighboring cells of Purkinje. These collaterals are believed to occasion a certain co-ordination in the action of those cells of Purkinje which are near each other.

The stellate cells of the molecular layer (Fig. 226, *b, e*) possess neurites, which lie in the same plane with the arborescent dendrites of the cells of Purkinje, and send collaterals to end in a basket-work of teleneurites applied to the bodies of the cells of Purkinje. The terminal teleneurites of these stellate cells also end in the same situation. Other smaller collaterals extend toward the surface of the cerebellar lamina.

The granular layer of the gray matter also contains two varieties of nerve-cells: the "small stellate cells," which are most numerous, and the "large stellate cells."

Fig. 227.

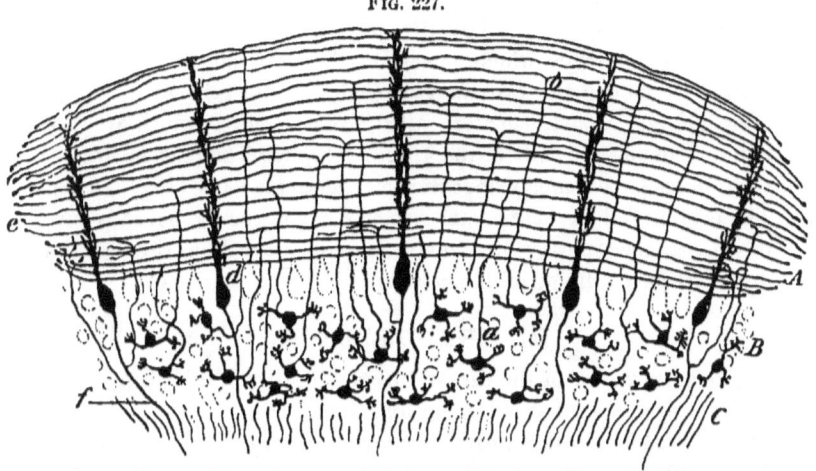

Section of a cerebellar lamina parallel to its axis. (R. y Cajal.) *A*, molecular layer of the gray matter; *B*, granular layer; *C*, white substance; *a*, small stellate cell of the granular layer, from which a neurite enters the molecular layer, where it bifurcates, sending branches throughout the length of the lamina; *b*, bifurcation of one of these neurites; *c*, slightly bulbous termination of one of the neuritic branches; *d*, body of a cell of Purkinje seen in profile; *f*, neurite of a cell of Purkinje.

The small stellate cells (Fig. 226, *g*, and Fig. 227, *a*) are scattered throughout the granular layer, and it is owing to the abundance of their nuclei that this layer has received that name. Their dendrites are few in number and short, but their neurites are very long. They extend perpendicularly into the molecular layer, where they bifurcate, the branches lying parallel with the axis of the cerebellar lamina and its surface. These fibres appear to run the whole length of the lamina, and to come in contact with the teledendrites of the cells of Purkinje, to the planes of which they run perpendicularly. They are thought to coördinate the action of a long series of the cells of Purkinje.

The large stellate cells of the granular layer lie near its external margin, whence they send their dendrites into a large area of the molecular layer, while their neurites are distributed in the granular layer, where they must come into relations with the dendrites of the small stellate cells (Fig. 226, $f$).

The distribution of the cells and their processes in the cerebellum indicates a very complex interchange of nervous impulses and an extraordinary coördination in the action of the various neurons.

This complication is still further increased by the presence of centripetal neurites, which enter the cerebellum through the white matter and are distributed in the gray matter. These are of two sorts: first, neurites which penetrate the granular layer and are distributed among the proximal dendrites of the cells of Purkinje (Fig. 226, $n$); second, neurites, called "moss" fibres, which are distributed among the cells of the granular layer. The teleneurites of these fibres have a mossy appearance, whence the name (Fig. 226, $h$). The origin of these centripetal neurites is not known, but it is surmised that the "moss" fibres may enter the cerebellum through the direct cerebellar tracts of the cord.

## III. THE CEREBRUM.

The gray matter of the cerebral cortex has been divided into four layers: first, an external molecular layer; second, the layer of small pyramidal cells; third, the layer of large pyramidal cells; and, fourth, an internal layer of irregular or stellate cells. Of these layers, the second and third are not clearly distinguishable from each other (Fig. 228).

The molecular layer contains three sorts of nerve-cells, two of which are closely related to each other, differing only in the form of the cell-bodies, which are small in both varieties (Fig. 229, $A$, $B$, and $C$); while the cell-bodies of the third variety are large and polygonal (Fig. 229, $D$). The small cells ($A$, $B$, $C$, Fig. 229) possess two or three tapering processes, which at first resemble protoplasmic processes, but soon assume the characters of neurites or axiscylinders. These neurons, then, resemble the type depicted in Fig. 217, III. Their neurites run parallel to the surface of the convolution in which they are situated, sending off numerous perpendicular collaterals, and finally end in teleneurites within the molecular layer. The collateral and terminal teleneurites are probably in relations with the dendrites of the pyramidal cells of the underlying layers, which form arborescent expansions in the molecular layer, similar to those of the cells of Purkinje in the cerebellum, extending to the surface of the gray matter.

The large stellate cells of the molecular layer (Fig. 229, *D*) send their dendrites in various directions into the molecular layer and the layer of small pyramidal cells lying beneath it. The neurite is distributed in the molecular and upper portions of the underlying layers, but is never extended into the white matter. The dendrites of these cells come into relations with the neurites of the other cells of this layer and with those that proceed upward from some of the cells in the deeper layers.

The small spindle- and stellate cells (*A, B, C*, Fig. 229) are considered to be the autochthonous cells of the cerebral cortex—*i. e.*, the cells of the brain in which the highest order of nervous impulses find their origin. The small spindle-shaped cells, with their peculiar neurites, are extremely abundant and fill the molecular layer with a mass of interwoven filaments.

The second and third layers of the cerebral gray matter are characterized by the presence of pyramidal nerve-cells of various sizes, the smaller being relatively more abundant in the second

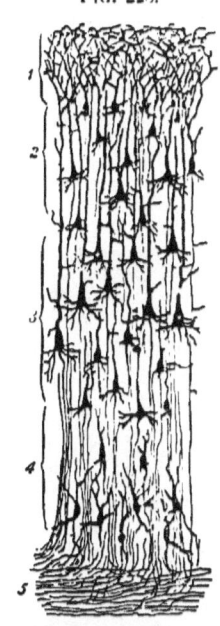

FIG. 228.

Vertical section of the cerebral cortex, showing its layers. (R. y Cajal.) *1*, molecular layer; *2*, layer of the small pyramidal cells; *3*, layer of the large pyramidal cells; *4*, layer of polymorphic cells; *5*, white matter.

layer and the larger in the third layer. From the apex of the pyramidal cell a stout, "primordial" dendrite passes vertically into the molecular layer, where, as well as during its course to the molecular layer, it gives off numerous branches, and finally ends in a brush of teledendrites extending to the surface of the gray matter (Fig. 230, *A, B*). Other and shorter dendrites are given off from the body of the cell, which ramify and end in the second, third, or fourth layer of the gray matter. The neurites from the bases of the pyramidal cells pass vertically downward into the white substance, where they may bifurcate, giving axis-cylinders to two nerve-fibres. While within

FIG. 229.

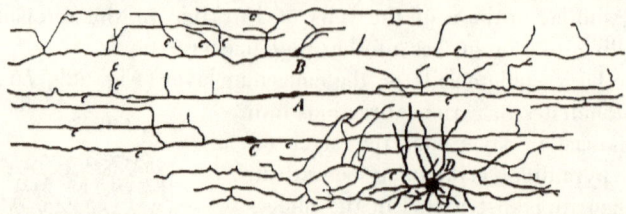

Cells of the molecular layer of the cerebral cortex. (R. y Cajal.) *A, C,* small spindle-shaped cells; *B,* small stellate cell; *D,* large stellate cell. The branches marked *c* are neurites.

FIG. 230.    FIG. 231.

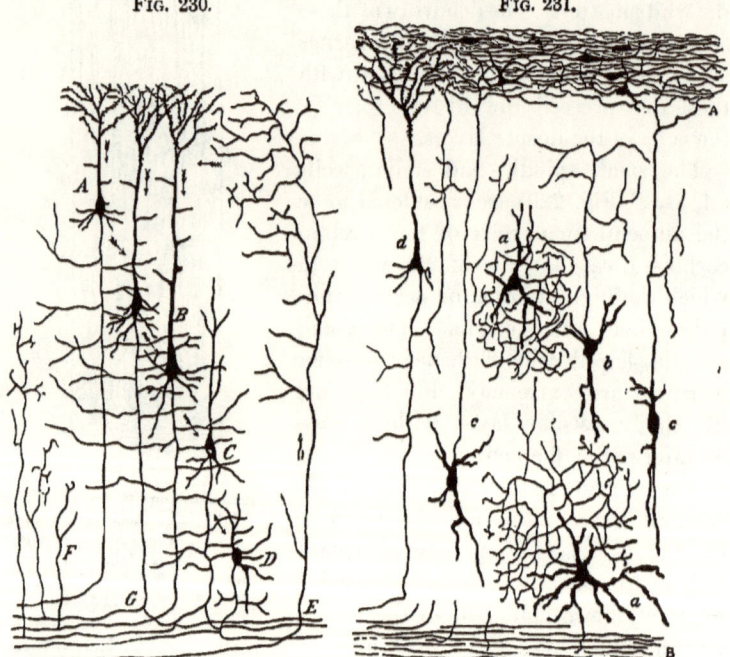

Fig. 230.—Diagrammatic section through the cerebral cortex. (R. y Cajal.) *A,* small pyramidal cell in the second layer; *B,* two large pyramidal cells in the third layer; *C, D,* polymorphic cells in the fourth layer; *E,* centripetal neurite from distant nerve-centres; *F,* collaterals from the white substance; *G,* bifurcation of a neurite in the white substance. The arrows indicate the centripetal and centrifugal courses of nerve-impulses, but it is probable that centripetal impulses have to pass through other neurons (perhaps the spindle-cells of the molecular layer) before they are translated into centrifugal impulses.

Fig. 231.—Cells with short neurites in the cerebral cortex. (R. y Cajal.) *A,* molecular layer; *B,* white substance; *a,* cells with neurites, which speedily divide into numerous teleneurites in the neighborhood of the cell belonging to the same neuron; *b,* cell with a neurite extending vertically toward, but not entering, the molecular layer; *c,* cell with a neurite distributed within the molecular layer; *d,* small pyramidal cell.

the gray matter, and after their entrance into the white matter, these neurites give off collaterals, which branch and end in terminal bulbous expansions without breaking up into a set of teleneurites.

The irregular cells of the fourth layer (Fig. 230, C, D) do not send their dendrites into the molecular layer, but distribute them within the deeper layers of the gray matter. Their neurities, like those of the pyramidal cells, enter the white matter, where they may or may not bifurcate.

Besides the cells in the deeper layers of the gray matter hitherto described, those layers contain cells with short neurites, which are divisible into two classes: first, spindle-shaped or stellate cells, sending their neurites into the molecular layer (Fig. 231, c) or into the second layer of the gray matter (Fig. 231, b); second, polymorphic cells with radiating dendrites and copiously branching neurites, both of which are distributed within a short distance of the cell. These cells are believed to distribute nervous impulses to the neurons in their vicinity.

The gray matter of the cortex also receives centripetal neurites from the white matter, which give off numerous collaterals and terminate in the molecular layer.

The white matter of the cerebrum contains fibres that may be divided into four groups: first, centrifugal or "projection" fibres; second, "commissure-fibres," which bring the two sides of the brain into coördination (these lie in the corpus callosum and in the anterior commissure); third, "association-fibres," which coördinate the different regions of the cerebral cortex on the same side; fourth, centripetal fibres, reaching the cortex from the peripheral nervous system or cord.

The centrifugal or projection-fibres arise from all parts of the cortex, springing from the pyramidal and, perhaps, also from the irregular cells. Many of these fibres give off a collateral, which passes into the corpus callosum, to be distributed in the cortex of the opposite side, commissural collaterals, and then pass on to the corpus striatum, to the gray matter of which further collaterals may be given off, after which the main neurite probably passes into the pyramidal tracts of the cord through the cerebral crus (Fig. 232, a).

The commissure-fibres (Fig. 232, b, c) also arise from the pyramidal cells of the cortex, mostly from the smaller variety, and pass into the corpus callosum or the anterior commissure, to be distributed in the gray matter of the cortex of the opposite hemisphere, but not necessarily to the corresponding region. These commissural

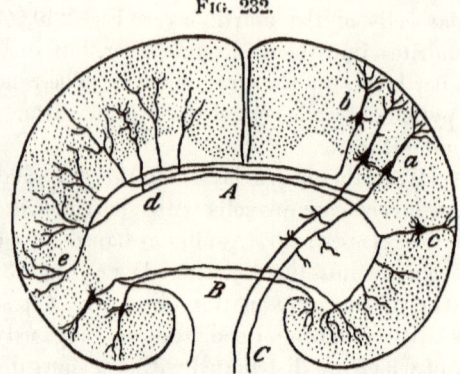

Fig. 232.

Centrifugal and commissural fibres of the cerebrum. (R. y Cajal.) *A*, corpus callosum; *B*, anterior commissure; *C*, pyramidal tract; *a*, large pyramidal cell, with a neurite sending a large collateral into the corpus callosum and then entering the pyramidal tract. Between *a* and *b* is a second similar cell, the neurite from which contributes no branch to the corpus callosum. *b*, small pyramidal cell giving rise to a commissural neurite; *c*, a similar cell, the neurite of which divides into a commissural and an association branch; *d*, collateral entering the gray matter of the opposite hemisphere; *e*, terminal teleneurites of a commissural fibre.

fibres give off collaterals, which also end in the gray matter, and are accompanied by collaterals from the centrifugal fibres, which likewise end in, and send collaterals to, the gray matter.

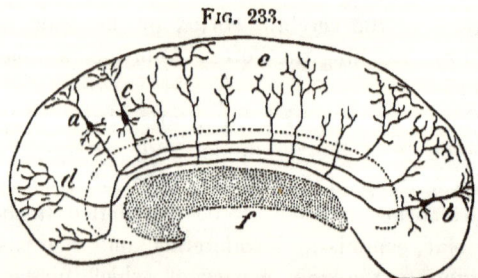

Fig. 233.

Association-fibres of the cerebrum. (R. y Cajal.) The figure represents, diagrammatically, a sagittal section through one of the cerebral hemispheres. *a*, pyramidal cell, with neurite giving off collaterals to, and ending in, the gray matter of the same side; *b*, a similar cell; *c*, cell with a branching neurite passing to different parts of the hemisphere; *d*, teleneurites; *e*, terminal collateral twigs.

The origin, course, and general distribution of the association-fibres are indicated in Fig. 233. They are so numerous that they form the great bulk of the white substance, where they are inextricably interwoven with the other fibres there present.

Besides the centripetal neurites of the association and commissural neurons, their collaterals and those of the projection-fibres, the gray matter of the cortex receives terminal neurites from larger fibres that are probably derived from the cerebellum and cord (Fig. 230, *E*). These give off numerous collaterals and teleneurites, which are distributed to the small pyramidal cells of the second layer, and probably also penetrate into the molecular layer, where they end in numerous teleneurites among the cells of that layer.

In the diagrammatic figure 230 the probable course of nervous stimuli to and from the cerebral cortex is indicated. The possibilities of transmission within a structure of such marvellous complexity are incalculable.

The above structural details of the central nervous system are chiefly taken from the publications of Ramon y Cajal. They are the result of researches carried on by the application of the methods devised by Golgi to the nervous structures of the lower vertebrates and embryos. Such details cannot be observed when specimens have been hardened and stained by methods used for the study of other structures. In such specimens the nuclei of the nerve-cells and those of the neuroglia are stained and become prominent. But the multitude of nervous filaments lying between the cells and the processes of the neuroglia-cells are not differentiated, but appear as an indefinite, finely granular material, in which the cell-bodies apparently lie. Where the cells are sparse or small, as in the first layer of the cerebral gray matter, the tissue appears finely molecular. Where the cells are numerous but small, their stained nuclei give the tissue a granular appearance, as, for example, in the second layer of the cerebellar cortex.

The brain and spinal cord are invested by a membrane of areolar tissue, called the "pia mater." Extensions of this areolar tissue penetrate the substance of the cord and brain, giving support to bloodvessels and their accompanying lymphatics. This areolar tissue also extends into the ventricles of the brain, where it receives an external covering of epithelium continuous with that lining the ventricles, which is ciliated. Externally, the areolar tissue is condensed to form a thin superficial layer.

# CHAPTER XIX.

## THE ORGANS OF THE SPECIAL SENSES.

1. **Touch.**—The nervous filaments distributed among the cells of stratified epithelium have already been depicted in Fig. 93. Similar filaments occur in the human epidermis, and it is probable that some of them are the teledendrites of spinal ganglion-cells, while others are centrifugal teleneurites subserving the functions

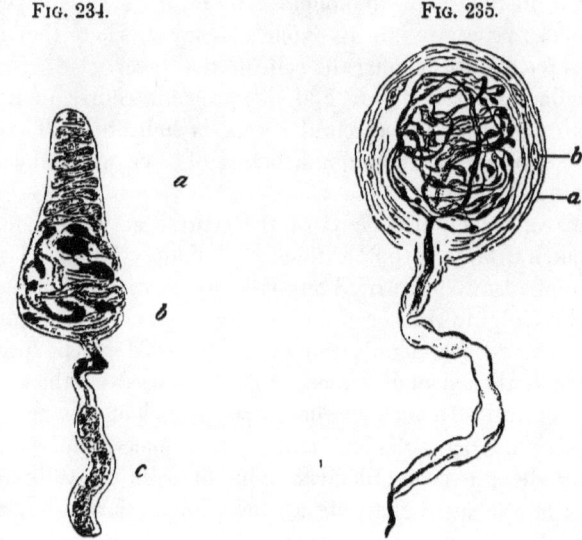

Fig. 234.    Fig. 235.

Tactile corpuscles.

Fig. 234.—Meissner's corpuscle, from the human corium. (Böhm and Davidoff.) *a*, upper portion, in which the epithelial cells alone are represented. The nuclei of those cells are in the broader peripheral portion of the cytoplasm; *b*, nerve-dendrite coiled about the epithelial cells; *c*, nerve-fibre.

Fig. 235.—Krause's corpuscle, from the human conjunctiva. (Dogiel.) *a*, endothelial envelope; *b*, nucleus of connective-tissue cell within the fibrous capsule; *c*, nerve-fibre.

of nutrition, etc., or the teledendrons of neurons belonging to other than the spinal system of nerves.

Besides these nervous terminations the skin possesses certain bodies, which are called "tactile corpuscles" and "Pacinian bodies."

These are situated in the corium, the former lying in some of the papillæ projecting into the rete mucosum.

The tactile corpuscles are of two forms, differing slightly from each other in structure: first, those of Meissner, and, second, those of Krause.

The tactile corpuscles of Meissner (Fig. 234) consist of a group of epithelial cells closely associated with the teledendrites of a nerve-fibre. The cells are closely compacted together to form an ellipsoid body. The nervous dendrite, with its medullary sheath, enters this body at one of its ends, and, after making one or two spiral turns around the mass of epithelial cells, loses its medullary sheath and breaks up into a number of teledendrons, which are distributed among the epithelial cells. The neurilemma and the endoneurium of the fibre are continued over the corpuscle, constituting a species of capsule.

The tactile corpuscles of Krause (Fig. 235) possess a capsule composed of delicate fibrous tissue, covered and lined with endothelial cells. The dendrite of the nerve-fibre loses its medullary sheath upon penetrating this capsule, and then breaks up into teledendrites, that form a complex tangle within the cavity of the corpuscle. There appear to be no cells among the teledendrites, the interstices being occupied by lymph. These corpuscles are especially abundant in the conjunctivæ and the edges of the eyelids, but occur also in the lip, large intestine, posterior surface of the epiglottis, and the glans penis and clitoris. They may receive dendrites from more than one nerve. Those of Meissner are found throughout the skin, being most abundant where the tactile sense is most acute.

The Pacinian corpuscles (Fig. 236) are large oval bodies, composed of a number of concentric cellular lamellæ, surrounding a central, almost cylindrical cavity, and covered externally with a layer of endothelioid cells, which appear to be continuous with the delicate endoneurium of the fibre. The latter enters the corpuscle at one of its ends, soon loses its medullary sheath, and is finally subdivided into a number of teledendrites within the central cavity.

The "genital corpuscles" which are found in the glans of the penis and that of the clitoris are similar in structure to the Pacinian corpuscles, but the lamellar envelope of the latter is here reduced to one or two ill-developed lamellæ.

The nervous impulses inaugurated in the tactile and Pacinian corpuscles are probably transmitted to the sensorium in the manner indicated in Fig. 225.

Pacinian corpuscles are found in the palms and soles, on the nerves of the joints and periosteum, in the pericardium, and in the pancreas.

2. **Taste.**—The special organs of taste appear to be the taste-

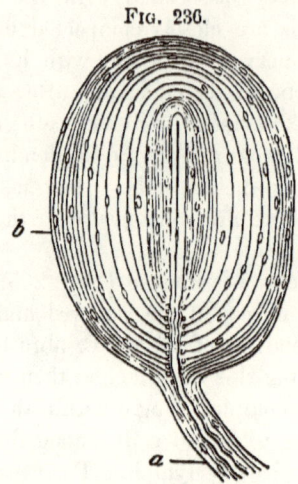

FIG. 236.

Pacinian corpuscle, from the mesentery of the cat. (Klein.) *a*, nerve-fibre; *b*, concentric capsule. The nature of the cells in this capsule is a matter of doubt; analogy would suggest their epithelial nature.

buds, situated in the walls of the sulci surrounding the circumvallate papillæ of the tongue (see Fig. 109).

The taste-buds are bulb-shaped groups of epithelial and nervous cells, situated within the stratified epithelium lining the sulci. The cells composing these buds are spindle-shaped or tapering, and their ends are grouped together at the base of the bud and converge at its apex, where they occupy a "pore" in the stratified epithelium. The epithelial cells do not appear to be active in the inauguration of nervous impulses, but the more spindle-shaped cells lying among them seem to be endowed with nervous functions. They may, possibly, be regarded as peculiar neurons; their distal processes, which receive stimuli at the pore, being the dendrite, while the proximal process is the neurite. The latter divides into a number of minute branches, which, from this point of view, might be regarded as teleneurites. Be this as it may, these branches come into close relations with the teledendrites of nerve-fibres supplied to the taste-bud (Fig. 237). The stratified epithelium surrounding the taste-buds, as elsewhere, contains teledendrites from sensory nerves.

3. **Smell.**—The olfactory organ occupies a small area at the top of the nasal vault, and extends for a short distance upon the septum and external wall. Its exposed surface is about equal to that

Fig. 237.

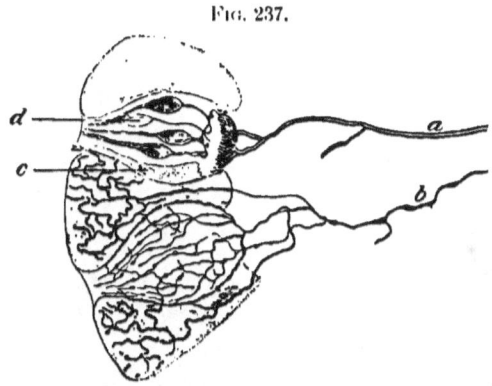

Diagram of a taste-bud and its nervous supply. (Dogiel.) *a*, radicle of the gustatory nerve; *b*, radicle of a sensory nerve; *c*, epithelial cell; *d*, nerve-cell. The shaded part of the figure represents the stratified epithelium lining the sulcus of the circumvallate papilla. Only one of the epithelial or supporting cells of the upper bud is represented in the figure; the others are omitted. The structure of the lower bud is not shown.

of a five-cent piece. It is a modified portion of the mucous membrane of the nose, which may be divided into this, the olfactory portion, and the general or respiratory portion.

The respiratory portion of the nasal mucous membrane is covered with a stratified, columnar, ciliated epithelium, with occasional mucigenous goblet-cells, resting upon a basement-membrane. Beneath this is the membrana propria, resembling that of the small intestine in being rich in lymphadenoid tissue, which may, here and there, be condensed into solitary follicles. Beneath the membrana propria is a richly vascularized submucous areolar tissue, containing compound tubular glands, the glands of Bowman, which open upon the surface of the mucous membrane. These glands secrete both mucus and a serous fluid.

In the olfactory region the columnar epithelial cells are devoid of cilia, but possess a thin cuticle, and the epithelium rests directly upon the lymphadenoid tissue, without the intermediation of a basement-membrane (Fig. 238). Between these epithelial cells are the nervous cells, which constitute the receptive elements of the olfactory nervous tract. These are cells with large nuclei and cylindrical distal bodies, which terminate at the surface of the epithelial layer in several delicate hairs projecting from the surface (Figs. 239 and 240). The proximal ends of the cells rapidly taper to a delicate

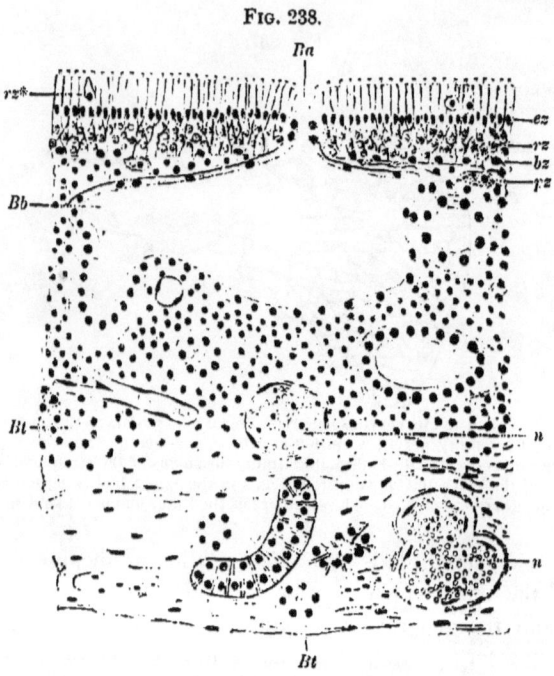

FIG. 238.

Vertical section through the olfactory mucous membrane of the human nose. (Brunn.) *ez*, nuclei of the columnar epithelial cells; *rz*, nuclei of the nervous or olfactory cells lying among those of the epithelium; *bz*, nuclei of basal pyramidal epithelial cells lying among the branching proximal ends of the columnar epithelial cells and tapering ends of the nervous cells; *pz*, pigmented cell in the layer of lymphadenoid tissues beneath the epithelium; *Ba*, duct of a gland of Bowman; *Bb*, dilated subepithelial portion of the duct, receiving several of the tubular acini, *Bt*. The connection between the duct and tubes is not shown. *n*, *n*, branches of the olfactory nerve; *rz\**, atypical nervous cell.

filament, which extends through the subepithelial tissue and becomes associated with others to form the olfactory nerve. The distal ends of the nerve-cells represent the dendrites of neurons, the neurites of which form the axis-cylinders in the olfactory nerve.

The neurites in the olfactory nerve pass through the cribriform plate of the ethmoid bone to the olfactory bulb of the brain, where

Fig. 239.

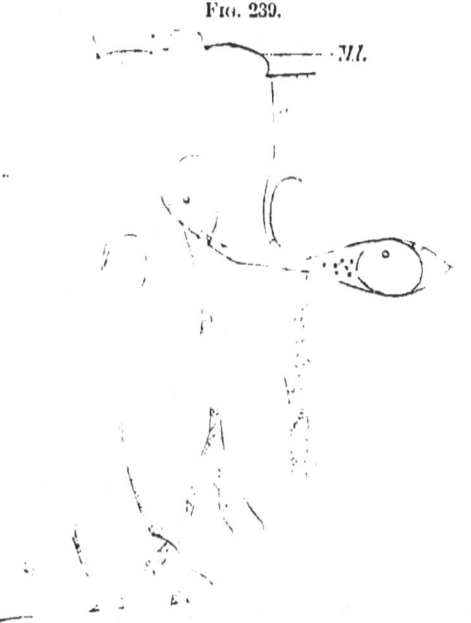

Epithelial layer of the human olfactory mucous membrane. (Brunn.) Isolated elements. Three epithelial cells, with forked proximal ends, are represented, together with a nervous cell bent out of position and the distal end of a second nervous cell. $M.l$, cuticle of the columnar epithelium, which is not continued over the end of the nervous cell. The cuticle of neighboring cells unites at the edges to form a species of membrane, which appears to be perforated for the exit of the distal ends of the nervous cells. A similar cuticle is found in the retina, where it has received the name "limiting membrane."

Fig. 240.

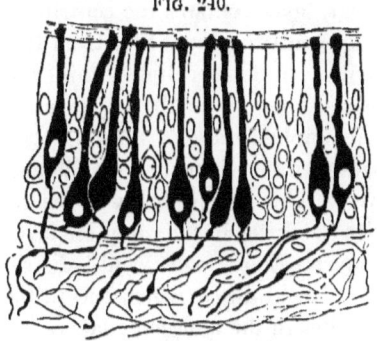

Vertical section of the epithelium, showing the arrangements of its elements. The nervous cells, with their neurites, are black.

they terminate in telencurites within little globular structures, called the "glomeruli of the bulb."

17

The olfactory bulb may be divided into five layers: first, the layer of peripheral nerves, containing the neurites of the olfactory nerve; second, the layer containing the olfactory glomeruli; third, the molecular layer; fourth, the layer of the mitral cells; fifth, the granular layer.

The first layer is, as already stated, occupied by the neurites from the nervous cells in the olfactory mucous membrane. These neurites constitute the axis-cylinders of the olfactory nerve.

The glomeruli of the second layer are small globular masses formed by the closely associated teleneurites of the olfactory nerves and teledendrites from the mitral cells of the fourth layer, the dendrites from which pass through the third or molecular layer. A few cells of neurogliar nature may be associated with these nervous terminations, but the chief mass of each glomerulus is composed of interwoven teleneurites and teledendrites.

The third, or molecular, layer contains small spindle-shaped nerve-cells, which send dendrites to the glomeruli of the second layer and neurites into the granular (fifth) layer, where they turn and take a centripetal direction toward the cerebrum.

The fourth layer is characterized by the presence of large triangular nerve-cells, the mitral cells, the dendrites from which pass through the molecular layer, to end in teledendrites within the glomeruli. A single mitral cell sends dendrites to more than one glomerulus. The neurites from these cells pass, centripetally, to the olfactory centre of the cerebrum.

The fifth, granular, layer contains the centripetal neurites of the mitral cells, and also centrifugal neurites from the cerebrum. The latter are distributed in teleneurites within the granular layer itself. This layer also contains small polygonal nerve-cells of two sorts: first, cells resembling those of the third type represented in Fig. 217, the processes from which are distributed in the granular and molecular layers. They are probably association-cells. Second, cells (Fig. 241) with dendrites in the granular layer and teleneurites in the molecular layer. These cells would distribute impulses received from the centrifugal fibres, which end in the granular layer, among the teledendrites in the molecular layer.

The sense of smell, then, is aroused by stimulations of the distal ends of the nervous cells in the olfactory mucous membrane (Fig. 241), which are transmitted to the glomeruli, where they leave the first neuron, being communicated to the second, represented by the mitral cells and their processes, by which they are conveyed to the cerebral cortex. In its passage through this tract numerous collat-

Fig. 241.

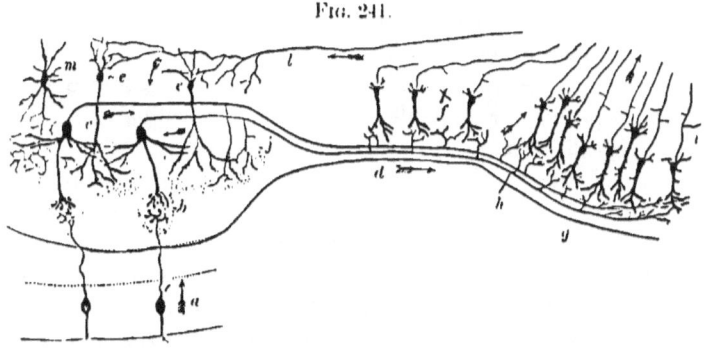

Diagram of the nervous mechanism of the olfactory apparatus. (R. y Cajal.) *a*, olfactory portion of the nasal mucous membrane; *b*, second or glomerular layer of the olfactory bulb *j*, at the right edge of the molecular layer, which is dotted. The cells of this layer are omitted. *c*, fourth layer of the bulb, the layer of the mitral cells, two of which are represented; *e, m*, cells of the fifth or granular layer; *d*, olfactory tract; *g*, cerebral cortex; *h*, neurite from a mitral cell, giving off a collateral to the dendrites of a pyramidal cell in the gray matter of the brain; *f*, pyramidal cells of the olfactory tract; *j*, collateral from a mitral neurite passing, recurrently, into the molecular layer; *l*, centrifugal neurite from the cerebrum.

eral and association-tracts may be influenced in a manner too complicated to be readily followed.

Fig. 242.

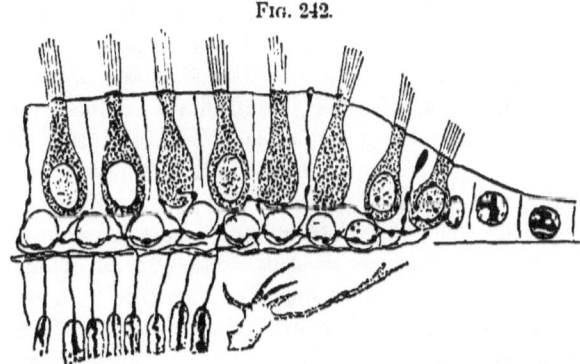

Diagram of the distribution of the auditory nerve within the mucous membrane of the crista acustica. (Niemack.) The bodies of the hair-cells are dotted. Between them are the cells of Deiters, the nuclei of which are shown below the hair-cells. The nervous filaments are distributed between these cells.

4. **Hearing.**—The acoustic nervous apparatus resembles somewhat that which subserves the sense of touch. The receptive portion consists

of a layer of epithelium containing two sorts of cells: first, ciliated cells, which are somewhat flask-shaped and are called "hair-cells"; second, epithelial cells, the "cells of Deiters," which surround and enclose the hair-cells, except at their free ends, and reach the surface of the mucous membrane, where their ends are cuticularized. These cells of Deiters extend from the surface of the membrane to the basement-membrane, while the hair-cells extend only for a portion of that distance.

The dendrites of the auditory nerve are distributed among these cells, but are not in organic union with them (Fig. 242). In this respect the auditory apparatus differs from the olfactory and resembles the tactile. The nervous dendrites are processes of bipolar ganglion-cells situated in the ganglia on the branches of the auditory nerve. The neurites from those cells presumably carry the nervous stimuli to the cerebrum. The bipolar cells are, therefore, analogous to the posterior root ganglion-cells of the spinal nerves. Whether this single neuron carries the nervous stimulus directly to the cerebral cortex cannot be stated, but it is probable that there is an intermediate neuron in the tract of transmission, perhaps in the medulla oblongata.

5. **Sight.**—The receptive nervous organ of vision is the retina. This has an extremely complicated structure, which may be divided into the following nine layers:

1. The layer of pigmented epithelium, which lies next to the choroid coat of the eye, and is, therefore, the most deeply situated coat of the retina; 2, the layer of rods and cones; 3, the external limiting membrane; 4, the outer granular layer; 5, the outer molecular layer; 6, the inner granular layer; 7, the inner molecular layer; 8, the ganglionic layer; 9, the layer of nerve-fibres.

Internal to the ninth layer is the internal limiting membrane, which separates the retinal structures from the vitreous humor occupying the cavity of the eyeball. The general character and associations of these layers are shown in Fig. 243.

1. The layer of pigmented epithelium is made up of hexagonal cells, which are separated from each other by a homogeneous cement and form a single continuous layer upon the external surface of the retina. They are in contact with the rods and cones of the next layer, and send filamentous prolongations between those structures. The pigment lies within these filamentous processes and the portion of cytoplasm continuous with them, but its position

varies with the functional activities of the organ. When the eye has been exposed to light the pigment is found lying deeply between the rods. When the eye has been at rest for some time the pigment is retracted in greater or less degree within the body of the cell.

2. The rods and cones are the terminal structures of cells which extend from the fifth layer to the first. The nuclei of these cells

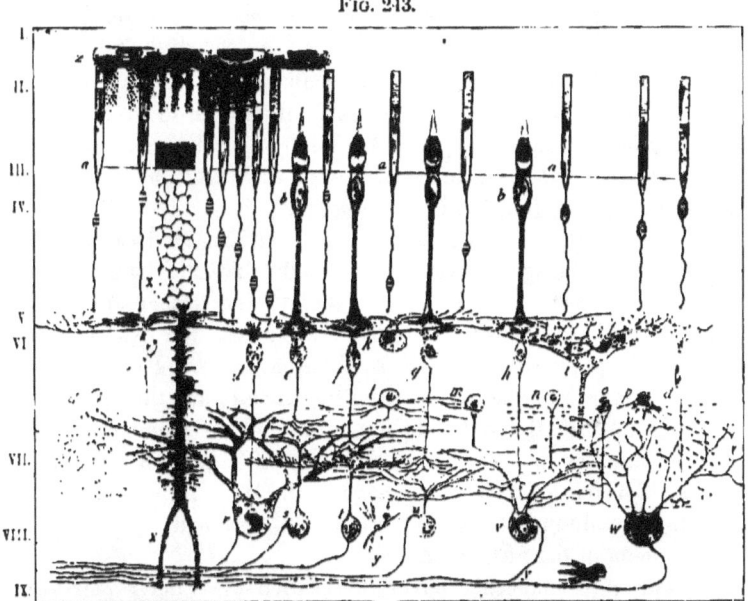

Fig. 243.

Diagram of the retina. (Kallius.) I., pigmented epithelial layer; II., layer of the rods and cones; III., external limiting membrane; IV., outer granular layer; V., outer molecular layer; VI., inner granular layer; VII., inner molecular layer; VIII., ganglionic layer; IX., layer of nerve-fibres. z, pigmented epithelial cells; a, at the bottom of the external limiting membrane, rods; b, cone cells; c-h, ganglion-cells of the sixth layer connecting the fourth layer with the eighth; i, horizontal cell sending a process into the seventh layer; k-q, "spongioblasts," or neurons of the third type (Fig. 217); r-w, ganglion-cells of the eighth layer; x, sustentacular cell of Müller, with striated upper end forming a part of the external limiting membrane; y, y, neuroglia-cells. It should be borne in mind that in sections of the retina numerous elements of the various sorts here represented are crowded together to form a compact tissue. The centrifugal fibres which reach the retina from the cerebrum are omitted from this diagram. They are distributed in the inner granular or sixth layer. The light entering the eye passes through the layers represented in the lower part of this figure before it can affect the rods and cones.

lie within the fourth layer, to which they give a granular appearance (Fig. 243).

3. The external limiting membrane is formed by the cuticularized outer ends of certain sustentacular epithelial cells, the "cells of

Müller" (Fig. 243, *c*), which extend from this layer to the internal limiting membrane and serve to support the various elements of the retina. The nuclei of these cells lie in the seventh layer, to the granular character of which they contribute. The portion of the cell which lies in the fourth layer of the retina is indented with numerous oval depressions receiving the nuclei of the cells carrying the rods and cones, which they both support and isolate from each other. The filamentous cell-bodies of those elements are also separated by the cells of Müller. In the sixth and seventh layers delicate processes from these cells serve a similar purpose, and in the eighth layer their deep extremities fork to give support to the ganglion-cells. Beyond the ninth layer the ends of these forks expand and come in contact with each other at their edges to form the "internal limiting membrane."

4. The fourth, or outer granular layer contains, as already stated, the nuclei and elongated bodies of the cells that carry the rods and cones of the second layer. The bodies of the former are almost filamentous in character, but expand to enclose the oval nucleus, which lies at various depths in different cells. The cell-body expands again near the external limiting membrane, through which it passes to form the rod. At the other end the filamentous cell-body terminates in a minute knob in the fifth layer of the retina. The cells which form the cones have nuclei lying near the external limiting membrane and cylindrical bodies terminating in a brush of filaments in the fifth layer.

5. The outer molecular layer, also called the "outer plexiform layer," owes its appearance to a multitude of filaments, part of which have been described as the terminations of the cells bearing the rods and cones, the rest being the terminations of nerve-processes springing from the cells of the sixth layer.

6. The sixth layer has a granular appearance, because of the presence within it of the cells of a great number of short neurons. These are of two sorts: first, those belonging to the first type, represented in Fig. 217, which have dendrites in relation in the fifth layer with the filaments of the cells bearing the rods and cones, and neurites that come into relation in the seventh layer with the dendrites of ganglion-cells lying in the eighth layer; second, neurons of the third type, shown in Fig. 217, which, in this situation have been called "spongioblasts." These, which we may regard as association-neurons, form two groups: first, those which send

processes into the fifth layer; and, second, those which send their processes into the seventh layer; but, aside from the neurons included in these two groups, there are certain cells (Fig. 243, *i*) which send processes into both the fifth and the seventh layers.

7. The seventh, inner molecular or "inner plexiform" layer owes its delicate structure to the fact that it is here that the teleneurites of the cells in the sixth layer come into relations with the teledendrites of the ganglion-cells of the eighth layer.

8. The eighth layer contains those ganglion-cells whose teledendrites receive impressions from the teleneurites derived from the sixth layer, and send their neurites into the optic nerve. These neurites form the chief constituent of the ninth layer of the retina.

It will be observed in Fig. 243 that the basal expansions of the cells bearing the cones are mostly in relation with the teledendrites of a single neuron of the sixth layer, and that this neuron is, again, in close relations with the teledendrites of but one ganglion-cell of the eighth layer. This arrangement would not favor a diffusion of

Fig. 244.

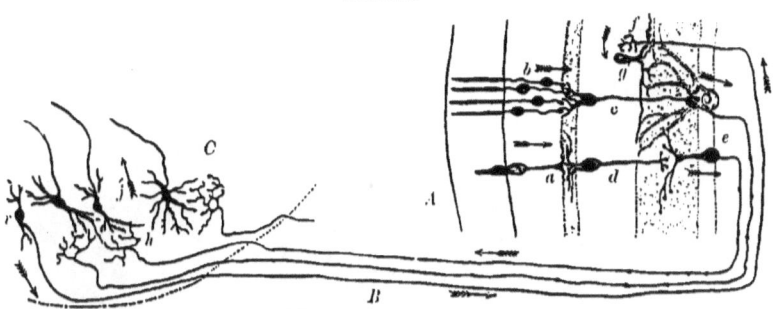

Diagram of the nervous mechanism of vision. (R. y Cajal.) *A*, retina; *B*, optic nerve; *C*, corpus geniculatum. *a*, cone; *b*, rod; *c, d*, bipolar nerve-cells of the outer granular layer; *e*, ganglion-cell; *f*, centrifugal teleneurites; *g*, "spongioblast"; *h*, teleneurites from optic nerve; *j*, neuron receiving and further transmitting the nervous impulse; *r*, cell transmitting the centrifugal impression. The courses of nervous impressions are indicated by the arrows.

the impressions inaugurated in the cones. The arrangement is quite different in the case of the cells bearing the rods.

The probable course of nervous impressions to and from the retinal elements is represented in Fig. 244.

# PART II.

## HISTOLOGY OF THE MORBID PROCESSES.

### CHAPTER XX.

#### DEGENERATIONS AND INFILTRATIONS.

As the result of disturbances in the internal economy of the cell, a variety of changes, called degenerations or infiltrations, are occasioned, some of which are accompanied by visible alterations in the structure of the cell or of the intercellular substances. We are so ignorant of the exact nature of the normal processes carried on by the cell that it is impossible for us to furnish an explanation of most of these changes due to abnormal conditions. We can only describe and group the results according to their apparent likenesses until such time as an increased knowledge permits a more enlightened conception of their significance.

The degenerations are changes in which one of the results is the conversion of a part of the normal structure into some other substance. They imply a loss on the part of the tissue-elements suffering the change.

The infiltrations are departures from the normal in that material from without is deposited either within or between the tissue-elements in an abnormal form or degree. They imply a gain of material, but not necessarily an advantageous gain, on the part of the tissues affected.

Such general statements of an obscure subject must inevitably be vague. They are largely based upon theoretical considerations, and it becomes difficult in many cases to decide definitely whether a given condition is due to degenerative changes or is the result of infiltration, or whether both processes may not have contributed toward producing the abnormal appearances which are observed.

It must be borne in mind that changes which are morbid in a given part of the body may be included in perfectly normal processes carried on in other parts, and are, therefore, not beyond the pale of possible normal cellular activity. In fact, most of the morbid processes observed find parallels in the physiological activities of some portion of the body.

In bone, for example, it is a pathological condition when the intercellular substance fails to be impregnated with earthly salts; but if such salts are deposited in the somewhat similar fibrous intercellular substance of the closely related tissue forming a ligament, the process is then morbid. The two tissues are closely related in structure and are built up by cells having a common, not very remote, ancestry: yet the uses the cells made of the materials brought to them are, to us, very different, and, as yet, inexplicable.

Nor do we know much concerning the way in which, or the extent to which, normal conditions must be modified in order to occasion visible morbid changes in the tissues. We do know that apparently very slight alterations in those conditions may cause profound tissue-changes, as is exemplified in the cachexia following extirpation of the thyroid gland (see p. 183). The amount of thyroid secretion allotted to individual cells of the body must be almost infinitesimal, but its importance is strikingly demonstrated when the cells are deprived of that supply.

In this case we have at least an inkling of how slight an abnormal condition may suffice to work profound alterations in the cellular economy. When, therefore, we meet with evidences of a marked disturbance of the processes within the cells of a tissue, or of their formative activities, we need feel no surprise if an explanation of the causes underlying those morbid manifestations is incomplete or even entirely wanting.

1. **Albuminoid and Fatty Degenerations.**—These two forms of degeneration are frequently associated with each other, and have so much in common that they may well be considered together. They both affect the cells of the parenchymatous organs, such as the kidney, liver, and other secreting glands, the heart and other muscles.

Albuminoid, or "parenchymatous," degeneration results in a swelling of the cells, with an increased granulation of their cytoplasm. The granules are rendered invisible when acted upon by weak acids or alkalies, and are considered to be of albuminoid nature. They are formed at the expense of the cytoplasm, or, at any rate, the cytoplasm disappears as they accumulate.

If the change be only moderate in degree, it is possible for the cell to return to its normal condition. The granules then disappear, the cell recovers its original size, and there is no trace of the morbid condition left. But the degeneration may be too extensive to permit of recovery. The cell then suffers disintegration; the granules become more abundant, the normal cytoplasm disappears,

Fig. 245.

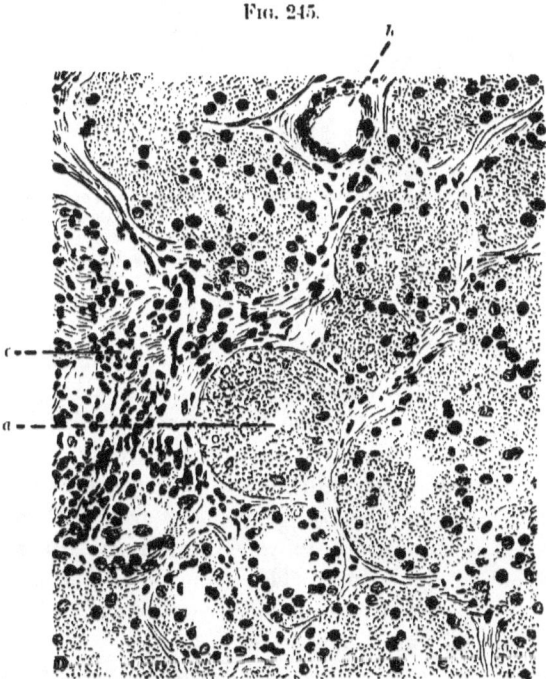

Parenchymatous nephritis. *a*, cross-section of a convoluted tubule of the kidney, the lining epithelium of which is the seat of albuminoid degeneration. The cells are swollen and their bodies filled with abnormally coarse granules. The cells to the left are so far disintegrated that the nuclei have lost most of their chromatin. Such cells cannot recover. The cells to the right are less profoundly altered and their nuclei retain sufficient chromatin to stain slightly. These cells might, perhaps, recover. Other convoluted tubules, similarly affected, are represented in oblique section. *b*, tubule with low, unaffected epithelium, the nuclei of which stain deeply; *c*, round-cell infiltration of the interstitial tissue in the neighborhood of a Malpighian body, the edge of which is just above the line *c*. Section stained with haematoxylin and eosin.

and the nucleus falls into fragments ("karyolysis"), the whole cell being reduced to a granular débris exhibiting no evidence of organization (Fig. 245).

In fatty degeneration the process is similar to that already described as taking place in albuminoid degeneration; but here the albuminoid granules are replaced by globules of fat. These vary in size from mere granules of minute dimensions to distinct globules of considerable diameter (Fig. 246). The fat is left

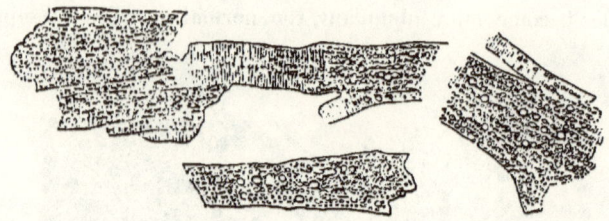

Fatty degeneration of the cardiac muscle. (Israel.) In some portions of the preparation the cross-striations of the contractile substance are retained. In these portions the fatty metamorphosis has not taken place. In other places the contractile substance has been destroyed and the cells are charged with minute granules and with small globules of fat. The preparation is unstained, so that the nuclei are not prominent. They have been omitted from the figure. Specimen prepared by teasing the fresh tissue.

unchanged upon treatment with weak acids or alkalies, and is stained a dark brown or black by solutions of osmic acid (see Fig. 186), reactions which distinguish fatty from albuminoid granules. They are, furthermore, dissolved by ether or strong alcohol, which leave albuminoid granules undissolved. In specimens which have been hardened in alcohol the fat is removed from the cells, which then contain little clear spaces in which the fat was situated in the fresh condition of the tissues. This removal of the fat is likely to be still more perfect if the specimen has been embedded in celloidin, solutions of which contain ether.

Albuminoid degeneration occurs in acute diseases, such as the exanthemata, typhoid fever, septicæmia, etc., which are all characterized by fever. It also occurs in cases of damage to the tissues, insufficient immediately to kill the cells, but great enough to induce inflammation. Because of this frequent association with inflammatory changes in other tissue-elements albuminoid degeneration has been termed "acute parenchymatous inflammation." The damage may be the result of some externally applied injury, or it may be occasioned by a sudden diminution, but not complete arrest, of the nutrient supply; e. g., by the incomplete plugging of a bloodvessel by an embolus. Albuminoid degeneration may also be the result of toxic conditions that are not accompanied by rise of temperature.

In all the foregoing cases the cause is of an acute nature, acting rapidly on the cells. If that action be moderate in degree and persistent, the albuminoid degeneration passes into fatty degeneration. Hence the latter has been called "chronic parenchymatous inflammation."

But fatty degeneration is not always preceded by albuminoid degeneration. It is found widely distributed in the cells of the body in anæmia (Fig. 247), leucæmia, and phthisis, and in many

Fig. 247.

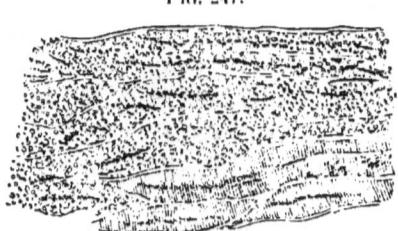

Localized fatty degeneration of the cardiac muscle in a case of pernicious anæmia. (Birch-Hirschfeld.) The three or four fibres at the bottom of the figure are nearly, if not quite, normal. The rest of the fibres are the seat of an extensive fatty degeneration, resulting in a complete obliteration of the normal striations of the contractile substance. Section of the fresh, unstained muscle. The nuclei, being unstained, are but faintly visible in such sections, and are not represented.

toxic conditions that are of a subacute or chronic character. In a more localized form it follows those diseases of the bloodvessels which interfere with a normally abundant supply of blood to the parts in which they are distributed. It appears, again, in parts the functional activity of which is markedly increased without a corresponding increase in the nutrient supply. For example, in stenosis of an orifice of the heart, when extra work is thrown on the cardiac muscle and the nutrient supply is insufficient to permit of hypertrophy, the muscle-cells suffer fatty degeneration, and the consequent weakening of its walls results in dilatation of that particular cavity which is subjected to the difficult task of urging the blood through the narrowed orifice.

If we examine these various conditions with a view to determining their effects upon the cells, we shall find that they have one common feature. There is in all of them a lack of balance between the nutrient supply of which the cells can avail themselves and the consumption of material made necessary by the work required of them.

Under these circumstances the cells appear, first, to utilize the food-materials which they already contain as an accumulated stock (metaplasm); but when these are exhausted they are forced to draw upon those materials which exist as a part of their own organized structure, if they are to maintain their functional activities. They thus sacrifice the integrity of that structure in order to do the work that has been assigned to them in the organization of the whole body.

Now, there is a difference in the immediate availability of the various classes of foods. The carbohydrates appear to be the most susceptible of rapid utilization; the proteids come next, and the fats last. We may imagine, then, that in a sudden emergency the cells will first consume the greater part of their store of carbohydrates, then the proteids, and lastly the fats. If the condition be an acute one, so that a part of the organized proteids are utilized as food, this utilization is not complete, but the proteids are split up into a portion that can be most readily oxidized and turned to account, and a residual portion, which appears in granular form within the cytoplasm.

We may also imagine that, in its efforts to obtain adequate nourishment, the cell imbibes an excessive amount of fluid from its surroundings.

If the adverse circumstances are extreme, the nucleus is also overworked and relatively starved, and suffers in its integrity (karyolysis). When the nucleus is destroyed, or when there is no longer sufficient cytoplasm to aid it in its assimilative function, a recovery of the cell becomes impossible.

Let us now consider how this conception of albuminoid degeneration may serve to explain its occurrence in the various conditions in which it is found.

In fevers the rise of temperature is evidence of an increased metabolism within the body—i. e., the cells of the body are more active in bringing about chemical changes. The amount of urea eliminated from the body is also increased, showing that those chemical changes involve an additional consumption of proteids.

In febrile conditions, then, the cells are unusually active and consume an increased amount of proteids. Let us next inquire what conditions exist which are likely to interfere with their nutrient supply.

The source of all nourishment, which is not gaseous, being the food taken into the system, it is evident that any condition interfering with digestion and absorption must influence the general nutrient supply. In fevers the glands of the alimentary tract, as well as the cells of other organs, are affected with albuminoid degeneration. Their secretions are diminished or altered, the digestion arrested in greater or less degree, and the appetite lost or perverted. For these reasons the diet must be adjusted, not only to the needs of the patient, but also to his powers of digestion. But this state is established only after the degenerative changes have been inaugurated, and does not explain the way in which they start.

If we bear in mind that the febrile condition is the result of a toxic state of the blood and nutrient fluids, and that the poisons present are probably obnoxious to the cells, we shall find no difficulty in understanding that the cells might reject a nutrient supply so vitiated. Where we can observe the action of cells, we know that they are repelled by certain substances, and it appears reasonable to suppose that cells which we cannot directly study during life possess similar powers of rejection. If this view be correct, the very condition which induces fever would also interfere with the proper nutrition of the cells.

The causation of fever, according to this argument, is to be sought in the toxic condition of the blood and other nutrient fluids, the poisons disturbing the action of the thermo-regulating mechanism of the nervous system and also interfering with the nutrition of the cells of the body. As soon as fever begins, its influence upon the cells is to stimulate their activities, for we know that a moderate elevation of temperature causes an increased metabolism in those cells that we can study while alive. It is, consequently, not necessary that a direct functional demand should bear upon the cells in order that the chemical changes within them be augmented. The rise of temperature is sufficient to account for increased metabolism, which, in turn, implies a liberation of heat, and, therefore, an aggravation of the morbid condition. The increase of noxious waste-products of cellular activity, which enter the circulating fluids, may also add to its toxicity.

But, in addition to this thermal cause of increased metabolism, the toxæmia throws extra work upon those cells that are charged with the function of maintaining the quality of the blood or lymph. The kidney contains such cells, and is one of the organs most likely to be severely affected with albuminoid degeneration (acute parenchymatous nephritis, Fig. 245). The spleen and lymphatic glands are also exposed to an increased functional demand, and respond in an increase of their active tissues, which may pass into degenerative conditions if the task be greater than they are able to cope with successfully.

In the other conditions in which albuminoid degeneration is found the factors determining its causation appear less complicated than in the fevers. Many of the acute inflammatory processes are accompanied by a rise of temperature, due to the absorption of poisons from the seat of the inflammation, and then the degeneration will be more widely distributed than in those cases in which the general reaction is less marked or entirely absent. But the tissues immediately involved in the inflammatory process will suffer in their nutrition, whether toxæmia be present or not, and in certain of them the result will be a degeneration, while in others it will be necrosis or death. In the case of albuminoid degeneration following incomplete embolism the explanation is even simpler; for here the nutrition is directly reduced by the mechanical effect of a partial plugging of a bloodvessel.

In all the cases in which albuminoid degeneration occurs in a comparatively pure form the cause is an acute one—*i. e.*, the cells are called upon to meet a sudden change of condition in their activities and nutrition : the former being, as a rule, increased ; the latter, probably always diminished.

The explanation which can be offered of the way in which fatty degeneration is brought about is very similar to that already given for albuminoid degeneration.

In fatty degeneration the emergency which the cells have to meet is less sudden than in albuminoid degeneration. The adverse conditions to which they are subjected are more slowly developed, though not necessarily less serious. The cells appear to be able to accommodate themselves to a considerable extent to the abnormal circumstances, but eventually their powers of metabolism are disturbed and they are incapable of utilizing the less readily available food-materials. When the organized proteids are then drawn upon their nitrogen appears to be completely used, so that no residual albuminoid substances are deposited in granular form, but a remnant of the cytoplasm, free from nitrogen and taking the form of fat, the least readily oxidized form of food, is left. If, now, the cells continue to appropriate and utilize albuminoid food-material, this fatty residue would accumulate within the cytoplasm. Fatty foods would, of course, be little, if at all, utilized.

This leads to the inference that one of the chief features in the disturbed metabolism of the cell is an inability to bring about the complete oxidations that normally take place in the cytoplasm, and when we examine the conditions in which fatty degeneration occurs we notice that a group of them are such as would involve a diminished amount of oxygen in the blood. This is manifest in cases of anæmia, advanced phthisis, and poisoning with carbonic oxide, which destroys the respiratory value of the hæmoglobin.

In the subacute and chronic toxic conditions—*e. g.*, such cases of poisoning by phosphorus or arsenic in which the patient survives for a considerable time—the blood probably contains a sufficiently abundant supply of oxygen for the needs of the tissues. But intracellular respiration is a complicated process; not a simple and direct burning of substances occasioned by their immediate conversion into fully oxidized compounds when brought into relations with free oxygen. The food-materials are split up within the cell into compounds of simpler constitution, some of which receive a sufficient amount of oxygen, from the original material of which they are derivatives, to satisfy their affinities, and are, therefore, stable; while others are organic substances in a chemically reduced state, which unite with the free oxygen that may be accessible. The oxidation is not caused by the presence of free oxygen, but is an incident in the chemical changes carried on by the cell.

In the toxic conditions leading to fatty degeneration this intracellular oxidation is probably interfered with through the action of the poisons upon the cytoplasm, and, as a result, the least easily oxidizable substance, fat, remains as an unutilized residue. The poisons at the same time probably interfere with the nutrition of the cell, which draws upon its organized proteids for a supply of nitrogen, leaving again a remnant of unavailable fat.

It is easily comprehensible that relative overwork may have the same effect upon the cell as relative innutrition. The fatty degeneration of the heart-muscle as the result of stenosis or of valvular insufficiency at one of its orifices would, therefore, be explained as an example of a lack of balance between the supply and consumption of food in the economy of the cardiac cells. Relative overwork of the heart is also one of the effects of marked anæmia. The anæmic condition involves a diminished supply of oxygen, from which the heart, as well as the other tissues, suffers. But the demand for oxygen on the part of the general economy requires an acceleration of the circulation; this throws extra work upon a relatively starved heart.

It is evident, from the foregoing considerations, that albuminoid and fatty degenerations must be very common conditions in the cells of the body. Their close etiological similarity makes it obvious, also, that they must very frequently be associated with each other, either in the same cell or in different cells of the same organ. The fact that fatty degeneration is often a sequel of albuminoid degeneration may be explained as the result of a toxic or other condition, which has been sudden in its onset, but has declined in intensity with the lapse of time. Or it may be possible that the cells are able gradually to adapt themselves, in a measure, to the new conditions under which they must do their work, and that they become able to utilize more completely the foods they receive; leaving a fatty, instead of an albuminoid, residue.

Fatty degeneration, like albuminoid degeneration, may lead to a total destruction of the cell, leaving the fatty globules free, or recovery may take place on the subsidence of the cause.

2. **Cheesy degeneration** is a term applied to an association of albuminoid and fatty degenerations with necrosis, in which the detritus of the tissues forms a dry material, somewhat resembling the softer varieties of cheese. Under the microscope this cheesy material has a finely granular appearance, with here and there small fragments of nuclear chromoplasm which still retain their affinity for nuclear dyes.

3. **Fatty Infiltration.**—Essentially different from fatty degeneration is an accumulation of fat in cells as the result of their overfeeding. It may be due to an excessive reception of fat by the cells, but this is not necessarily the case. A supply of any form of food that is in considerable excess of the needs of the body may result in a fatty infiltration of its cells, for fat is the least readily consumed variety of food, and where the other varieties are in great abundance it may be guarded against destruction and remain in the tissues. Furthermore, a part of the excess of other food-materials may be converted into fat within the cells and be retained by them.

Fatty infiltration is a normal condition of many cells. Those which form the characteristic element in adipose tissue (Fig. 65) are connective-tissue cells that have undergone extensive fatty infiltration. A transitory fatty infiltration is also normal in the cells of the liver (Fig. 248).

Fig. 248

Cells from the human liver, normal. (Orth.) *a*, cells free from fat. The isolated cell to the right contains two nuclei and three or four granules of pigment. The three lower cells, *b*, are infiltrated with globules of fat. It will be noticed that these three cells contain as much cytoplasm as the two contiguous cells, *a*. This is taken as an indication that the fat is superadded to the cytoplasm, and has not been produced at the expense of part of the *organized* substance of the cell. This does not imply that the fat was necessarily taken into the cell *as such*, for it may have been produced within the cell from food-materials; but it has not been produced at a sacrifice of the organized materials forming an essential part of the living cell.

The globules of fat form a part of the metaplasm of the cells in which they are situated; *i. e.*, they do not constitute an integral part of the cytoplasm, but lie within it, leaving it intact, unless the accumulation is so great that the functions of the cell are interfered with. Then the cytoplasm may suffer atrophy and its usefulness be diminished.

It is not possible to lay down any practical rule for distinguishing between fatty infiltration and fatty degeneration when cells are examined under the microscope, beyond the general statement that in degeneration there is a corresponding destruction of the cytoplasm as the fat accumulates. In fatty infiltration the globules of fat are rather more apt to coalesce with each other than in fatty degeneration, so that the globules appear larger. This is not invariably the case, however, the behavior of the fat in this respect differing in different kinds of cell.

4. **Glycogenic Infiltration.**—This is a condition analogous to fatty infiltration, but the stored excess of food-material in this case belongs to the class of carbohydrates. The condition is found in the cells of the convoluted renal tubules in cases of diabetes mellitus, sometimes in the leucocytes in inflammatory foci, and occasionally in the cells of tumors where the functional activities of the cells are in abeyance and only their formative powers call for a consumption of food.

The glycogen occurs either in granules or in small, irregular masses within the cytoplasm (Fig. 249). It is soluble in water, and its detection is a matter of difficulty unless special methods of prep-

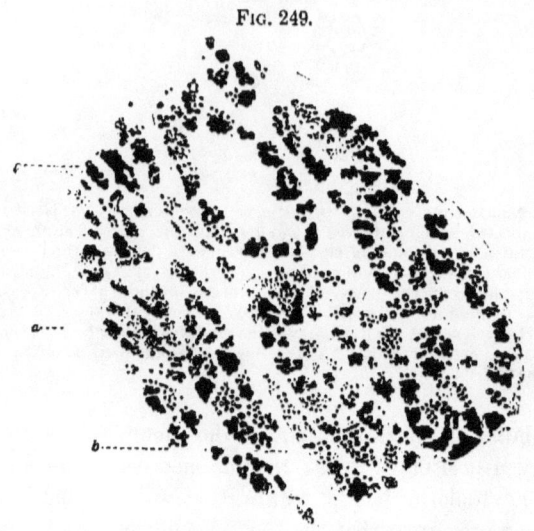

FIG. 249.

Glycogenic infiltration of the cells in an endothelioma. (Driessen.) *a*, cell crowded with granular masses of glycogen; *b*, fibrous tissue forming the stroma of the tumor; *c*, space within the growth containing blood. Section from an endothelioma of bone, stained with a solution of iodine and gum-arabic in water. Iodine stains glycogen brown. The nuclei and cytoplasm of the cells are not represented. A section from the same tumor after the extraction of the glycogen and staining with nuclear dyes is shown in Fig. 222.

aration are employed to retain it *in situ* and so facilitate its recognition. When it is dissolved from the cytoplasm it leaves small, clear, empty spaces behind.

Glycogenic infiltration is a normal condition in the cells of the liver and in muscular fibres. In the latter situation it serves as a store of rapidly available energy, which can be drawn upon during the functional activity of the cells. In the liver it serves a similar purpose for the whole body.

5. **Serous Infiltration.**—In œdematous conditions of the tissues their cells sometimes imbibe fluid from their surroundings, which appears as clear drops or vacuoles within the cytoplasm (Fig. 250). The condition may subsequently subside, or it may lead to a disintegration of the cytoplasm and nucleus. The cell then undergoes a form of destruction very closely resembling that in albuminoid

degeneration. Serous infiltration, more or less complicated with albuminoid degeneration, also occurs in inflammations when the serous constituent of the exudate is prominent.

6. **Mucous Degeneration.**—This form of degeneration has its normal analogue in the elaboration of mucus by the epithelial cells covering many of the mucous membranes or lining mucous glands.

Fig. 250.

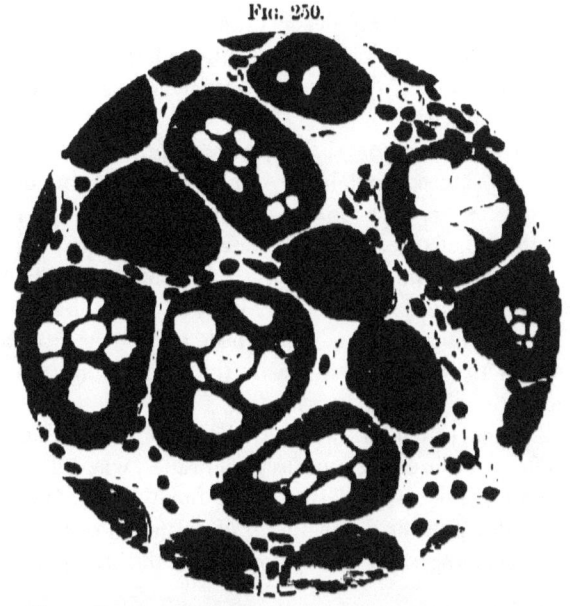

Vacuolation of striated muscle. (Volkmann.) The specimen is from the rectus abdominis muscle from a case of typhoid fever. The cross-sections of the muscle-fibres contain spaces within the contractile substance, which are filled with a clear, fluid serum. The fibres so infiltrated are larger than those containing no such vacuoles. The cavities are, therefore, not produced at the expense of the contractile substance. Between the fibres is the intermuscular, vascularized fibrous tissue, forming the interstitial tissue of the muscular organ.

But the elaboration of mucin is not confined to epithelium. It may be produced by the cells of the connective tissues, appearing among the intercellular substances. This is most marked in mucous tissue, where the general character of the tissue is determined by the mucus in the intercellular substance. There is also a comparatively small amount of mucus in other forms of connective tissue, especially in the fibrous varieties.

Under morbid conditions, which we are not able exactly to define, this production of mucus is increased. In epithelial and other cells its production may involve a destruction of the cytoplasm, which appears to be sacrificed. A similar transformation or replacement of the normal intercellular substances may also occur in the connective tissues, such as bone, cartilage, fat, or fibrous tissue, which then contain more than the normal proportion of mucin. This proportion may be so great as to alter the physical properties of the tissue. In these cases the cells may undergo mucous degeneration, or they may ultimately suffer a fatty degeneration. It is a question to what extent the cells are active in the substitution of mucous for the usual intercellular substances, the manner in which it is produced being as yet undetermined.

The mucus is a clear, viscid fluid, which appears to be a mixture of various substances containing either mucin or pseudomucin. These substances are precipitated by alcohol, so that in hardened specimens the mucus becomes granular or is streaked with linear coagula. Hæmatoxylin usually stains the whole mass a faint blue; the granules and streaks a little more intensely than the clearer portions. This staining serves to distinguish the mucus from a serous fluid, which is also made granular by the coagulating influence of alcohol upon the albumin it contains.

Mucous degeneration of the epithelia is a frequent accompaniment of inflammation of the mucous membranes, where it appears to be due to an excessive stimulation of the functional activities of the cells. A similar mucous degeneration of epithelial cells is also very common in tumors; *e. g.*, the cystomata of the ovary and colloid cancer.

7. **Colloid Degeneration.**—This is a form of degeneration in which the substance of cells is converted into a clear, homogeneous, gelatinous material of greater consistency than mucus, and, unlike the latter, is not precipitated by alcohol, so that in hardened specimens it retains its homogeneous appearance.

The production of colloid seems to be normal in the thyroid gland after the attainment of a certain age. In this situation the colloid material is formed in the cells of the alveoli and then discharged into their lumina, where it forms a mass that may completely fill its cavity (Fig. 154); but the cells of the thyroid not infrequently suffer destruction in the elaboration of the colloid material, so that even here the process partakes of a degenerative character.

The material forming the hyaline casts in various kinds of

nephritis appears to be colloid elaborated by the cells lining the renal tubules, but those casts may not always owe their origin to this form of degeneration.

Fig. 251.

Fig. 252.

Hyaline degeneration. (Ernst.)

Fig. 251.—Hyaline degeneration of cells in the choroid plexus. In this case the hyaline material appears to be derived from the cytoplasm of the cells, the process constituting a true degeneration. Transitional conditions from the unchanged cells to masses of hyaline without traces of cellular structure are found in the specimen.

Fig. 252.—Hyaline degeneration of the capillary walls in a psammoma of the dura mater. Here the endothelial lining of the capillaries is intact, the hyaline material being outside of it. This disposition of the hyaline would lead to the inference that in this case it was the result of infiltration.

It is probable that the composition of colloid is not always the same. It is identified by the facts that it is a clear, structureless substance, derived from cells and not presenting the characteristics of mucus. The causes and mode of its production are unknown.

**8. Hyaline Degeneration.**—This term is used to designate the occurrence of a material similar to colloid, which appears chiefly in the intercellular substances or in the interstices of the tissues, and is apparently not immediately derived from the substance of cells. It is a question whether it should, in such cases, be regarded as a degeneration—*i. e.*, the result of a transformation of pre-existent normal structures—or whether it is not a form of infiltration, the material being simply deposited between the normal structures, which may atrophy and disappear in consequence of its presence. Its most common site is beneath the endothelial linings of the bloodvessels, where it forms a homogeneous layer, greatly thickening the vascular wall and often causing a narrowing of the lumen of the vessel (Figs. 251 and 252). It may also affect the fibrous tissues, replacing the intercellular substances with hyaline material, made up of an agglomeration of little masses, or appearing quite homogeneous. The cells of the tissues gradually undergo atrophy and disappear, but do not seem in most cases to suffer a transformation into hyaline substance. In some instances, however, the cytoplasm of the cells appears to undergo a hyaline transformation (Fig. 251).

A hyaline transformation sometimes affects thrombi, which lose their fibrinous character and become homogeneous.

Hyaline material may take a faint bluish tint when treated with hæmatoxylin, or it may remain colorless.

Various attempts have been made to define more clearly the conceptions of colloid and hyaline substances, and to distinguish them by means of reactions with different staining-fluids. These attempts have not led to satisfactory results, probably because the colloid and hyaline substances are mixtures of various chemical compounds; the whole subject awaits further investigation.

**9. Keratoid Degeneration.**—This form of degeneration is a transformation of the cytoplasm into a substance called keratin, which gives to horn, the nails, etc., their peculiar character. It is normally produced in the epidermis, where this degenerative process is not pathological. The transformation appears to involve the preliminary formation of a substance called eleidin (Fig. 175), the chemical nature of which is unknown, which subsequently changes into keratin. These two substances may be distinguished by the facts that eleidin is deeply stained by carmine and not by fuchsin, while keratin is readily stained by the latter dye.

The cells in the epithelial pearls of epitheliomata often undergo these degenerative changes, producing large masses of eleidin or keratin. The change in these cases may be considered as due to a retention of this normal tendency by the epidermal epithelium under the abnormal conditions in which it is placed in the tumor. In those cases of metaplasia in which columnar epithelium becomes converted into the stratified variety the susceptibility to keratoid degeneration is an acquired character, columnar epithelium under normal conditions never suffering this change.

10. **Amyloid Infiltration.**—The change in the tissues known by this name, or that of amyloid degeneration, has many resemblances to hyaline degeneration (or infiltration). Amyloid differs, however, from the hyaline substances in being recognizable by means of a number of characteristic reactions, although they vary considerably

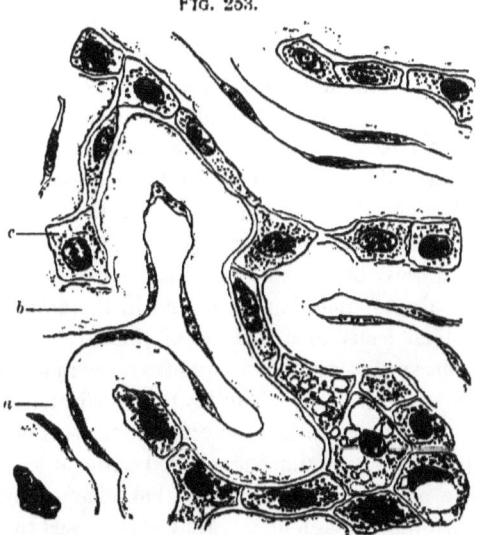

FIG. 253.

Amyloid infiltration in the liver. (Thoma.) *a*, lumen of an intralobular capillary, surrounded by the endothelial wall of the vessel; *b*, amyloid substance immediately beneath the endothelium; *c*, epithelial cells of the hepatic parenchyma, some of which show a fatty infiltration.

in sharpness in different cases, and give rise to the suspicion that the amyloid substance is not always of constant chemical composition, or that it may be transformed into other substances of similar physical and optical properties.

Amyloid is a nitrogenous material, which is stained a dark brown by aqueous solutions of iodine, while the normal tissues acquire a yellow color. Under the microscope the brown color has a marked reddish tinge. Solutions of methyl-violet give amyloid a red color and stain the rest of the tissues blue or bluish-violet. It is upon these reactions, and not upon the optical appearance of the material when unstained, that the recognition of amyloid depends.

Its most frequent situation is in the walls of the smaller blood-vessels, where it lies in the deeper layers of the intima or in the muscular coat. It may also be deposited around the endothelial walls of the capillaries (Fig. 253).

Amyloid infiltration occurs in syphilis, advanced tuberculosis (especially of bone), long-continued suppuration, and similar conditions in which there is profound cachexia. It evidently depends upon conditions of marked malnutrition or chronic toxic conditions, and it is believed that its occurrence depends upon the inability of the tissue-cells to utilize the proteids that are present in the interstitial serum. These are thought to accumulate and gradually become transformed into amyloid. The deposition of amyloid, according to this hypothesis, would depend primarily upon a lack of power to assimilate proteids on the part of the cells.

The presence of amyloid between the cellular elements of the tissue interferes with their nutrition, and they suffer atrophy.

11. **Calcareous Infiltration** (Figs. 254 and 255).—There appears to be a marked affinity between necrosed tissues, or tissues of low vitality, and the salts of lime that are found in the circulating fluids of the body, which leads to a deposit of the latter within those tissues. The cheesy material that results from tubercular or other processes is prone to this form of infiltration. Cicatricial tissue, when abundant and poorly nourished, may also be the seat of lime-deposits. Similar deposits are sometimes associated with those of urates in the inflammatory nodules of low vitality that characterize gout. Bits of organic or other foreign matter that are exposed to fluids containing salts of lime are liable to become encrusted with a coating of calcareous material. This is the origin of many renal and other calculi and of the vein-stones that form around small thrombi of occasional occurrence where the circulation is very sluggish; e. g., in the venous plexuses within the pelvis, or behind the valves that occur in the course of most of the veins. Calcification of cartilage is also common after the individual has attained a certain age. Tumors in which the tissues are of low vitality or have degenerated are

also liable to calcareous infiltration. That infiltration appears, then, to be always secondary to some morbid process lowering the vitality of the tissues.

Calcareous infiltration may serve as a type of infiltrations with other materials, such as urates, and of the formation of concretions; for example, gall-stones. These and other concretions contain a nucleus of organic or other nature, upon which the salts are deposited from their solutions very much as sugar crystallizes upon threads suspended in a syrup.

**12. Degeneration of Nerves.**—If a nerve-fibre be severed from its connection with the ganglion-cell of which it is a process, it suffers disintegration. The medullary sheath breaks up into a number of globular masses, which are subdivided and eventually absorbed. The axis-cylinder becomes swollen, granular, and also disappears. If the ganglion-cell retains its vitality.

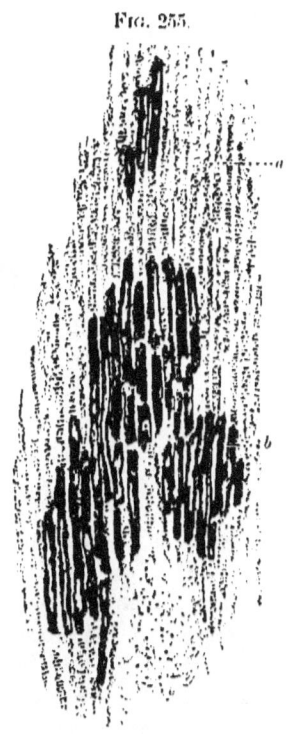

Fig. 255.

Fig. 254.

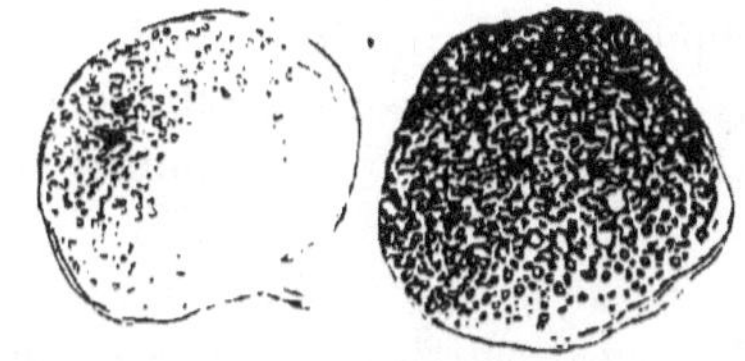

Fig. 254.—Calcareous infiltration of renal glomeruli, secondary to hyaline degeneration of the capillary walls, obliteration of the vascular lumen, and death of the tissue. The glomerulus to the left shows a slight granular deposit of calcareous material in the hyaline glomerulus. The figure to the right shows the organic base almost completely obscured by calcareous granules. (Ribbert.)

Fig. 255.—Calcareous infiltration of the cardiac muscle. (Langerhans.) a, degenerated cardiac muscle; b, muscular fibres impregnated with lime-salts. The specimen was taken from a case of chronic lead-poisoning. The cells which are the seat of the calcareous infiltration must have been dead for a considerable time before the death of the individual.

it may regenerate the nerve by the development of a new process. If, however, the ganglion-cell has been destroyed, regeneration does not take place. This exemplifies the statement, made in the chapter on the cell, that portions of cells which were devoid of a nucleus could not continue their existence.

# CHAPTER XXI.
## ATROPHY.

ATROPHY is a diminution in the size of a part, due to a deficient nutrition of its constituents, which is neither so rapid nor so destructive as to cause necrotic, degenerative, or inflammatory changes. The tissue-elements appear comparatively normal under the microscope, but are either all or in part diminished in size. This diminution in size is frequently accompanied by an increased depth of the usual coloring of the tissue-elements, or with the appearance of granules of pigment (Fig. 256).

FIG. 256.

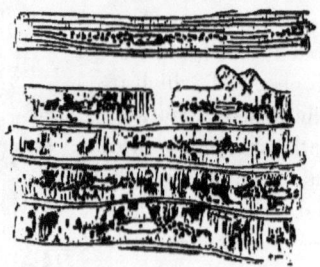

Brown or senile atrophy of the heart. (Ribbert.) The muscle-fibres are reduced in diameter. At the ends of the nuclei are collections of pigment-granules.

The cause of atrophy may operate almost directly upon the cells involved, or it may indirectly influence the nutrition of the cells through lesions in the circulatory or nervous system, or through an interference with the processes of general nutrition maintaining the whole body.

1. **Functional Atrophy.**—It appears to be a general principle governing living organisms that functional activity, within a certain normal range, is necessary to the maintenance of the normal nutrition of a part. When the required degree of functional activity is not called forth, the nutrition of the part suffers and it undergoes atrophy. Paralyzed muscles lose their normal size through innutrition following their disuse. Secreting glands may also suffer atrophy

when there is no longer an adequate call for their functional activities.

This form of atrophy is probably attributable in some measure to a diminished flow of blood to the part, for in health, when the functional activity of an organ is called into play, there is an increased volume of blood conveyed to that organ. But this element in the innutrition does not account for the whole process. The intracellular metabolism also falls below the normal level, and this appears to reduce the state of nutrition of the cellular constituents.

2. **Pressure-atrophy** (Figs. 257 and 258).—When a part is subjected to moderate but constant, or oft-repeated pressure, it undergoes atrophy through a disturbance in its nutrition. This may be

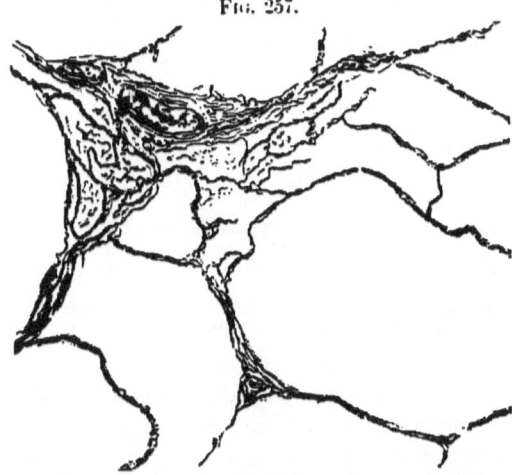

Fig. 257.

Section from an emphysematous lung. (Ribbert.) The pulmonary alveoli are enlarged; their walls are stretched and thinned; atrophied because of repeated excessive air-pressure within the alveoli. In more extreme cases of emphysema the atrophy of the alveolar walls may lead to their total destruction in places, so that the cavities of neighboring alveoli communicate. (Compare with Fig. 150.)

partly due to a direct influence exerted by the pressure upon the processes carried on in the cells of the tissue, but it is probable that interference with the circulation, including the lymph-currents, has a greater influence in bringing about the lack of nourishment. Examples of this form of atrophy are furnished by cases in which a contracting cicatricial tissue is formed between the parenchymatous cells of an organ, as the result of a chronic interstitial inflammation. These cells then undergo atrophy and may eventually disappear (Fig.

288). In passive hyperæmia of the liver the cells situated around the central veins of the lobules suffer atrophy. This is due in part to the pressure exerted upon them, in part to an interruption of the lymphatic circulation, and in part to the fact that the blood reaches them last in its course through the organ and is probably less richly provided with oxygen and other nutritive materials than when it

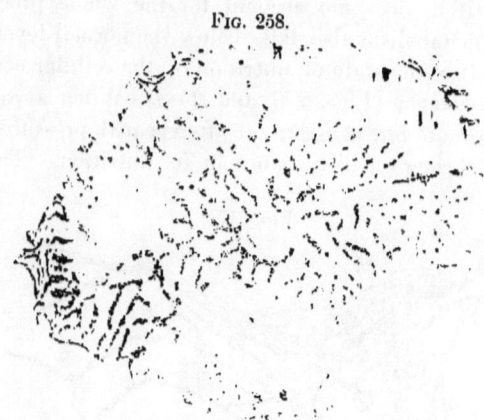

FIG. 258.

Lobule of the liver, showing atrophy from chronic passive congestion. (Ribbert.) In the centre is the central vein, with slightly thickened walls. Surrounding this are the dilated capillaries, forming the intralobular vessels, between which are the atrophic liver-cells containing pigment. This pigment is probably of biliary origin. The pressure upon the cells must interfere with the discharge of the bile through the bile-capillaries (Figs. 127 and 128), and lead to an accumulation of its constituents within the cells, where the pigment collects.

passed through the other parts of the vascular system within the liver. The capillaries are enlarged around the central vein; the hepatic cells between them are diminished in size and pigmented (Fig. 258).

The growth of tumors may exert a pressure upon neighboring parts, causing their atrophy, the explanation of which is similar to that of atrophy of the liver as the result of passive hyperæmia. Pressure upon a tissue does not always, however, occasion atrophy. If the function of a part be to resist pressure, an increase of pressure may lead to hypertrophy, provided the nutrient supply be sufficient. Thus pressure upon the walls of a bloodvessel may cause them to increase in thickness.

Aside from the two forms already mentioned, atrophy may be the result of a diminution in the nutritive supply: local, as the result of disease in the vessels of a part; general, when all the vessels are affected with disease, or when the general nutrition of the body is reduced. Both these causes operate in the general condition known as "senile atrophy."

More obscure forms of atrophy are those which appear to be occasioned by lesions of trophic nerves, or are caused by toxic conditions; *e.g.*, lead-poisoning.

## CHAPTER XXII.

### HYPERTROPHY AND HYPERPLASIA.

By hypertrophy is meant an increase in the size of the elements composing a tissue; by hyperplasia, an increase in their number. Both conditions usually lead to an enlargement of the organ in which they are found, but this is not necessarily the case, for all the elements in the organ need not participate in the increase; some may diminish in bulk.

**1. Functional Hypertrophy.**—This process, like that of functional atrophy, depends upon the activity of the part undergoing the change. In this case the parenchyma of the part is increased to meet a gradually increasing demand for the work it is fitted to perform. This increase may take the form of hypertrophy or that of hyperplasia. The muscular tissues meet the demand by an increase in the size of the muscle-cells. This is illustrated in the hypertrophy of the heart in valvular lesions, which throw extra work upon the muscle; in the enlargement of the uterus during gestation, fitting it for the strong contractions during labor; and in the enlargement of the voluntary muscles by exercise.

In glandular organs an additional demand for work results in hyperplasia, in which the epithelial cells of the parenchyma multiply (Fig. 259).

Functional hypertrophy, or hyperplasia, takes place only under certain favorable conditions. The demand for extra functional activity must not be too great, otherwise degenerative changes ensue. The same result would follow were the nutritive supply insufficient to meet the loss of material and force sustained by the cells in doing the increased work. It is evident, then, that the condition occasioning the hypertrophy or hyperplasia must develop gradually, and not interfere with the supply of nutrition. The nature of the tissue also influences the result. In general, it may be stated that tissues of high specialization are less capable of either hypertrophy or hyperplasia than those less specialized, and that hypertrophy is the rule in tissues of higher function, while hyperplasia is more common in those of lower function, where the formative powers of the cells are less in abeyance.

COMPENSATORY HYPERTROPHY is a term applied to functional hypertrophy or hyperplasia following the destruction of an organ or part of an organ. This leads to an increase of the work demanded of other parts capable of performing the function normally carried on by the part destroyed, or capable of assisting the function that has

FIG. 259.

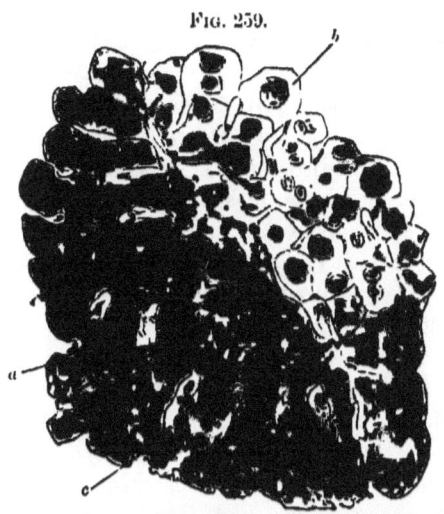

Necrosis of part of an hepatic lobule. (v. Meister.) *a*, necrosed cells, the nuclei of which have lost their affinity for dyes; *b*, hypertrophic cells with large nuclei; *c*, detritus of blood-corpuscles in the capillaries. Section taken eighteen hours after removal of a portion of the liver in a rabbit. The section is taken at the margin between that tissue which is affected with necrosis and that which retains life, but is stimulated to proliferation by the irritative effects of the amputation. After a while the hypertrophied epithelial cells will divide by karyokinesis and attempt a restitution of the lost tissue—a species of compensatory hyperplasia.

suffered diminution. Thus, disease of one kidney may indirectly occasion hypertrophy of the other kidney, or, more properly, hyperplasia of its functional epithelium, or chronic interstitial nephritis affecting both kidneys may lead to hypertrophy of the heart by throwing more labor upon that organ in order that the remaining renal parenchyma may perform the work demanded of the kidneys. In like manner the auxiliary muscles of respiration may become hypertrophic in cases of embarrassed respiration.[1]

Functional hypertrophy may also find expression among the con-

[1] Attention has already been called to the hypertrophies of the hypophysis and parathyroids in cases of thyroidectomy or disease of the thyroid gland (see p. 191).

nective tissues of the body, in which the usefulness of the tissue resides in its physical properties. In muscular individuals the bony ridges giving attachment to the tendons are more strongly accentuated than in those whose muscles are less highly developed.

A very familiar illustration of functional hyperplasia is furnished by the skin of the palms. Manual labor that is *habitual* occasions a thickening of the epidermis due to hyperplasia; *exceptional* overwork causes damage leading to inflammation, blisters.

2. **Developmental Hypertrophy.**—Hypertrophy of a part occasionally arises without assignable cause and apparently as a mere anomaly in development. Such structures as horns and warts are examples of this form of hypertrophy, which are not readily separated from the group of growths called tumors. When the growth is limited and not progressive it may in most cases be attributed to this form of hypertrophy; when apparently unlimited, progressive, and atypical in structure, it must be classed among the tumors.

3. **Inflammatory Hypertrophy.**—Under the influence of damaging agents which act with such mitigated intensity that their effect upon the cells amounts merely to a decided irritation, the formative powers of the cells may be stimulated and an enlargement of the part be brought about, either as the result of hypertrophy or of hyperplasia of its elements. This form of hypertrophy is nearly, if not quite, equivalent to the results of chronic productive inflammations, for an account of which the student is referred to another chapter. In cases where the evidences of damage are inappreciable the process may be considered as irritative hypertrophy or hyperplasia; where they are at all marked, it must be regarded as inflammatory.

The microscopical evidence of hypertrophy is found in an increase of size in the elements composing the tissue. It is not a simple matter to decide from a microscopical examination whether hyperplasia exists or not, for the microscopical appearances are almost, if not quite, normal. It is often necessary to consider the changes in the gross appearances of the part in order to determine whether its constituent elements have increased in number or not.

## CHAPTER XXIII.

### METAPLASIA.

WHEN a fully developed tissue becomes modified in its structure to resemble another form of adult tissue, without passing through an intermediate stage of indifferent or more embryonic tissue, the process is known as "metaplasia." It differs from the inflammatory process in that the rejuvenescence of the tissue is not obvious, and it is unlike the development of a tumor because the tissue-change is a *conversion* of one form of tissue into another, and not the production of a new tissue within another.

Metaplasia only results in the formation of a tissue closely allied to that in which it takes place. It is most commonly met with in the connective tissues, where a change in the character of the intercellular substances and in the form of the cells, which all spring from the same original source, the mesoderm, is all that is necessary to convert one form of connective tissue into another variety of the same group. We must attribute the change to a modification in the functional activity of the cells, the reasons for which are in most cases very obscure. We may, perhaps, in some cases, seek the explanation in conditions that lead to an altered functional demand on the part. Thus, for example, it has been noticed that bone sometimes develops in the fibrous tissues of the thigh or shoulder in soldiers that are obliged to ride or carry a musket for a long time. It may be that the fibrous tissue becomes reinforced in these cases with bone, because it is better calculated to withstand the pressure; but the fact that such cases are exceptional shows that this response on the part of the tissues is by no means constant and that the explanation is incomplete.

Metaplasia may result in the conversion of fibrous tissue into mucous or osseous tissue; hyaline cartilage into fibro-cartilage, or into fibrous, mucous, or osseous tissue; adipose tissue into mucous tissue, etc. The metaplastic tissue is usually not typical; that is, it differs somewhat from the normally developed tissue in the finer details of its structure. Thus, the bone that is produced by metaplasia from fibrous tissue lacks the elaborate system of canaliculi that is found in normally developed osseous tissue, although in its essential features it is virtually bone, the intercellular substances being impregnated with calcareous matter and yielding gelatin on boiling.

Epithelial tissues may also be the seat of metaplasia. Under the influence of moderate but repeated damage, columnar epithelium may become modified into a stratified variety. In such cases the cause may, presumably, be traced to a change of conditions, which calls for an unusual exercise of the protective function of the epithelium. The uterine cavity and the respiratory tract are the most common situations in which this transformation of epithelium is met with. A similar conversion of transitional epithelium into true stratified epithelium is occasionally met with in the bladder and renal pelvis, as the result of a calculus not causing sufficient damage to induce an active inflammation.

Metaplasia appears to result from a change in the functional activities of the cells, which lose their accustomed form of specialization and acquire new ones of closely related character.

# CHAPTER XXIV.

## STRUCTURAL CHANGES DUE TO AND FOLLOWING DAMAGE.

### I. NECROSIS.

THE term necrosis designates a local death of tissue during the life of the individual.

In our study of the normal tissues under the microscope we are obliged to use methods of preparation which, in nearly all cases, kill the tissues before they come under observation. When we examine them with a view to determining their structure, they are nearly always necrotic, if we may use that term in this connection. Our standards of the normal appearances are, therefore, largely based upon what we learn from recently killed tissues.

In some instances it is possible, however, to examine even highly developed tissues while still living. If, for example, the superficial layer of a frog's cornea be stripped off and mounted in a drop of serum, the cells composing it may be readily seen under the microscope. While such a preparation is quite recent it is difficult to distinguish clearly the nuclei within the cells, their refractive indices being nearly the same as that of the surrounding cytoplasm; but in a short time the nuclei suddenly become very distinct, as though they had undergone a sort of crystallization. This is probably an indication of the death of the nuclei, the substances composing them having suffered a coagulation which increases their powers of refracting light and, in consequence, the distinctness with which they are seen. This conclusion is strengthened by the fact that the change may be hastened by the application of reagents, such as acetic acid.

The modern methods of preparation used in histological studies aim at bringing about a sudden death of the cells and such a coagulation of the tissue-elements as shall prevent further changes of structure before the tissues can be studied. For, if the tissues are allowed to die spontaneously, their elements suffer changes that greatly alter their appearance. When they die and remain within the living body, as is the case in necrosis, those changes in structure are more diverse and more marked than those incident to spontaneous death resulting from removal. This has led to the distinction of several varieties of necrosis, characterized by different structural changes in the dead tissue, which are dependent upon the conditions obtaining in the tissue at the time of death or after death has taken place.

Among the most striking changes incident to necrosis are those affecting the nucleus. This may retain its form in great measure, but lose its affinity for the nuclear dyes ("chromolysis," Fig. 262), or the chromoplasmic substances may retain that affinity, but be broken up into fragments, thus destroying the form of the nucleus ("karyolysis," Figs. 260 and 261). Both of these changes are indicative of the death of the nucleus and assure the death of all parts of the cell.

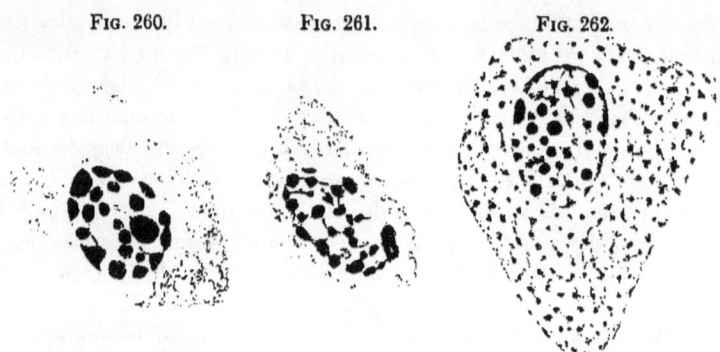

Fig. 260. Fig. 261. Fig. 262.

Changes in the nuclei of renal epithelial cells incident to necrosis. (Schmaus.)
Fig. 260.—Destruction of the chromatic reticulum and condensation of the chromatin in masses of various sizes; early stage of karyolysis. Nuclear membrane nearly gone.
Fig. 261.—More advanced stage of nuclear destruction. The nuclear fragments lie free in the cytoplasm; later stage of karyolysis.
Fig. 262.—Disintegration and disappearance of the chromatin without a coincident disintegration of the form of the nucleus-chromolysis.

1. **Coagulation-necrosis.**—When the tissues that have suffered death liberate fibrinoplastic substances and fibrin-ferment these interact with the fibrinogen in the lymph and occasion a coagulation of the necrosed tissue analogous to the production of fibrin. These coagulated materials may appear as fine granules or as hyaline masses of a dense, glassy character. This form of necrosis is illustrated in the formation of the "membrane" in diphtheria, which is the superficial portion of the affected part that has under-

gone coagulation-necrosis (Fig. 263). When the granular form of coagulation-necrosis is associated with albuminoid and fatty degeneration the result is a cheese-like mass, and the process is known as cheesy degeneration (p. 274).

2. **Colliquative Necrosis** (Fig. 281).—This form of necrosis is followed by an imbibition of fluid, occasioning a disintegration of the

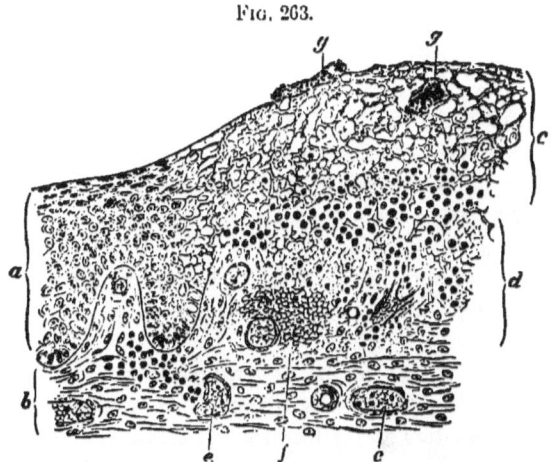

Fig. 263.

Edge of a diphtheritic membrane. Section from the human uvula. (Ziegler.) *a*, normal stratified epithelium; *b*, subepithelial fibrous tissue of the mucous membrane; *c*, epithelium that has undergone coagulation-necrosis. Only remnants of cells remain in the coarse fibrinous meshwork. *d*, œdematous subepithelial fibrous tissue containing fibrin and leucocytes; *e*, bloodvessels; *f*, hæmorrhage; *g, g*, groups of the bacteria causing the necrosis.

tissue-elements, which are broken up into a granular detritus suspended in the fluid.

The foregoing two forms of necrosis may be associated with each other, or one may follow the other.

The fate of the necrosed tissue depends upon a variety of circumstances. The presence of dead tissue excites an inflammation in the living tissue surrounding it, and the character of this inflammation often determines the fate of the necrosed mass. (See article on inflammation.) The situation of the dead tissue also affects the result. The following examples will serve to illustrate these variations:

1. ABSORPTION.—The necrosed tissue-elements become disintegrated, and the débris either dissolved or carried away through the lymphatic channels by the currents of fluid, or through the

agency of leucocytes, which incorporate them and then pass out of the necrotic area. This disintegration appears to be due partly to a simple maceration or separation of the particles of the tissue, partly to a solvent action exerted by the fluids in the tissues upon dead organic matter. While absorption is going on there is an inflammatory reaction in the surrounding tissues that still retain life, which results in the formation of cicatricial tissue. This may ultimately occupy the site of the necrosed tissue, or it may form a capsule around a collection of fluid occupying that site, the result being a cyst with a fibrous wall.

2. ENCAPSULATION.—The necrosed tissues may remain unabsorbed, or be only partly absorbed, and eventually become enclosed in a capsule of new-formed fibrous tissue arising through the inflammatory process mentioned above. In this case the necrosed mass becomes desiccated through absorption of its fluid constituents, and may eventually be infiltrated with lime-salts, calcified.

3. GANGRENE.—This occurs in two forms, distinguished as dry and moist gangrene.

Dry gangrene is due to the desiccation of dead tissues that are exposed to the air. The tissues become discolored, owing to changes in the coloring-matter of the blood, and shrink, the skin assuming the appearance of parchment. After a time the dead mass is cast off by the formation of granulation-tissue from the neighboring living tissues.

Moist gangrene is the result of putrefactive changes in dead tissue, due to infection with bacteria causing decomposition. The parts are discolored, swollen, moist, and often contain bubbles of gas having a foul odor. The gangrenous part may here also be cast off as the result of the formation of granulations, but the gangrenous process may spread before it can be checked by an inflammatory demarcation, the products of decomposition having a poisonous effect upon the neighboring tissues that leads to necrosis and prevents the development of granulation-tissue.

4. SUPPURATION.—If the dead matter contain pyogenic microorganisms, they exert a peptonizing action upon the necrotic mass, causing it to liquefy. At the same time they excite a purulent inflammation in the surrounding tissues which leads to the formation of an abscess or an ulcer.

In those cases of necrosis in which the necrosed tissues are not speedily absorbed the dead mass is known as a "sequestrum," and the zone of inflammation separating it from the living tissues is called the line or plane of demarcation. (For a fuller explanation of the process of demarcation and of the tissue-changes that lead to encapsulation, the student is referred to the article on inflammation.)

## II. INFLAMMATION.

It is difficult to frame an accurate definition of inflammation, for the reason that the term includes a number of different conceptions that cannot be readily expressed in concise form. In general, it may be stated that inflammation is a process of repair following a limited damage to the tissues. The injurious agent acting upon a part must inflict a certain amount of damage in order to bring about inflammation; if its action be slight, it will cause only an evanescent irritation which does not pass into inflammation; if, on the other hand, its action be severe, it occasions necrosis or degenerative changes at the point of its application, and only in remoter parts of the tissue, where its action is moderate, will inflammatory changes be manifested. The nature of the damaging cause and that of the tissues affected both influence the character of the inflammatory process. It therefore manifests many variations under different circumstances, and in order to understand the underlying principles of the process it will be best to select some particular example for a somewhat close study, and then to consider some of the circumstances that modify the phenomena presented by that example. A severe burn, the effects of which extend deeply enough to destroy a part of the true skin, will serve this purpose, as affording an example of acute inflammation of a vascularized part following a cause that has acted for only a short time and has then been removed.

In considering this example we must distinguish between those destructive effects that are due to the damaging cause, and the reparative processes that follow in the tissue-elements that have been less seriously affected. It will make the example clearer if we also separately consider the phenomena presented by the vascular system from those taking place in the fixed tissues of the part exclusive of the bloodvessels.

Those tissues which have come into the closest contact with the source of heat will have been quickly killed and, perhaps, charred. Beyond this point of complete destruction the tissues may be roughly divided into zones, in which the direct damage is successively less marked. In the first zone necrosis will have taken place; in the tissues that are more remote, degenerative changes will be occasioned; and still farther away from the seat of injury the tissues will show a vital reaction to the stimulation or irritation they have received, which will reveal itself in a growth, eventually leading to a repair or patching of the defect in the tissues occasioned by the damage.

1. **The Bloodvessels and the Circulation.**—The vessels most seriously damaged, together with the blood they contained, will have been completely destroyed; in those less affected the circulation will have been arrested and the blood coagulated. But beyond the zones in which the function of the circulation has been abolished the first marked effect is an increase in the volume and rapidity of the current of blood. This increased flow of blood to the part is attributed to the action of the injury upon the vaso-motor system of nerves, causing a relaxation of the walls of the arteries supplying the part which has been damaged. A similar increase in circulation follows slighter stimulation of the skin, as, *e. g.*, rubbing, so that this determination of blood to the part as the result of vasomotor disturbance is comparable with entirely normal hyperæmias; but it is greater in degree when the irritation of the parts is great enough to cause damage.

After an interval the velocity of the circulation in the part which is becoming inflamed is reduced, without any diminution in the calibre of the vessels, and the slackening of the current may pass into complete stasis. This is probably due to two causes: first, to the extension of the vaso-motor disturbance beyond the area of the injured part, so that collateral branches of the main arteries are dilated; this would diminish the pressure of blood going to the inflamed part. Second, to alterations in the walls of the smaller vessels in the inflamed part, especially the capillaries and small veins. These become more pervious, probably as the result of the damage they have sustained in common with the other tissues, allowing a greater amount of fluid to pass through them than when they were in the normal condition. This comparatively rapid extraction of its watery constituent increases the viscosity of the blood, and that increased viscosity, together with the changes in the walls of the vessels, increases the friction between the two, impeding the circulation.

Thus, two influences appear to check the flow of the blood after the inflammatory process has been inaugurated: (1) a diminution of the pressure urging the blood forward, and (2) an increase in the resistance offered to the passage of the blood through the smaller vessels. To these, another factor increasing the resistance is added as soon as the current has become slowed beyond a certain point. During the normally rapid flow of the blood the corpuscles it contains, being heavier than the serum, form a column in the axis of the vessels, with a clear zone of serum around it (Fig. 264). This is in accordance with the physical laws governing the behavior of suspended particles in fluids circulating in a tube; but if the rate of flow be diminished beyond a certain point, the suspended particles

Positions of the corpuscles in circulating blood. (Eberth and Schimmelbusch.)

Fig. 264.—Appearance when the velocity of the circulation is normal: *a*, axial column of corpuscles, both red and white, in such rapid movement that individual corpuscles cannot be distinguished. Occasionally a white corpuscle is thrown from the axial mass and appears in the plasmic zone, *b*.

Fig. 265.—Appearance when the velocity of the circulation is moderately reduced. The zone *b* contains numerous leucocytes.

Fig. 266.—Appearance when the current of blood is sluggish; *a*, red corpuscles, still in the axis; *b*, peripheral zone, containing leucocytes, *d*, and blood-plates, *c*.

When stasis is fully established the red corpuscles also invade the peripheral zone.

The figures are from observations made on the vessels of a dog's omentum during life.

invade the fluid zone at the periphery of the current, those which are specifically most nearly of the same weight as the fluid passing most freely into it. In the case of the blood those particles are the leucocytes, which are lighter than the red corpuscles, and, as the current slackens, it is these which first make their way into the clear serum at the periphery of the stream and soon come in contact with the vascular wall (Figs. 265 and 266). Here, by virtue of their adhesiveness, they cling to the endothelium, and must materially increase the difficulty with which the blood is forced forward and promote stasis.

While the blood is circulating freely in the vessels the leucocytes it contains are subjected to repeated mechanical shocks through contact with other corpuscles or with the walls of the vessels where these branch or form sharp curves. These blows cause the cytoplasm to contract, maintaining the globular form of the corpuscle; but when they come to rest upon the surface of the vascular wall, as may occasionally happen under normal circumstances, and is always the case in acute inflammations, the leucocytes have an opportunity to execute the movements which have been called "amœboid," from their resemblance to those displayed by the amœba. The leucocytes send out pseudopodial processes and creep along the surface of the vessel-wall. We must bear in mind that at this time the capillary vessels are dilated, and that the cement between the endothelial cells is somewhat stretched and thinned. The passage of the pseudopodia of the leucocytes through the cement is facilitated by these circumstances, so that soon after the circulation has become slowed there is a passage of leucocytes through the walls of the vessels into the spaces in the surrounding tissues. This escape of the leucocytes is called their "emigration" (Fig. 267). The number

FIG. 267.

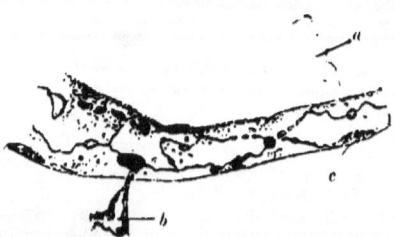

Emigration of leucocytes through a capillary wall. (Engelmann.) a, leucocyte just leaving one of the pseudostomata between the endothelial cells of the capillary wall; b, leucocyte partly within and partly outside of the capillary; c, nucleus of an endothelial cell of the capillary wall.

of leucocytes that escape from the blood in the manner described is variable. In some varieties of inflammation the tissues outside of the vessels contain substances that have an attraction for the leucocytes. This is particularly the case when the cause of the inflammation is an infection with bacteria. Under these circumstances the leucocytes that emigrate from the blood accumulate in great numbers in the tissues around the site of infection.

The leucocytes, by their passage through the cement between the endothelia, open minute channels through which the red corpuscles of the blood may be pressed into the surrounding tissues, when they come in contact with the vascular wall after stasis (complete arrest of the circulation) has become established. These corpuscles are soft, and can be forced through orifices much smaller than their normal diameters; but the number that escape from the vessels varies greatly in different cases of inflammation, and it is probable that the integrity of the vascular wall is more affected when the number is great than when it is slight, and that the leucocytes prepare the way for only a portion of the red corpuscles that escape from the vessel in those cases in which large numbers pass into the surrounding tissues. The escape of red corpuscles from a vessel without obvious rupture of its walls is called "diapedesis."

As a result of the processes already described, it will be observed that three of its constituents pass from the blood into the surrounding tissues: (1) serum, (2) leucocytes, and (3) red blood-corpuscles. These constitute what is known as the "exudate." But to these three a fourth constituent is soon added, namely, fibrin. The formation of fibrin is still awaiting a perfectly clear explanation, but it is usually assumed to be the result of the interaction of three substances: (1) fibrinogen, derived from the plasma of the blood; (2) fibrinoplastin and (3) fibrin-ferment, both of which may come from the bodies of cells. In the exudate of acute inflammation all of these elements necessary for the formation of fibrin are present in greater or less amount. (See explanation of fibrin-formation on p. 127.) As found in the tissues, therefore, the exudate consists of serum, fibrin, leucocytes, and red corpuscles (Fig. 268). But in different cases their relative abundance differs, and the acute inflammations have been roughly classified according to the character of the exudate. Thus, the serous inflammations are those in which serum predominates in the exudate. In like manner inflammations are designated by the terms fibrinous, hæmorrhagic, and purulent (when the leucocytes predominate), or sero-fibrinous, sero-purulent, fibrino-purulent, etc. These terms are descriptive, and merely indicate variations in the proportions

302   HISTOLOGY OF THE MORBID PROCESSES.

of the different constituents in the exudate. The general nature of the process is the same in all cases.

We are now in a position to explain four of the cardinal symptoms of acute inflammation. The increase of temperature and the redness (calor and rubor) are attributable to the hyperæmia of the part and its surroundings. The swelling and pain (tumor and dolor) are caused, at least chiefly, by the presence of the exudate. The suspension of function, or fifth cardinal symptom of acute

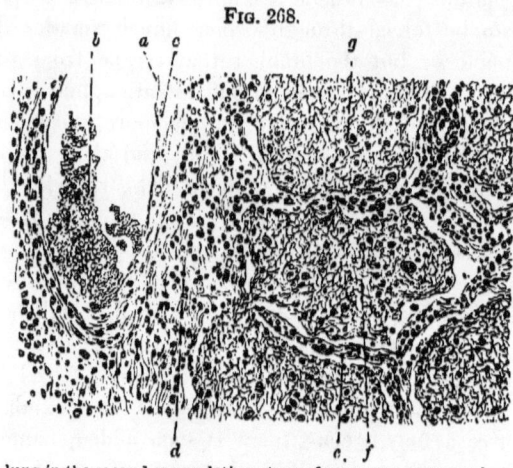

Fig. 268.

Section from lung in the second or exudative stage of croupous pneumonia: *a*, endothelial wall of a small vein; *b*, blood within the vein, unusually rich in leucocytes, which have collected during the slowing of the circulation. The line *b* points to the nucleus of a leucocyte. Part of the blood has fallen out of the section during its preparation. *c*, leucocytes beneath the endothelium of the vascular wall; *d*, œdematous fibrous tissue surrounding the vessel. The fibres of the tissue have been separated by the exuded serum. This tissue is also moderately infiltrated with leucocytes that may have passed through the walls of the vein, and contains a few red blood-corpuscles. *e*, wall separating two pulmonary alveoli. This is also somewhat infiltrated with leucocytes. *f*, exudate within an alveolus, consisting of serum, fibrin, leucocytes, and red blood-corpuscles; it also contains a few epithelial cells desquamated from the alveolar wall, *g*.

inflammation, may have a more complex causation. It may be due to the immediate effects of the injury that occasioned the inflammation, to disturbance of nutrition, to the presence of the exudate, or perhaps to an interruption of the normal nervous mechanism. All these disturbing factors are present, and may vary in their potency in different cases.

All the changes that have been hitherto described are the immediate or only slightly remote effects of the damage to the tissues, and have nothing to do with the process of repair. They may be regarded as constituting the *destructive phase* of acute inflammation.

2. **The Fixed Elements of the Tissues.**—It is evident that the cause of damage itself, or the disturbances of nutrition resulting from the changes in the circulation, must either cause rapid death, necrosis, or that slower form of death entailed by a relatively insufficient supply of nourishment, which has been described in the chapter on the degenerations. The cells are either killed at once, or are starved within a certain radius of the point at which the cause of the inflammation was applied. Beyond this radius these changes give place to those that bring about repair. But the susceptibility of the different tissue-elements varies: an injury that would kill some might hardly affect others; a given degree of innutrition might cause degeneration in some and not in others, so that the depth to which those changes are felt will depend upon the nature of the tissues present. In general, it may be stated that those tissues which are highly specialized and those which carry on functions requiring active intracellular metabolism are the ones most deeply affected by damaging influences.

**Repair.**—The view was at one time strongly upheld that emigrated leucocytes were active in the formation of the new tissues that developed during inflammation. These corpuscles were regarded as of indifferent character, capable of differentiation into the various forms of connective tissue. This view has not been supported by the results of experimental study, and is now abandoned, giving place to a revival of the earlier belief that the cells of the fixed tissues are the active elements in the reparative process which results in the formation of new tissues.

Since the significance of the mitotic figures during karyokinesis has been learned, it has become possible to ascertain positively that the fixed cells multiply beyond the zone of destruction in acute inflammations. The cells which have suffered neither destruction nor degeneration beyond their powers of recuperation undergo a species of rejuvenescence, returning to a comparatively undifferentiated condition, in which their powers of reproduction and tissue-formation are revived. It is as though they reverted, under the influence of strong irritation, to the condition in which their progenitors existed at an earlier stage of tissue-development. The process of repair depends upon this capacity for rejuvenescence on the part of the cells of the tissues, but that power varies greatly in the cells of different tissues, being, roughly, inversely proportional to the degree of specialization to which they have attained. Those tissues whose functional activities in the adult are chiefly formative possess this capacity for rejuvenescence in a high degree. In fact, epithelium in many situations—e. g., upon the skin—merely requires a little stimulation of its normal activities to produce new tissue. The case is different with tissues of higher function, in which the cells have become greatly specialized at a sacrifice of their formative activities. In these the capacity for rejuvenescence is always comparatively slight, and may be entirely lost; as, for example, in the ganglion-cells of the central nervous system. Such parenchymatous cells of high function are also more vulnerable than cells of a lower type of specialization, because they are more dependent for their functional activity upon a maintenance of the normal conditions of nutrition.

The foregoing considerations explain why the more highly specialized cells are damaged for a greater distance from the point of injury than are the connective-tissue cells, and also why they play a less prominent part in the restorative processes that follow those which have been destructive. The result is that the zone of connective tissue capable of rejuvenescence is nearer to the site of injury than the zone which includes undegenerated cells of higher function, and from this it follows that the defects in the tissues are made good by a proliferation of connective tissue, accompanied in only slight degree by a proliferation or restitution of the tissues of greater specialization. The process of repair is more a patching of the defect than a restoration of the normal structure. It results in a permanent scar, and not the perfect replacement of lost tissues by others of the same structure and function.

During rejuvenescence the cells of the connective tissues enlarge and become more cytoplasmic, and their nuclei become richer in chromatin. They then divide by the indirect process, giving rise to a number of spheroidal cells, which, together with newly developed loops of capillary bloodvessels, constitute an undifferentiated tissue, called "granulation-tissue." During its formation at least a part of the original fibrous intercellular substance appears to be removed by absorption. This may be brought about by maceration in the fluids present, or through the agency of the leucocytes that have emigrated from the vessels and play the part of phagocytes (Fig. 269).

The young vascular loops that supply the granulation-tissue are

Fig. 269.

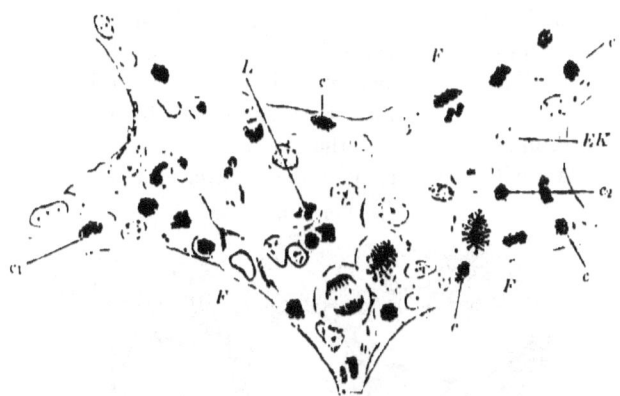

Section from adipose tissue in the neighborhood of a phlegmonous inflammation due to infection with streptococci. (Grawitz.) $F$, the boundaries of fat-cells, the tissue represented being the connective tissue between those cells. Four large karyokinetic figures are seen in that tissue; these are in the rejuvenescent cells of the fibrous tissue. The section also contains leucocytes that have wandered into the tissue from the neighboring focus of exudation. These are designated by the letters $L$ and $c$. $c_1$ and $c_2$ are connective-tissue cells undergoing destruction, their nuclei showing chromolysis. Other connective-tissue cells show a swelling of the nucleus (karyolysis), and the interstitial tissue is the seat of a moderate œdema.

produced through a similar rejuvenescence of the endothelial cells of the older capillaries. These cells become richer in cytoplasm, and acquire a strong resemblance to epithelial cells (Fig. 270). They then multiply, forming little collections of cells in contact at

Fig. 270.

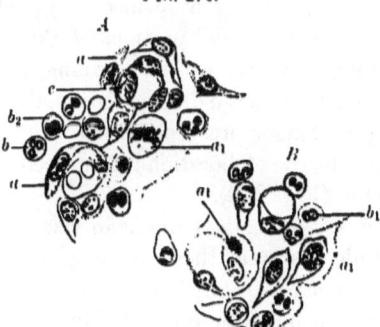

Sections from granulations forty-eight hours old. (Nikiforoff.) In both $A$ and $B$ two capillaries are represented. $a$, young connective-tissue cell; $a_1$, karyokinetic figures in such cells; $b$, $b_1$, $b_2$, leucocytes with single, polymorphic, or fragmented nuclei, the latter suffering karyolysis and, consequently, death; $c$, endothelial cell with nucleus in spirem stage of karyokinesis, demonstrating the proliferation of those cells.

20

one point with the walls of the capillaries and reaching out in columns or bands among the cells of the granulation-tissue. Here they may become united with each other, forming loops that spring from the same capillary vessel, or connect it with other capillaries. Subsequently these solid columns or bands of cells become channelled, the cells forming the walls of the new vessels, the lumina of which communicate with those of the parent capillaries (Fig. 271).

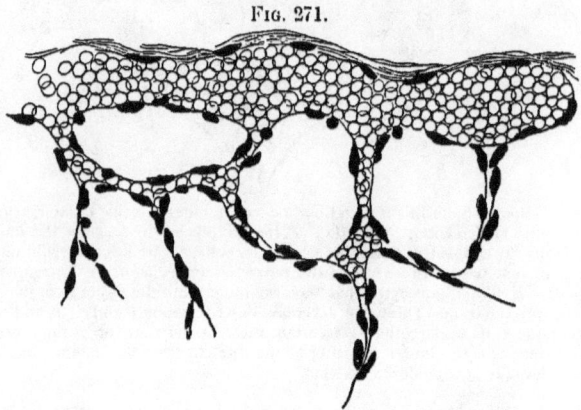

Fig. 271.

New-formation of bloodvessels in granulation-tissue. (Birch-Hirschfeld.)

The granulation-tissue thus formed is continuous with the adjacent uninjured fibrous tissues, and serves to separate the tissues that have been killed or have undergone irrevocable degeneration from the living tissues that lie beneath it. The dead mass is finally loosened and cast off, leaving a surface of growing granulations. While the cells in the superficial portions of this granulation-tissue continue to multiply and produce fresh, young, undifferentiated tissue, the deeper portions undergo differentiation, the formative powers of the cells being no longer preoccupied with the production of new cells, but diverted to the elaboration of intercellular substances of a fibrous character (Fig. 272).

During this process the cells dwindle in size as the intercellular substances accumulate between them, and may suffer complete extinction. This may be due to atrophy in consequence of pressure exerted by the fibrous constituent of the intercellular substances, which has a marked tendency to shrink as it becomes older. Another probable reason for the disappearance of many of the cells may be the lack of a well-defined lymphatic circulation in the granulation-tissue and the young cicatrix, which, if it existed, would serve to assist in the nutrition of the tissue. There is a manifest advantage to the whole organism in this absence of lymphatics in granulation-tissue, for the absorption of injurious substances from the region beyond the granulations is hindered. But the nutrition of the granulations themselves is impoverished and the fibrous tissue

FIG. 272.

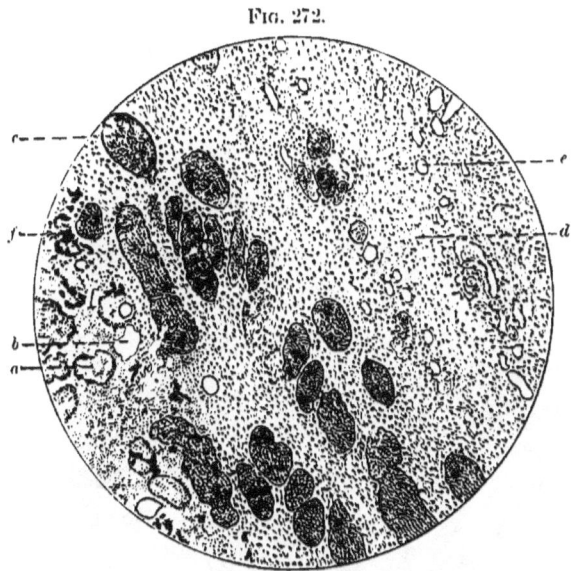

Newly formed fibrous tissue from a case of pleurisy: *a*, pulmonary alveolus filled with an exudate largely composed of leucocytes (pneumonia; stage of gray hepatization passing into resolution); *b*, alveolus, from which the disintegrated exudate has fallen out. Before the alterations in structure due to inflammation took place this alveolus, and the one above it, lay immediately beneath the pleura. The thin pleuritic membrane has now been destroyed and its place taken by the fibrous tissue of inflammatory production, which fills nearly the whole field of vision. *c*, thin-walled bloodvessel in that fibrous tissue. This and those like it form a part of the older portion of the granulation-tissue which has replaced the fibrinous exudate at first covering the lung (see p. 313). The granulation-tissue between these vessels has organized into a young fibrous tissue. *d*, younger granulation-tissue; *e*, recently formed bloodvessel in the latter; *f*, masses of carbon deposited in the tissues by leucocytes, which have transported it thither from the air-passages. These deposits existed before the acute inflammation began. This form of pigmentation is called "anthracosis."

that results from its differentiation is of comparatively low vitality. While the tissue is young, succulent, and highly vascularized by capillaries, this deficiency in its organization may not be apparent; but as the intercellular substances contract they compress the vessels and cause obliteration of many of them, with atrophy and disappearance of their cellular walls (Fig. 273).

308    HISTOLOGY OF THE MORBID PROCESSES.

When, as in the example originally chosen, the injury affects tissues that are normally covered with epithelium, the cells of that tissue proliferate at the edges of the granulations until a layer of epithelium completely covering them is produced. The whole process of repair comes to an end with the formation of a dense fibrous tissue that is only slightly vascularized by thin-walled bloodvessels and is poor in cells. This is the scar, composed of "cicatricial" tissue (Fig. 273). Upon the skin it is covered with epithelium;

FIG. 273.

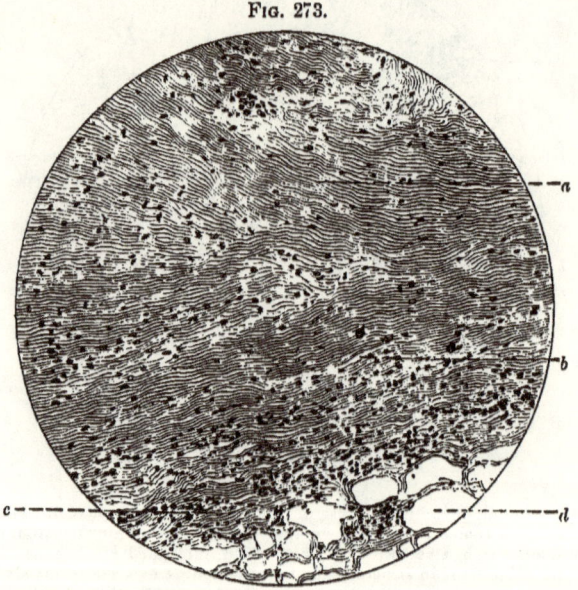

Dense fibrous tissue, or cicatricial tissue resulting from pericarditis: a, fibrous tissue, almost devoid of nuclei and vessels derived from granulation-tissue; b, lumen of a small remaining vessel; c, moderate round-cell infiltration in the deeper portion of the fibrous tissue, resulting from an immigration of leucocytes, and, perhaps, also from a slight irritative proliferation of the fixed cells of the tissue; d, subpericardial adipose tissue.

but there are no papillæ beneath this covering, and the epithelium is as poorly nourished as the cicatricial tissue beneath it.

The cells of higher function in the damaged part which have not been irremediably injured pass through the changes that will presently be described in the section on regeneration.

The course of a simple acute inflammation, as outlined above, may be modified and complicated by a number of circumstances to such an extent that these variations must be briefly described.

1. **The Healing of Fractures.**—When a bone is broken the rejuv-

enescence affects the tissues of the periosteum and endosteum, as well as the surrounding connective tissue of the fibrous type. In the subsequent differentiation of the granulation-tissue, which in this case is called the "callus," those cells which have been derived from the periosteum and endosteum produce bone, which becomes continuous with the osseous tissue of the fragments and restores the continuity of the broken bone. It is evident that in this case the rejuvenescence of the bone-forming cells has not caused a reversion to an entirely unspecialized type of connective-tissue cell. It is equally evident that in the production of cicatricial tissue the cells of fibrous tissue retain their special formative powers after rejuvenescence.

2. **Suppuration.**—This is occasioned by the persistent action of a damaging cause which is accompanied by the presence of substances exerting a "positive chemotactic influence" upon leucocytes (*i. e.*, attracts those cells) and at the same time effecting solution of the tissue-elements. In clinical experience nearly all cases of suppuration are due to infection with bacteria; but purulent inflammations of very limited extent may be caused experimentally by chemical substances free from micro-organisms.

Suppuration does not, however, always follow infection, even by pyogenic bacteria. Sometimes the virulence of the bacteria is too slight for the production of chemotactic substances in sufficient quantity to attract large numbers of leucocytes. Sometimes it is so great that the chemotactic influence becomes "negative" (*i. e.*, repels leucocytes), or the leucocytes are killed before they can collect in sufficient numbers to form pus. The relations between the leucocytes and the chemotactic substances are quantitative: if the substances be present in too great dilution, they fail to attract leucocytes; if in too great concentration, they repel them. Nor are bacteria and their products the only substances that attract leucocytes. Bits of dead tissue may do the same, a fact which would promote their absorption through the agency of the leucocytes.

These points will be made clearer if illustrated by an example, for which purpose an infection of the kidney through the vascular system may be selected. If a section be made through the organ so as to include a focus of infection, the bacteria will be found in the bloodvessels. The appearance of the tissues surrounding the vessel will depend upon a number of circumstances; among others, the length of time that has elapsed since the bacteria were brought to the part. In one case the walls of the obliterated vessel and the tissues in the vicinity may show chiefly necrotic changes; the tissue will be diffusely stained, the nuclei either unstained, only faintly tinged, or broken into fragments that take the dye in various intensities (Fig. 274). Around this necrosed tissue there

FIG. 274.

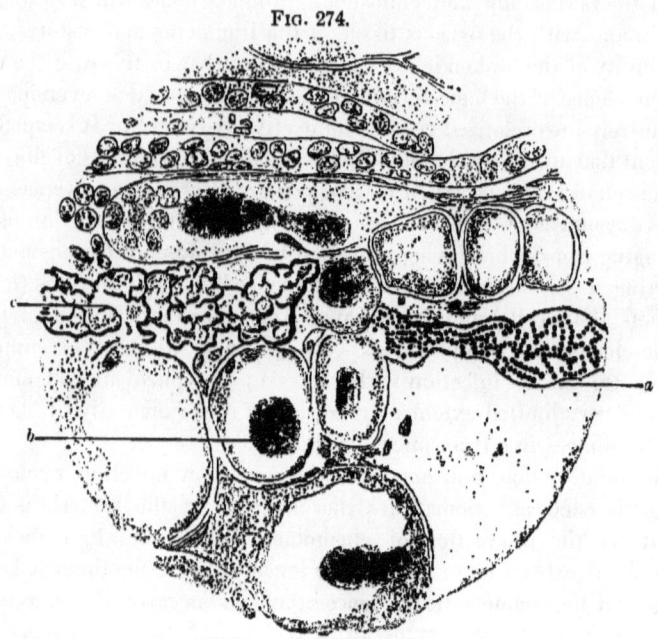

Secondary infection of the kidney in a case of erysipelas. (Faulhaber.) *a*, capillary containing streptococci; *b*, renal tubule containing a hyaline cast; *c*, renal tubule filled by a deposit of calcareous material. In the neighborhood of the capillary containing the bacteria the tissues have been necrosed, and have become reduced to a granular detritus through the peptonizing action of products formed by the bacteria. More remotely, at the upper left, the cells in the renal tubules are in a state of albuminoid degeneration. In this case the bacteria are evidently of great virulence; probably capable of destroying leucocytes that wandered into their neighborhood, through concentration of the poisons produced; for the section contains no evidence of a round-cell infiltration with emigrated leucocytes.

may be a ring of leucocytes, easily identified by their irregularly shaped or fragmented nuclei, which, unless necrosis has taken place, are more deeply stained than the normal nuclei of the surrounding kidney. The central necrosis is due to the poisons that have accompanied the bacteria at the time of infection or have been subsequently produced by them. Having killed a portion of the tissue through the action of these poisons, the bacteria thrive upon the dead matter and produce fresh poisons, which increase the area of necrotic

## STRUCTURAL CHANGES DUE TO DAMAGE. 311

Fig. 275.

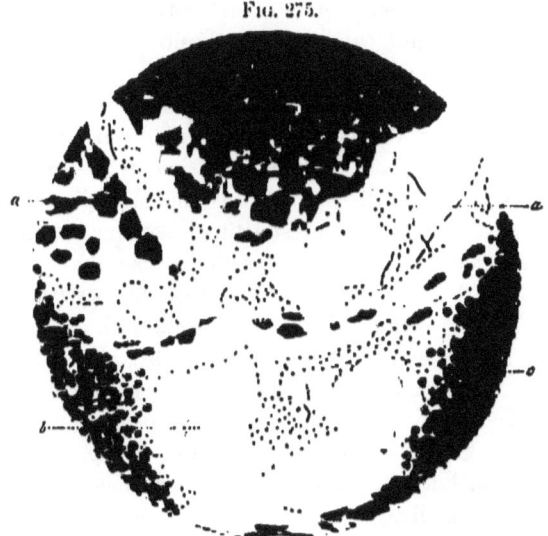

Beginning abscess-formation in the kidney. (Faulhaber.) The suppurative inflammation is due to secondary infection by bacilli carried to the kidney from a phlegmonous inflammation of the neck. *a, a*, bacilli in the capsule of a Malpighian body, the necrotic glomerulus of which is seen in the upper half of the figure; *b*, bacilli in the lumen of a convoluted tubule. The epithelial lining of that tubule has been destroyed and dissolved; only three nuclei, almost devoid of chromatin, remaining. The basement-membrane is also partially destroyed. *c*, beginning abscess-formation in the interstitial tissue between the convoluted tubules. These foci of suppuration are crowded with leucocytes, in some of which the nuclei have become poor in chromatin through the action of the poisons present. Among the leucocytes are a few bacilli, the virulence of which can only be moderate, since comparatively few of the leucocytes are necrotic.

Fig. 276.

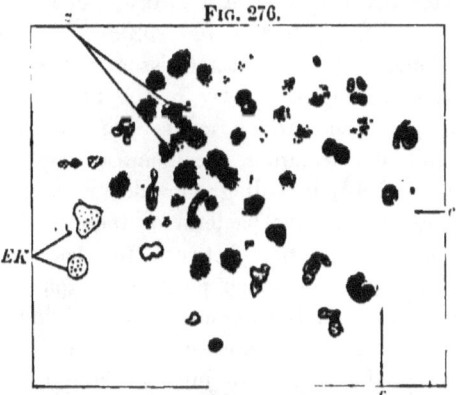

Pus from virulent abscess-formation. (Grawitz.) The leucocytes show marked necrotic changes, chromolysis. *c, c*, well-preserved leucocytes; *E. K.*, connective-tissue cells from the neighboring granulations; *z*, similar cells necrosed.

action. Toward the periphery of the inflammatory focus these poisons are more dilute, and exert a positive chemotactic influence upon the leucocytes, stimulating their emigration and progress toward the centre of the inflamed area. If they advance too far, however, or the accumulating poisons become too concentrated, they suffer necrosis or degeneration in the same manner as the tissues of the part. In this way the necrotic process may advance more rapidly than the restricting inflammatory process can cope with it. But to a certain extent the poisons they produce are injurious to the bacteria themselves, so that as they become more concentrated the growth of the bacteria is checked. The injurious influence of the bacteria upon the tissues is also, after a time, mitigated by the production within the body of chemical substances called "antitoxins," which neutralize the poisons produced by the bacteria. Other substances may also be produced which have a germicidal action. There will come a time, therefore, provided the individual lives, when the productive inflammatory process on the part of the tissues will predominate over the destructive action of the bacteria and confine the poisonous area within a zone of granulation-tissue. This demarcation does not take place in most cases until a collection of pus, an abscess, has been formed in and around the area of necrosis. The appearances are then different, and require a brief description.

An abscess or collection of pus within the tissues contains a fluid of serous character, in which there is such a great number of suspended leucocytes that they give it a milky or creamy appearance. This liquid is pus (Figs. 275, 276, and 292). The walls enclosing the pus are composed of granulation-tissue infiltrated with emigrated leucocytes making their way to the fluid contents. The liquefaction of the tissues which makes the central cavity possible is the result of maceration, the disintegrating action of the leucocytes, and, probably in still greater degree, is due to a peptonizing action exerted by the bacteria or their products. There is now an antagonistic action between the bacteria and their products and the tissues, in which possibly the phagocytic action of the leucocytes may aid the tissues. The activities of the tissues are directed to the formation of cicatricial tissue; the bacteria and their products tend to impede those activities or to destroy their results. If the destructive action predominates, the pus increases in amount and "burrows," following the direction of least resistance, until it is finally discharged along with some of the bacteria and poisons. This frequently brings relief, and the abscess becomes an open wound, which heals by granulations in the way already outlined.

In other cases the conflict between the bacteria and the tissues may be more evenly balanced and the pus confined by granulations, which are injuriously affected on the surface, but progress toward the formation of fibrous tissue in their deeper portions. Such a lining of granulation-tissue is called the "pyogenic membrane" of the abscess. Similar pyogenic membranes are formed on the walls of sinuses resulting from the discharge of an abscess when the infection is still sufficient to prevent the growth of healthy and vigorous granulation-tissue, or when the burrowing of the pus before its discharge has been so slow that the granulations surrounding the sinus have become organized in their deeper portions and are no longer capable of nourishing young and active tissues at the surface. In such a case curetting of the sinus-wall would remove this imperfectly nourished tissue and promote the development of vigorous granulations.

Still another variation of the process is possible when the infection becomes very greatly reduced in virulence or the bacteria die. In this case the granulations grow and obliterate the cavity in case its contents are absorbed, leaving a puckered scar, or its contents may become inspissated through absorption of the serum, and the leucocytes be converted into a cheesy mass by fatty degeneration combined with necrosis; in which case the resulting mass becomes encapsulated by cicatricial tissue. The resulting nodules are liable to subsequent calcareous infiltration.

3. **Fibrinous Inflammation.**—This frequently affects the serous membranes, the lung, etc. A case of lobar pneumonia may be selected as a typical example.

After a preliminary congestion of the vessels in the walls of the pulmonary alveoli an exudate, consisting of serum and red corpuscles, with a comparatively small number of leucocytes, is poured out into the alveoli. Here fibrin is formed, so that the exudate becomes solid (Fig. 268). This constitutes the stage of "red hepatization." This stage gradually passes into that of "gray hepatization," in consequence of an immigration of leucocytes into the fibrinous exudate, the red corpuscles meanwhile losing their coloring-matter, so that the red color due to them passes into a gray (Fig. 272, a). In favorable cases a stage of "resolution" follows that of gray hepatization; the fibrin disintegrates, and the exudate becomes softened (Fig. 272, b) and is expectorated. This is not the invariable outcome. Sometimes the fibrinous exudate is replaced by new-formed fibrous tissue, granulation-tissue, developing from the alveolar walls, and the alveoli become obliterated. The process in that case is similar to that which affects the pleura.

The pleural surface over the parts of the lung which are the seat of the pneumonia is usually also the seat of a similar inflammation; but here the course of the process is a little different. There are fewer red blood-corpuscles and less serum in the first exudate that is formed, probably because the proximity of the bloodvessels to the pleural surface is less immediate than the corresponding relations in the pulmonary tissue (Fig. 277). The exudate therefore

FIG. 277.

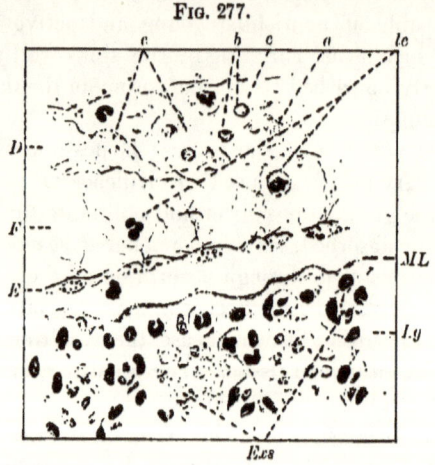

Fibrinous pleurisy, ten hours after its inception. (Abramow.) $Lg$, lung, in which three alveoli are shown in section. These contain an exudate, consisting chiefly of red blood-corpuscles and fibrin in somewhat granular form. In the alveolar walls are capillaries containing either red corpuscles or leucocytes. $ML$, membrana limitans of the subendothelial areolar tissue; $E$, endothelium with nuclear chromolysis; $F$, fibrin; $lc$, leucocytes; $D$, mass of red corpuscles, fibrin, and leucocytes, the latter with polymorphic nuclei; $a, b, c$, red corpuscles in various stages of decolorization and disintegration; $D$ and $F$ make up the exudate upon the pleural surface; $Exs$, exudate in the pulmonary alveoli.

first appears as a layer of fibrin upon the surface of the pleura. This may subsequently disintegrate and be absorbed, or granulation-tissue may develop from the pleura beneath it and grow into the fibrin, causing its gradual absorption and replacement with fibrous tissue.

In this way a fibrous thickening of the pleura is formed, which remains as an enduring evidence of the inflammation that caused it (Fig. 272). Again, it may happen that the inflammatory process is communicated to the costal pleura where it is in contact with the visceral layer. In this case fibrin is formed on both pleural surfaces, which become agglutinated in case they are in contact. When, in such cases, the interposed fibrin is replaced by cicatricial tissue, permanent fibrous adhesions between the lung and thoracic wall result. When the exudate contains sufficient serum to prevent the agglutination of the two pleural surfaces such adhesions do not take place, but each pleural surface receives a permanent layer of fibrous thickening.

Fibrinous inflammation may affect other tissues than those of the serous membranes (Figs. 278 and 279).

Fig. 278.

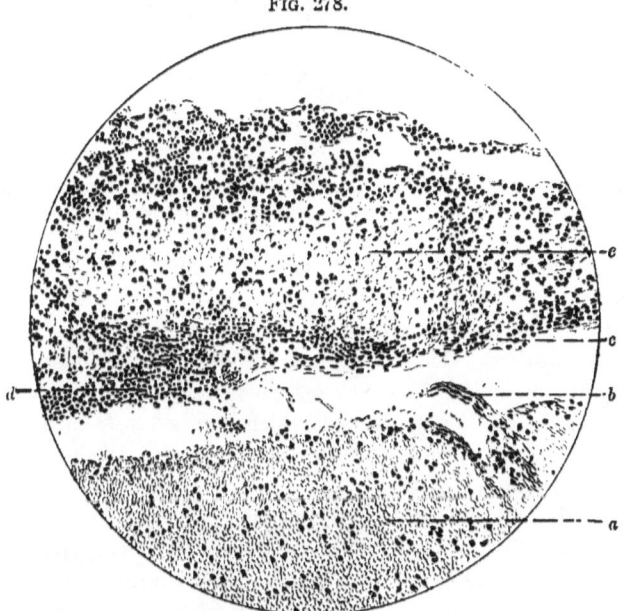

Fibrinous leptomeningitis; *a*, cerebral cortex; *b*, torn bloodvessel entering the brain from the pia mater; *c*, fibrous tissue of the pia mater; *d*, the same tissue infiltrated with emigrated leucocytes; *e*, fibrinous exudate in the wide-meshed areolar tissue of the pia mater.

**4. Serous Inflammations.**—Like the fibrinous, these inflammations are common affections of the serous membranes. Pleurisy is often an inflammation of this sort. The exudation is chiefly serous, of a light-straw color, and either quite clear or containing flakes of

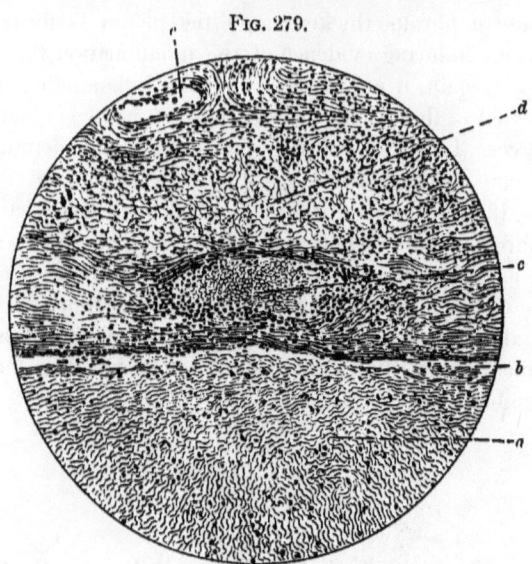

Fibrinous leptomeningitis: a, cerebral cortex; b, serum, with detritus, separating the brain from the pia mater; c, bloodvessel of the pia mater, the walls of which are infiltrated with emigrating leucocytes; d, fibrinous exudate; e, smaller vessel of the pia.

fibrin. Fibrin is also frequently deposited, or rather formed, upon the pleural surfaces; but agglutination of the opposed surfaces, with the formation of adhesions, is prevented by the fluid that keeps them apart. Another common site for serous inflammations is the skin, slight burns causing a serous exudation under or within the epidermis, the horny layer of which is raised to form the covering of a blister. Serous inflammations may also affect other portions of the body (Fig. 280).

Under the microscope a few leucocytes and blood-corpuscles can be detected in the serous exudate. Some of the leucocytes may be infiltrated with fat-globules, which they have appropriated from the débris of degenerated cells. These drops of fat may be so numerous as to obscure the nucleus and completely fill the cytoplasm, distending the cell to fully twice its normal size. These cells have received the name "compound granule-cells" (Fig. 195). When the inflammation affects a serous surface detached and swollen endothelial cells may also be present in the fluid.

5. **Catarrhal inflammations** are those which affect mucous membranes, with the production of a fluid exudate appearing upon their

surfaces. In the exudate, besides the usual constituents, there are desquamated epithelial cells and a variable amount of mucus. Mucus, it will be remembered, is a substance normally secreted upon the mucous membranes, where it serves to protect the underlying cells. When those membranes are irritated the supply of mucus is increased. In catarrhal inflammations it may be so abundant as to

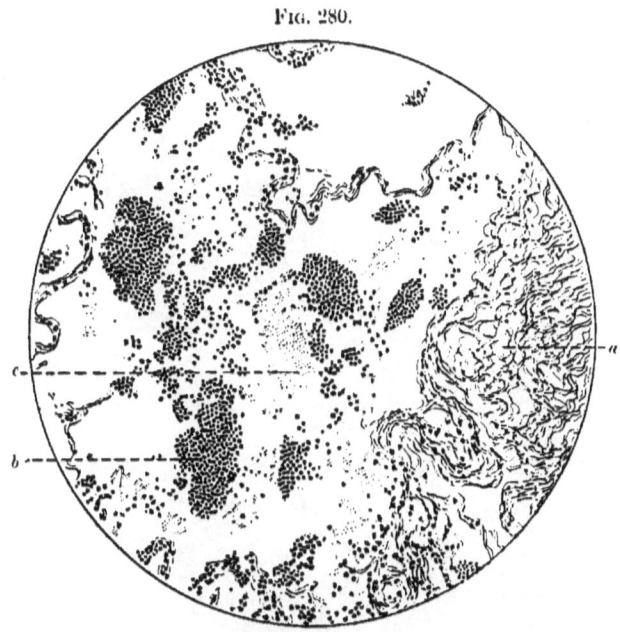

Fig. 280.

Serous leptomeningitis: *a*, œdematous fibrous tissue of the pia mater, the fibrous elements of the tissue being separated by the serous exudate; *b*, group of leucocytes, probably held together in part by fibrin; *c*, granular fibrin and detritus; *b* and *c*, and other similar masses, lie in the serum, which occupies the whole field between the visible elements.

predominate over the elements of the exudate, so that the fluid appearing on the surface of the membrane has a viscid character. In other cases the mixed secretion and exudate may be muco-serous or muco-purulent (Fig. 281).

In catarrhal or broncho-pneumonia the exudate appearing in the alveoli of the lung is of a serous character, with an admixture of desquamated cells from the alveolar walls and a variable number of leucocytes. These sometimes give the exudate an almost purulent appearance.

6. **Croupous inflammation** is an inflammation of a surface, char-

acterized by the formation upon it of a "pseudomembrane" composed chiefly of fibrin.

7. **Diphtheritic inflammation** is a term usually applied to inflammation affecting the tissues underlying a free surface. It is characterized by local death of the superficial portions of those tissues with an accompanying coagulation (Fig. 263). The result is the

FIG. 281.

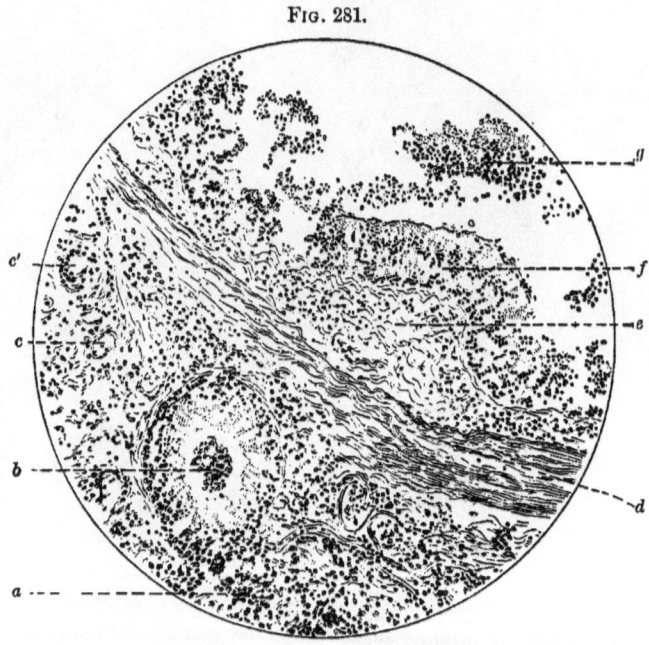

Catarrhal bronchitis: *a*, areolar tissue of the submucosa, infiltrated with serum and leucocytes; *b*, alveolus of a mucous gland, infiltrated at the periphery by leucocytes. The epithelium is undergoing colliquative necrosis, and in the centre of the lumen are a few leucocytes with fibrin. *c, c'*, bloodvessels. *c'* shows an infiltration of the wall by emigrating leucocytes. *d*, muscularis mucosæ; *e*, subepithelial areolar tissue of the mucous membrane, infiltrated with serum and leucocytes; *f*, columnar epithelium of the surface in a state of colliquative necrosis; *g*, exudate within the bronchus. In this portion of the bronchus the destructive processes are so acute that the epithelium is destroyed, instead of stimulated to the production of excessive mucus.

formation of a membranous mass of dead tissue closely adhering to the tissues beneath, a so-called "true membrane," in contradistinction to the "false membrane" of croupous inflammation. This membrane is subsequently separated from the underlying tissues by the formation of granulations, leaving an ulcer.

8. The "**infective granulomata**," such as tubercle, gumma, and the

nodules of leprosy and glanders, are forms of subacute inflammation which owe their peculiarities to the infections that occasion them. The tubercle, caused by the presence of the tubercle bacil-

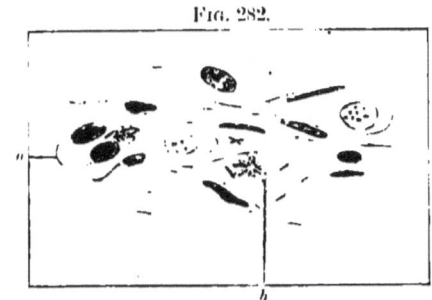

Fig. 282.

Early stage of experimental tuberculosis; cornea of rabbit. (Schieck.) Five days after inoculation. Rejuvenescence and beginning degeneration in fixed cells of the fibrous tissue. *a*, karyolysis in a cell affected by a group of tubercle bacilli within the cytoplasm; *b*, karyokinetic figure in another cell.

lus, is the most common of these inflammations and may be taken as a type of the whole group.

The tubercle bacillus does not always produce the little in-

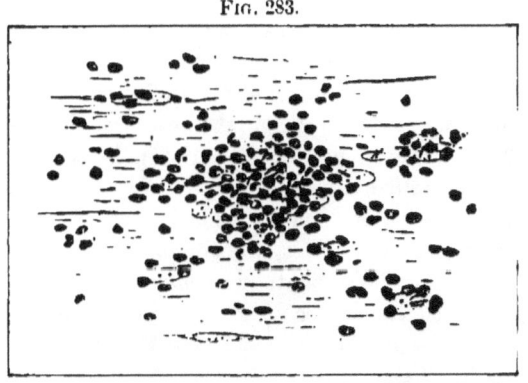

Fig. 283.

Early stage of experimental tuberculosis; cornea of rabbit. (Schieck.) Ten days after inoculation. Beginning of a tubercle. The "epithelioid" or young connective-tissue cells are masked by the presence of leucocytes with denser nuclei, which have been attracted by the chemotactic (positive chemotaxis) influence of the materials accumulating in the inflamed focus.

flammatory nodules called "tubercles." It sometimes occasions a suppurative inflammation of sluggish type, forming "cold abscesses," or purulent inflammations of mucous membranes. It may also cause sero-hæmorrhagic exudations from the serous membranes—*e. g.*, the pleura; but the most characteristic tissue-reaction due to its presence is the formation of the tubercle. This is the result of a rejuvenescence of the connective-tissue cells, without any preceding exudation, and an attempt at the production of granulation-tissue around the bacilli (Figs. 282 and 283). These multiply so slowly that they and their products exert merely an irritation on the cells of the tissue, stimulating them to reproduce, but they do not usually cause the growth of new bloodvessels, so that in the majority of cases the granulation-tissue is not vascularized. Furthermore, as they increase in number the bacteria cause degenerative and necrotic changes in the cells that have been produced, and, as their products increase in amount, the cells in the centre of the focus of inflammation are destroyed (cheesy degeneration, p. 274), while those at the periphery multiply, causing an increase in the size of the inflammatory nodule or tubercle. The multiplication of the cells is often hindered to a certain extent by the poisons present; the nuclei divide, but the protoplasm fails to undergo a corresponding division. In this way multinucleated cells, called "giant-cells," are produced.

As the result of these processes a developing tubercle presents the following appearances under the microscope. In the centre is a mass of cheesy matter, composed of fine granules of fat, albuminoid material, and fragments of nuclei, the result of degenerative and necrotic changes caused by the bacterial poisons. Around this mass is a zone of rather large "epithelioid" cells, which belong to the granulation-tissue, and among which there may be a variable number of emigrated leucocytes, probably attracted by the necrosed tissues in the centre. Also, near the centre or in the granulation-tissue, a few giant-cells may be present; but they are not invariably found, nor is their presence a conclusive sign that the process is tubercular (Fig. 284).

The ultimate outcome of the process varies in different cases. The inflammatory reaction may overcome the infection, encapsulating the nodule with a dense cicatricial tissue; or the infection may conquer; bits of the cheesy matter containing tubercle bacilli may then find entrance into the lymphatic circulation and be carried to the neighboring lymph-glands, establishing in them new foci of tubercular inflammation, or tubercle bacilli may get into the bloodvessels and carry the infection to all parts of the body, occasioning general tuberculosis.

The poisonous products of the tubercle bacilli are absorbed into the general system, producing disturbances of nutrition, emaciation, and fever. Old encapsulated tubercular products are prone to calcareous infiltration, but, even after prolonged encapsulation,

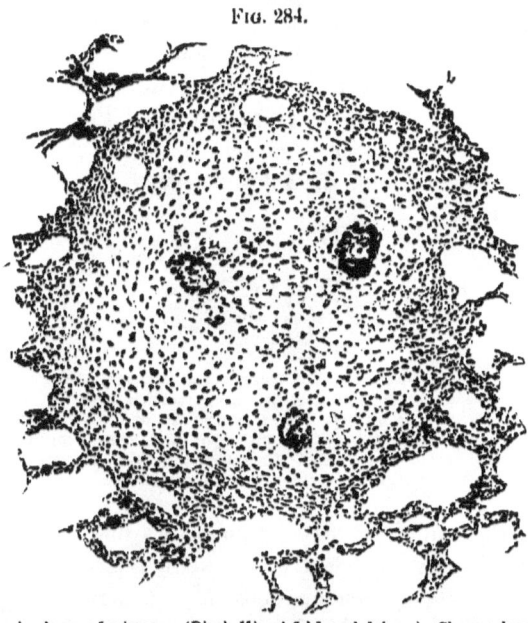

Fig. 284.

Miliary tubercle; lung of a horse. (Birch-Hirschfeld and Johne.) Cheesy degeneration has only just begun in the centre of the focus of inflammation, where the nuclei of epithelioid cells and leucocytes are still visible. At the periphery of the tubercle is a zone of round-cell or leucocytic infiltration. Three giant-cells, with peripheral nuclei, occupy intermediate positions; around the tubercle are the infiltrated walls of pulmonary alveoli.

the tubercle bacilli which have been imprisoned may retain their vitality, and, if for any reason the poorly nourished capsule suffers in its integrity, these old nodules may become the source of fresh infection. This is a not uncommon result of some acute disease like scarlet fever, typhoid fever, or influenza, convalescence from those diseases being followed by the development of tuberculosis springing from an old and long-dormant tubercular infection.

In the lungs the tubercles, as they increase in size, involve the walls of the alveoli or the bronchi, and when the cheesy matter escapes into the alveoli or bronchi cavities are produced. The process rarely remains a purely tubercular one in the lungs. The conditions there (exposure to inspired air) are favorable to a mixed infection with pyogenic bacteria, which hastens the destruction of the pulmonary tissues inaugurated by the tubercle bacillus.

Isolated tubercles, such as have been described, are not infrequently met with; but it is more usual to find a number of such nodules in close aggregation, each starting from a distinct focus of infection. As these enlarge, their peripheries coalesce, and finally their cheesy centres meet and blend. Meanwhile fresh young nodules are formed around the older mass, and thus the tubercular disintegration of the tissues spreads. It is for this reason that tubercular ulcers—*e. g.*, of the intestine—have swollen and undermined borders (Fig. 285).

Fig. 285.

Tubercular ulcer of the intestine. (Kaufmann.) The cavity of the ulcer was formed through disintegration and removal of the cheesy matter formed in the earlier tubercles. Now the base of the ulcer is formed by necrosed and cheesy material, beneath which eight or nine distinct tubercles are distinguishable, those in the centre extending into the muscular coat of the intestine. The infection has also extended into the lymphatics beneath the serous coat, where three tubercles can be seen.

The other granulomata have peculiarities due to their special causes, which are pretty clearly defined in typical cases; but, as in tuberculosis, these inflammations may in certain instances be structurally indistinguishable from those due to other causes.

## Chronic Inflammation.

A consideration of the infective granulomata makes the fact clear that inflammation may occur without the production of a distinct exudate, the damaging cause merely exciting the tissues to proliferation; but in that group of inflammations the excitation of the tissues was sufficiently intense to occasion the development of a tissue closely resembling the granulations of acute inflammation. For this reason they were designated as *subacute* inflammations.

There is another group of inflammations in which the irritation of the tissues is not sufficient to induce a rejuvenescence of the cells in such a pronounced degree as to cause their reversion to a comparatively undifferentiated condition. No granulations are, therefore, produced, but the cells are simply stimulated to a formative activity that is abnormal to the part. This is the group of chronic inflammations, of which three or four examples will be cited.

Chronic periosteal inflammation may be induced by a number of damaging causes of slight intensity, but repeated application. The response which the cells of the periosteum make to this irritation is a revival of their formative activity and the production of bone, which forms an "epiphyte," or other osseous excrescence, apparently springing from the surface of the older bone. Similar new-formations of bone may take their origin from the endosteum, forming

Fig. 286.

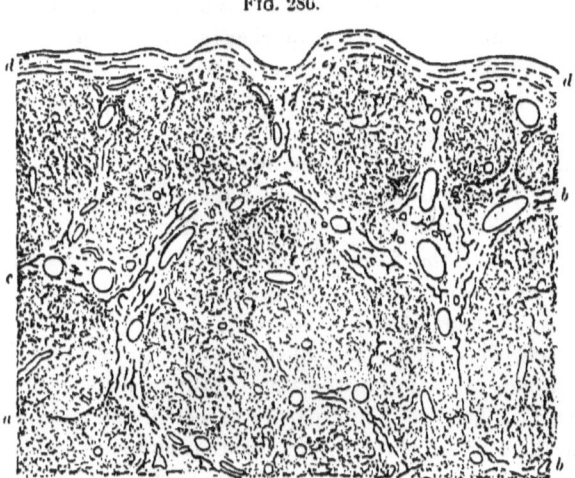

Cirrhosis of the liver; chronic interstitial hepatitis. (Kaufmann.) *a*, lobules of the liver; *b*, increased interstitial fibrous tissue, the result of the inflammatory process; *c*, collection of nuclei in the fibrous tissue, showing that the process is still in progress; *d*, thickened capsule of the liver.

layers that encroach upon the lumina of the Haversian canals or the medullary cavity of the bone. These deposits are more diffuse than those springing from the external surface of the bone, probably because they arise as the result of a more widespread irritation, such as the presence of some noxious substance in the circulation, and not from a localized point of irritation.

Another example of this group is presented by cirrhosis of the liver, selected from among the *chronic interstitial inflammations* that may affect any of the organs of the body. In hepatic cirrhosis there is a redundant production of fibrous tissue around the branches of the portal vein, and, therefore, appearing between the "lobules" of the liver (Fig. 286). This has the same tendency as other cicatricial tissue to contract, and that contraction causes atrophy of the hepatic cells through the pressure it exerts upon them. There may be another cause for this atrophy of the liver-cells, which will be more comprehensible after considering the probable etiology of the interstitial inflammation itself. This appears to be caused by the absorption of irritating substances from the digestive tract, which are carried in most concentrated form by the portal vein to the liver. Here they stimulate the cells of the connective tissue to produce fresh fibrous tissue around the branches of that vessel. But it is quite possible that those same substances may act injuriously upon the parenchymatous cells of the liver, impairing their nutrition and rendering them especially liable to atrophy under the increased pressure from the fibrous tissue in their neighborhood.

While the interstitial inflammation is in progress the connective tissue of Glisson's capsule appears not only increased in amount, but more highly cellular than normal. This is due in part to a multiplication of the fixed cells of the fibrous tissue, in part to a round-cell infiltration—*i. e.*, an immigration of leucocytes. This immigration is more abundant in some cases than in others, as would be expected, since the process must be subject to exacerbations, due to fluctuations in the amount of the irritating substances brought to the liver. In fact, we should hardly expect to find a sharp division between the slowest chronic inflammation and such inflammations as approach the character of a subacute manifestation of the same process.

A third example of the chronic inflammatory process may be found in the reaction of the tissues around the necrotic mass resulting from bland embolism. Suppose one of the vessels of the kidney to be plugged by an aseptic body. The tissues normally supplied with blood through that vessel will die (Fig. 293). But the presence of this dead tissue, although it contains no micro-organisms, acts as an irritant upon the surrounding tissues, which respond by the production of a capsule of fibrous tissue. The necrosed tissues may remain within this capsule, or they may be absorbed, in which case the capsule shrinks to a puckering mass of dense fibrous tissue.

In like manner a non-infectious foreign body may become encapsulated within any of the tissues of the body.

Still another example of chronic interstitial inflammation appears to be furnished by cases in which the parenchyma has suffered atrophy or some other form of destruction, and the loss is made good by the production of fibrous tissue without a precedent formation of granulations. In embolism of a branch of one of the coronary arteries supplying the heart-muscle the destruction of the muscle-fibres seems to stimulate the formative activities of the cells of the interstitial fibrous tissue. The deduction that the production of fibrous tissue is the direct result of a loss of parenchyma is, however, not quite clear, for the stimulus to tissue-production may

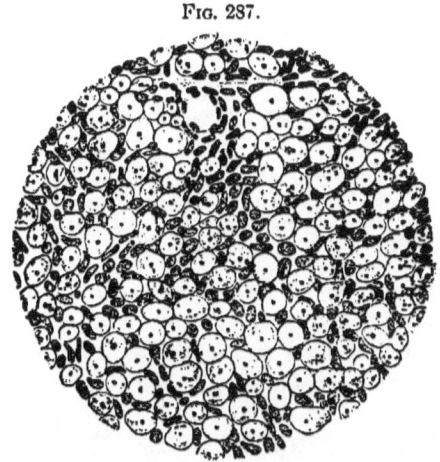

FIG. 287.

Chronic interstitial inflammation. Early stage of productive interstitial neuritis. (Nauwerck and Barth.) The section is from the anterior root of a lumbar nerve. It represents a number of apparently normal medullated nerve-fibres in cross-section, with proliferation of the cells of the endoneurium, as is evidenced by the abundance of nuclei in that tissue.

result from the unusual strain brought upon the part of the heart which is deprived of the usual support of muscular tissue. It may be that other cases in which a loss of parenchyma is replaced by fibrous tissue are also not due to stimulation of fibrous-tissue production because of that loss, but are to be explained in a manner analogous to the explanation of cirrhosis already offered.

Further examples of interstitial inflammations are shown in Figs. 287 and 288.

From the examples that have been given it will be noticed that, amid all its protean manifestations, the inflammatory process is fun-

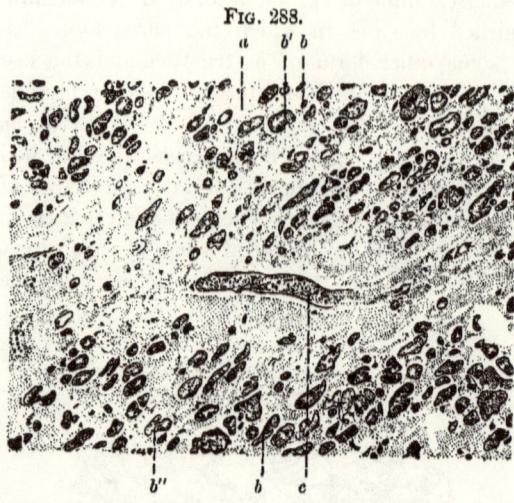

FIG. 288.

Chronic interstitial myocarditis, late stage: *a*, dense fibrous tissue, the final result of the interstitial inflammation; *b, b', b''*, atrophied cardiac muscle-cells; *b'*, vacuolation of a less atrophic cell; *b''*, section showing anastomotic branch joining two cells; *c*, partially obliterated bloodvessel.

damentally the same, but susceptible of many variations; and when the conditions are not too adverse it leads to a removal of the cause of an injury and to a more or less complete repair or patching of the tissues that have been damaged.

## III. INCIDENTAL CONSEQUENCES OF DAMAGE AND INFLAMMATION.

The damage and ensuing inflammation affecting a part of the body not only occasion changes in the structure of that part, but also, through those changes, very frequently cause morbid conditions in remote parts. It will be impossible to enumerate all the possibilities in this connection, but a few examples will suffice to show their importance. It is obvious that chronic interstitial hepatitis (Fig. 286) must affect the circulation in the portal system of vessels. The inflammatory fibrous tissue formed between the lobules of the liver, and, therefore, around the portal vessels within that organ, possesses the same tendency to contract after its formation that is manifested by cicatricial tissue of more acute inflammations, though perhaps in less degree. This contraction would suffice to compromise at least the smaller branches of the portal vein entering the lobules, so as to obstruct the current of blood flowing through them. The result is an increase of pressure in the portal circulation and the production of passive hyperæmia or congestion of the organs in which the portal radicles are situated.

This passive congestion results in a dilatation of the vessels in

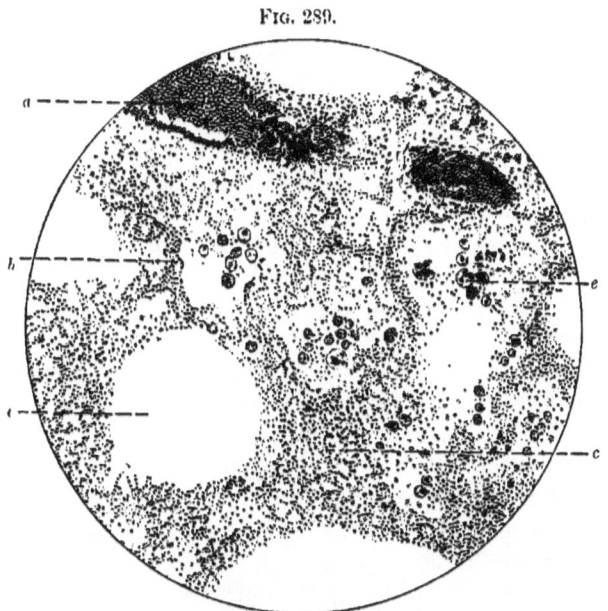

FIG. 289.

Brown induration of the lung, the result of chronic passive congestion caused by valvular disease of the heart: *a*, small radicle of the pulmonary vein, dilated and filled with blood; *b*, alveolar wall in cross-section, thickened and containing an abnormal number of nuclei (evidence of an increase of tissue, a chronic interstitial pneumonia); *c*, surface-view of an alveolar wall, showing similar abundance of nuclei and a dilatation of the capillaries, evidenced here and elsewhere in the section by a double row of corpuscles in a capillary; *d*, cavity of an alveolus; *e*, alveolus containing serum, red corpuscles, and leucocytes, and also large pigmented cells. These are chiefly leucocytes which have taken up pigment from the red corpuscles that have disintegrated—phagocytes. Some of these large cells may be desquamated epithelial cells from the alveolar walls, in a swollen and degenerated condition. The presence of serum is demonstrated by the fact that the cells in the alveolus are not lying against the alveolar walls. The escape of the blood-corpuscles from the capillaries is a result of the sluggish circulation.

those organs and a thickening of their walls, and also frequently induces a chronic interstitial inflammation. It may also so impede the lymphatic circulation and impair the nutrition of the vascular walls as to give rise to an excessive transudation of serum and occasion œdema and ascites.

Similar chronic passive hyperæmias may follow those inflammatory lesions in the valves of the heart which cause either agglutination and permanent adhesions of the valvular curtains, stenosis; or a contraction of one or more of those curtains, so that their proper closure is prevented, incompetency. In either case the circulation is impeded and the flow of blood from the organs behind the lesion interfered with (Fig. 289).

Hæmorrhage is another of the frequent results of damage. It may be recognized by the presence of blood outside of the vessels. This blood at first contains the red and white corpuscles in their normal proportions, but after a lapse of time the clot which forms becomes infiltrated with leucocytes as the expression of an inflammatory reaction induced by the extravasated blood. Subsequently the blood disintegrates, productive inflammation is induced, and the lesion heals, with the production of a scar. This is often colored brown or gray, from the presence of pigment derived from the hæmoglobin of the red blood-corpuscles. This pigment may be in the form of reddish-brown rhombic crystals, or granules, of hæmatoidin; or it may take the form of small granules of hæmosiderin. The latter substance contains iron, from which the former is free, and under the action of sulphuretted hydrogen produced by decomposition may give rise to sulphide of iron, changing its brown color to black, and the color of the pigmentation from a brown to some shade of gray.

Hæmorrhage may be among the direct results of damage to the tissues, or it may follow necrotic changes in the vascular wall. This is a not infrequent occurrence in virulent forms of infection, and results in the formation of small, punctiform hæmorrhages; for the vessels necrosed are usually of small calibre and surrounded by tissues sufficiently firm to check the flow of blood under the slight pressure within those vessels (Fig. 290). But more copious hæmorrhages may occur in the course of slowly progressing infections, notably in pulmonary tuberculosis. It will be remembered that the walls of the larger vessels are composed of a dense fibrous tissue rich in elastic fibres (Fig. 97). Such a tissue resists the necrosing action of tuberculosis for a longer time than the more succulent tissues of the lung. It therefore occasionally happens that a cavity may be formed by the destruction of the pulmonary tissue, and that through this cavity, or within its walls, a pervious vessel of considerable diameter may take its course. After a while the wall of this vessel may become sufficiently destroyed to yield before the pressure of the blood within it; rupture may then take place, with the effusion of considerable blood, hæmoptysis. In many cases, however, such a result is prevented by the formation of a clot (thrombus) within the vessel before erosion of its wall has gone far enough to threaten rupture.

**Thrombosis.**—This term is applied to the formation of fibrin within the circulatory system during life. It may take place when

Fig. 290.

Hæmorrhage in the kidney following general infection. (Tizzoni and Giovannini.) The hæmorrhage has taken place within the capsule of a Malpighian body and part of the extravasated blood has passed into the corresponding uriniferous tubule. The glomerulus has been compressed (to the right), an occurrence which probably checked the hæmorrhage. The tissues of the glomerulus and of the neighboring tubules are necrotic.

the circulation in a particular vessel or in a portion of the heart is sufficiently sluggish to permit leucocytes and, perhaps, blood-plates to collect and remain in one place long enough for their disintegration to begin. The elements required for fibrin-formation are then set free and thrombosis results. In this way thrombi may form between the columnae carneae in marantic conditions, behind the curtains of venous valves, or in the lumina of dilated veins within the pelvis. Thrombosis may also occur as the result of a roughening of the intima of a vessel or its mechanical destruction, as in the tying or crushing of a vessel.

Thrombosis may be the result of disease of the vessel-wall, caused by infection or malnutrition. The affection of the veins known as septic thrombophlebitis may be selected as one of the more important acute lesions of the vessels. This is caused by an infection of

330    HISTOLOGY OF THE MORBID PROCESSES.

the vascular wall, which eventually reaches the intima. Here a fibrinous inflammation, analogous to that of a serous membrane (p. 313), is induced. The roughness of the intima so occasioned induces the formation of a thrombus (Fig. 291). Meanwhile the

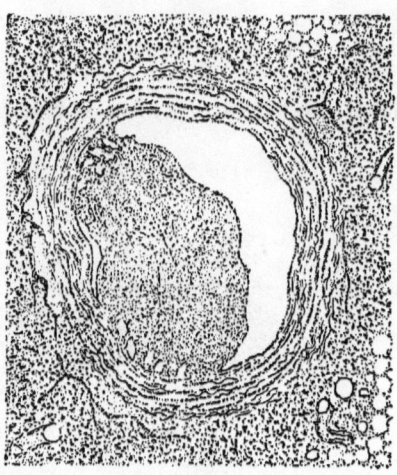

Fig. 291.

Thrombophlebitis, incident to erysipelas of the arm. (Kaufmann.) The thrombus occupies about two-thirds of the lumen of the vein, which is surrounded by areolar tissue infiltrated with serum and leucocytes.

septic process in the wall of the vessel progresses and extends into the thrombus, which is softened. The rate of softening may now exceed that of thrombus-formation, in which case the thrombus is broken up, and particles containing some of the bacteria occasioning the inflammation gain access to the venous circulation (see Embolism).

**Embolism.**—The obstruction of a vessel by a foreign body brought from a distance by the circulating blood is called embolism. The foreign body, or embolus, is usually a small mass of fibrin; but it may be air, fat (derived, for example, from the medulla of a fractured bone), a calcareous fragment, or a particle of tissue.

With the exception of the branches of the portal vein, the vessels obstructed by an embolus are arterial. The results of embolism will depend, first, upon the anatomical distribution of the vessel plugged, whether there are anastomotic branches of considerable calibre beyond the site of the obstruction; second, upon the nature

# STRUCTURAL CHANGES DUE TO DAMAGE. 331

of the embolus, whether it contain pathogenic bacteria or not. In the former case the embolus is called a septic, in the latter a bland, embolus.

In septic embolism an acute inflammation, similar to that at the

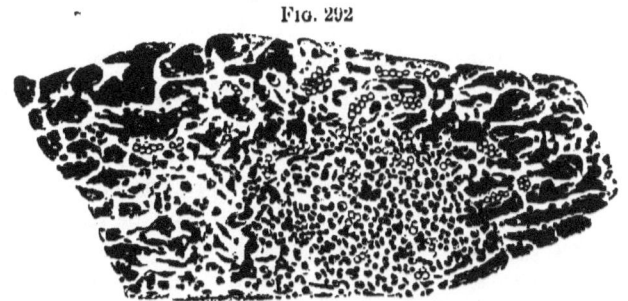

Fig. 292

Metastatic abscess in the heart, due to septic embolism. (Birch-Hirschfeld.) The abscess-cavity contains red blood-corpuscles and leucocytes with fragmented nuclei. The muscle-fibres within and near the cavity have been killed and many of them dissolved.

site of the original lesion, is induced by the bacteria brought with the embolus. If the original inflammation was suppurative, ab-

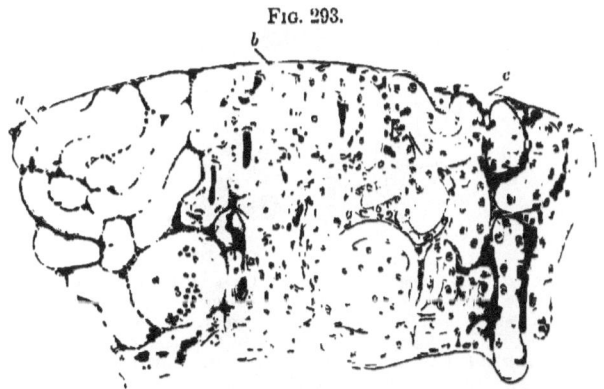

Fig. 293.

Experimental anæmic infarction of the kidney; rabbit. (Foa.) a, necrotic tissue formerly supplied by the artery obstructed; b, zone of affected tissue surrounding the infarct. In this zone the renal tubules contain hyaline casts, and their lining epithelium shows an evanescent tendency to proliferate, some of the cells containing karyokinetic figures. c, normal renal tissue.

scesses, called metastatic abscesses, are formed around each septic embolus (Fig. 292).

In bland embolism, when there are ample anastomoses between the vessel plugged and other vessels beyond the site of the embolus, no serious result follows. Thrombosis takes place around the embolus, but the circulation beyond it is maintained through the anastomotic vessels. If, however, the anastomoses are not sufficient to maintain the nutrition of the tissues normally supplied by the obstructed vessel, those tissues suffer necrosis (Fig. 293). Such a mass of necrosed tissue is called an "infarct."

Infarcts are divided into anæmic and hæmorrhagic infarcts. The former occur when the tissues are entirely deprived of blood by embolism (Fig. 293); the latter take place when, through innutrition of the vessels in the part affected by infarction, blood, derived from the veins or through capillary or other fine anastomoses, is permitted to pass into the interstices of the necrosed tissues. These then appear surcharged with blood. The most striking example of hæmorrhagic infarction is that following bland embolism of a branch of the pulmonary artery (Fig. 294).

FIG. 294.

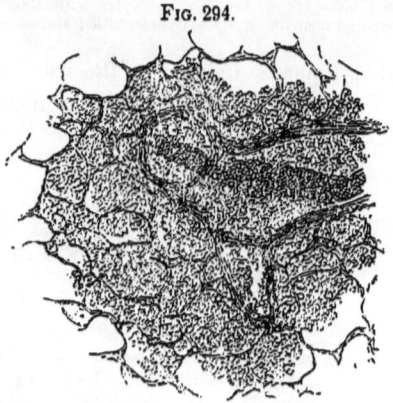

Hæmorrhagic infarct of the lung. (Kaufmann.) The section contains a portion of the plugged vessel beyond the site of the embolus. It and the pulmonary alveoli are filled with blood, which, in the latter, has passed through the capillary walls, rendered pervious by malnutrition. This blood may be derived from the pulmonary vein and also from the bronchial artery, which communicates with the capillaries of the alveolar walls.

**Phagocytosis.**—In the preceding pages incidental mention has been made of the ability of leucocytes and other amœboid cells to incorporate within their cytoplasm particles of foreign matter with which they may come in contact. Such cells within the body are called "phagocytes" (devouring cells). It was at one time thought that these cells had much to do with the killing and destruction of pathogenic bacteria and other organisms that might gain access to the system; but it is now believed that such is not the case.

Phagocytes do incorporate bacteria; but if those bacteria are virulent, the phagocyte either refuses to take them within its cytoplasm, or, after doing so, suffers degeneration or necrosis. It has no peculiar immunity against the action of the bacteria. On the other hand, it has been shown that the fluids of the body are capable of diminishing the virulence of bacteria or of killing them. It often takes some time for the production of the substances that have this effect, and their elaboration is frequently too tardy to check the destructive action of the bacteria. But upon the surface of granulations, from which absorption is slow or does not take place, the effects of the tissue-fluids have been studied and an attenuation of bacteria (decrease in their virulence) observed. These attenuated

FIG. 295.

Phagocytes from granulations infected with virulent anthrax bacilli. (Afanassieff.) a, thread of bacilli, partly within and partly outside of a phagocyte. Both portions show a vacuolation of the bacilli, indicative of their degeneration. d, thread almost entirely incorporated. Within the cell the incorporated bacilli lie in vacuoles in the cytoplasm; probably digestive vacuoles. In b and e similar appearances are presented. c, degenerating thread of bacilli from the fluid of the granulations. Vacuolation has also taken place in this thread, showing that the fluids of the granulations have a destructive influence upon the bacilli.

bacteria may be taken up by phagocytes with impunity and subsequently digested within their cytoplasm (Fig. 295).

The digestion and removal of degenerated or dead materials appear, then, to be the useful rôle played by phagocytes. They appear to be the active agents in the absorption of organic fragments, such as fibrin, macerated necrotic tissue, etc., which may be present in the tissues of the body (Fig. 296).

The majority of phagocytes are probably leucocytes, identical with

334    HISTOLOGY OF THE MORBID PROCESSES.

Fig. 296.

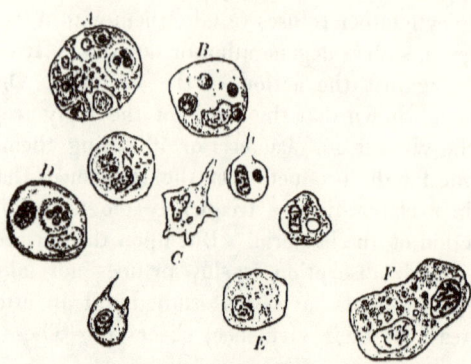

Phagocytes from aseptic granulations. (Nikiforoff.) C, phagocytes with pseudopodia; E, without pseudopodia; F, proliferating, the daughter-nuclei in the spirem phase of karyokinesis; A, B, D, with leucocytes, fragments of tissue, and red corpuscles in their cytoplasm.

those in the blood and lymph;[1] but it is possible that young connective-tissue cells, which are believed to possess the power of amœboid motion, may sometimes play the part of phagocytes.

## IV. REGENERATION OF THE TISSUES.

Frequent reference has been made to the power possessed by many cells to restore or regenerate structures that have been damaged by influences causing either necrosis or degeneration. The ability to effect this restoration varies greatly in the cells of different tissues, being, in general, inversely proportional to the degree of specialization to which they had attained at the time the damage took place. We must, therefore, consider this process in the different tissues separately, after taking a general survey of the facts that apply to all cases of regeneration.

It is needless to say that a cell which has once become necrotic is incapable of restoration; but if the nucleus be sufficiently preserved and enough cytoplasm be left after degenerative changes have come to an end, both these cellular constituents may take up nourishment and regenerate the parts destroyed. When whole masses of tissue have been killed, but some of the same form of tissue retains life and continuity with the necrosed portion, the dead tissue may be more or less completely replaced by tissue

[1] The polynuclear neutrophile leucocytes are those which most frequently act as phagocytes.

of new formation springing from the living portion. If this
takes place, the cells of the latter portion multiply and reassume
those formative activities that they possessed during the development of the tissues in earlier life. The division of the cells always takes place by the indirect method, that of karyokinesis. We
must not, however, assume that because the cells of a tissue may,
under the influence of damaging agents, contain karyokinetic figures,
they must necessarily possess the power of regenerating lost portions of tissue. More than mere observation of those figures is required to establish that fact. Such figures are occasionally met with
in the ganglion-cells of the central nervous system, and they show
that the nuclei of those cells retain, at least to a certain extent, the
power of division. But this by no means implies that new ganglion-
cells, capable of full functional activity, can be produced by the
division of an adult nerve-cell, and, as a fact, such an occurrence

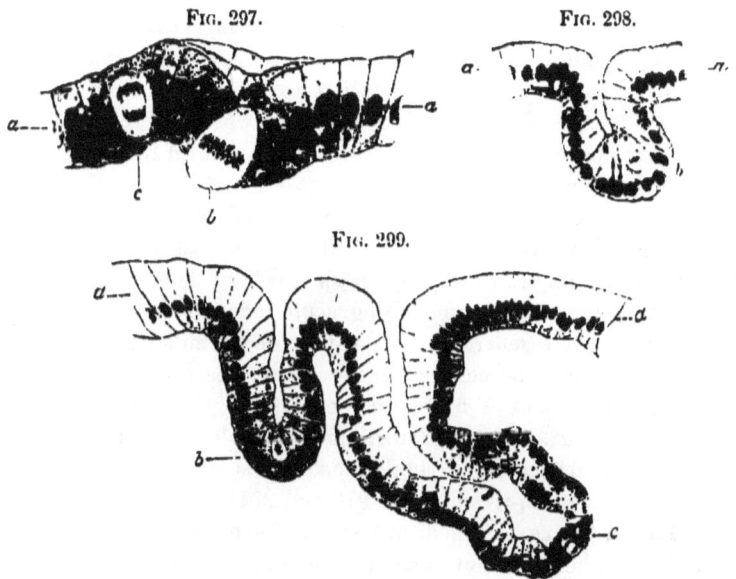

Fig. 297. Fig. 298.

Fig. 299.

Phases in the regeneration of the gastric mucous membrane; dog. (Griffini and Vassale.)
*a*, regenerated columnar epithelial cells covering the base of the wound; *b*, *c*, karyokinetic
figures indicative of proliferation.

does not appear to take place. In Fig. 293, zone *b*, karyokinetic
figures are seen in the renal epithelium; but it is doubtful whether
they signify the beginning formation of new renal tissue to replace

that killed in the anæmic infarct. Such a replacement does not take place in the kidney, but a scar of fibrous tissue is formed around or in place of the necrosed mass. The karyokinetic figures, then, simply demonstrate a tendency toward cell-division, and further observations are necessary in order to determine the significance of that tendency.

1. **Epithelium.**—The regenerations of which epithelium is capable are very extensive and perfect. In some forms of epithelium—*e. g.*, the stratified variety and that found in sebaceous glands—the regenerative process is a part of the functional activity of the tissue. After wounds of the skin the epithelium forming the epidermis regenerates a new epidermis for the injured area. In this case the epithelial layer, provided the wound be extensive, is relatively thin and of low vitality. This is not because the epithelial regeneration was imperfect, but because the nourishment it receives from the underlying cicatricial tissue is deficient. There is in this case a lack of coördinate development in the regenerations effected by the epithelium and underlying fibrous tissues. Remarkable examples of a more perfect coördination are exhibited in the regeneration of glands (Figs. 297, 298, and 299), where the regenerating epithelium and fibrous tissues appear to coöperate in the restitution of lost glandular structures.

The complicated glandular structure of the liver is also capable of regeneration when a portion of that organ has been removed under aseptic precautions (Fig. 300). Where, however, the destruction is due to damage exciting acute inflammation it is doubtful whether any regeneration is possible, owing either to the injurious action upon the cells, or to the hindrances interposed by the regenerating portions of fibrous tissue in the neighborhood.

2. **Endothelium.**—That endothelium is capable of regeneration is shown by the formation of young bloodvessels during the development of granulation-tissue (Figs. 270 and 271).

3. **Fibrous Tissue.**—A mode of regeneration of this tissue has been described in the article on inflammation, and is illustrated in Figs. 269 and 270. This tissue, when fully developed, differs from normal fibrous tissue in its density and freedom from bloodvessels (Fig. 273). The regeneration of a tendon severed under aseptic precautions results in a much more perfect restitution of the normal structures. Here the cut ends of the fibre show softening, swelling, and final disintegration of the intercellular substance. Some of the cells are

Section of regenerating liver. (v. Meister.)

also affected by a degenerative process; but others rejuvenate, mul-

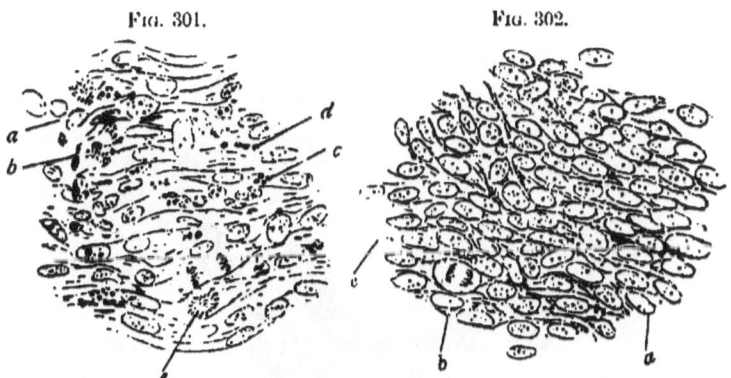

Phases in the regeneration of a tendon; guinea-pig. (Enderlen.)
Fig. 301.—Two days after section: *a*, swollen intercellular substance; *b*, karyolysis; *c*, *d*, leucocytes; *e*, karyokinesis.
Fig. 302.—Seven days after section: *a*, nucleus of young connective-tissue cell; *b*, karyokinesis; *c*, intercellular substance of new formation.

tiply, and eventually produce a highly cellular tissue, which develops into tendinous fibrous tissue (Figs. 301, 302, and 303).

**4. Bone.**—When a piece of bone dies fresh bone is produced through a rejuvenescence of the formative activities of the periosteum (or endosteum). While this new formation of bone is in progress the dead bone is removed by phagocytes, which are usually multi-

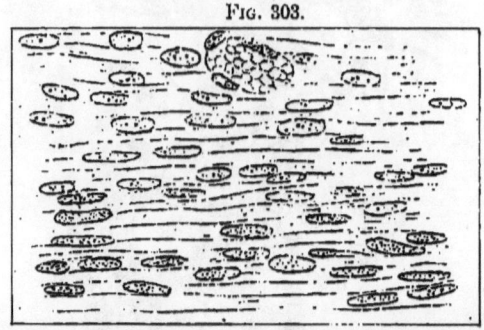

FIG. 303.

Phase in the regeneration of a tendon; guinea-pig. (Enderlen.) Seventy days after section. The tendon is still rather highly cellular, but its structure is, in the main, fully restored. At the top of the figure is the cross-section of a blood-vessel.

nucleated, and have received the name "osteoclasts" (bone-breakers), in contradistinction to the bone-forming cells of the periosteum, which are known as "osteoblasts" (bone-builders) (Fig. 304).

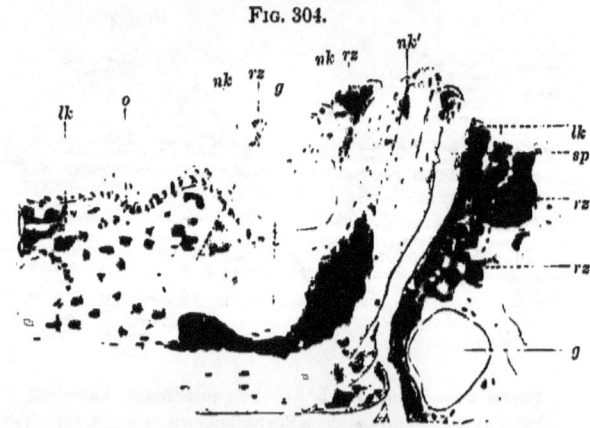

FIG. 304.

Regeneration of bone. (Barth.) *nk*, fragments of necrotic bone; *rz*, osteoclasts; *o*, osteoblasts; *lk*, bone of new formation; *g*, bloodvessels; *nk'*, lamina of dead bone. (*sp*, accidental crack in the section.)

**5. Cartilage.**—This tissue is capable of only a limited and imperfect regeneration. Defects in cartilage are usually made good by the development of fibrous tissue, which may become modified into adipose tissue, or by bone-production if the damage causes a rejuvenescence of periosteum or endosteum.

6. **Smooth Muscular Tissue.**—Non-striated muscle-cells are capable of multiplication, but in inflammatory conditions the tissue of the media of the vessels does not appear to keep pace with that of the intima in the production of new bloodvessels. The latter, therefore, usually lack a muscular coat and are thin-walled (Fig. 272). In the uterus and other situations smooth muscle-cells may multiply and occasion a hyperplasia of the tissue. This appears, however, to be in response to a functional demand, rather than one

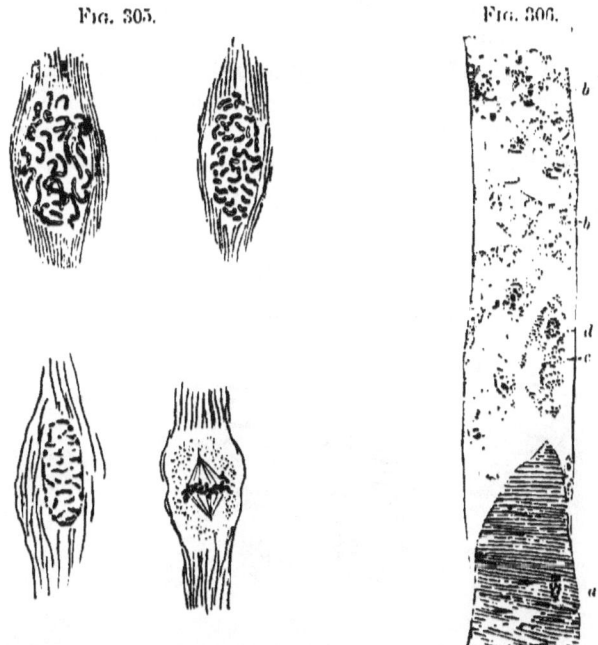

Fig. 305.—Karyokinetic figures in smooth muscular fibres. (Busachi.)
Fig. 306.—Regeneration of a striated muscle-fibre. (Kirby.) *a*, remains of the old contractile substance; *b*, rejuvenating cytoplasmic fragments, with their nuclei; *c*, similar fragment containing a bit of old contractile substance and a nucleus in karyokinesis, *d*.

of the results of damage : a functional hyperplasia. Karyokinetic figures have been observed in smooth muscle-cells after damage, but they do not lead to a restoration of the original tissue, which heals with the formation of a scar (Fig. 305).

7. **Striated Muscle.**—When a striated muscle-fibre undergoes

partial degeneration the cytoplasm around the nuclei that have been preserved may increase in amount, the nuclei may divide, and a multinucleated cytoplasmic mass result from the union of these rejuvenated portions. From this mass new contractile substance is then elaborated. This process results in regeneration of the particular fibre. It is still a question whether new striated muscle-fibres are produced in consequence of regenerative processes following damage. Wounds of voluntary muscles heal through the formation of a cicatrix (Fig. 306).

8. **Cardiac Muscle.**—Karyokinetic figures have been observed in the cells of the heart-muscle, but they do not appear to lead to regeneration of that tissue, which heals with the production of scar-tissue when wounded.

9. **The Nervous Tissues.**—Ganglion-cells have not been observed to rejuvenate so as to produce fresh nerve-cells; but if the cell-process forming part of a nerve is severed from the cell without serious damage to the cell-body, a new process or nerve-fibre is developed

FIG. 307.

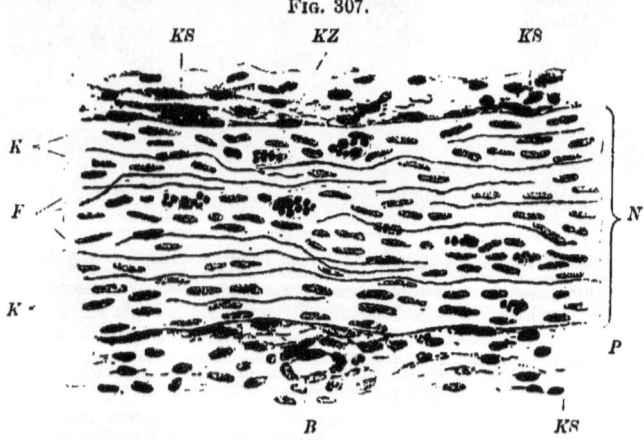

Longitudinal section of a regenerating nerve. (Stroebe.) N, nerve; P, perineurium, containing more cells than normally; KZ, phagocytes, containing globules of myelin from the medullary sheaths of degenerated fibres; K, nuclei of proliferated cells of the neurilemma; F, young axis-cylinders; KS, points showing the relations of the nuclei and young nerve-fibres; B, bloodvessel in the perineurium.

(Fig. 307). The cells of the neuroglia are, on the other hand, capable of regenerating that tissue. In this respect the neuroglia resembles the interstitial tissue of other organs than those of the central nervous system, often increasing in amount when there is a diminution in the bulk of the parenchyma, due to disease.

# CHAPTER XXV.

## TUMORS.

It will promote clearness of conception if the term tumor is restricted to abnormal masses of tissue produced without obvious reason and performing no function of use to the organism.

In the introductory chapter an attempt was made to show that under normal conditions the parts of the body develop in an orderly manner, which fits them for the performance of work useful to the whole organism, as well as for maintaining their own nutrition and structure. It was also pointed out that parts of the body, when occasion arises, frequently fulfil what appear to be their duties to the whole body, even if their own nutrition or structure suffers in consequence. From these observations we must conclude that throughout the life of the individual each part is controlled in its activities by influences having direct reference to the well-being of the whole body. Those influences control not only the functional activities of the tissues after the body has reached the adult state, but also control or guide the activities of the cells elaborating the body during development. The nature of those influences and the mechanism of their control are unknown to us. We are ignorant of any reason why the tissues of the body should develop to a certain point and then have their nutritive and formative activities restricted to a maintenance of the structures then existent. We attribute these phenomena to the force of heredity, but the explanation is incomplete, for that term merely expresses the fact that the offspring of an individual develops into a likeness to its parent.

In the development of tumors these guiding or controlling influences are in abeyance, sometimes in greater, sometimes in less degree. The tissues do not grow to meet a functional demand imposed upon them by the needs of the body, as appears to be invariably the case in the increase of tissue during the development of the individual. Instances of growth bringing about such adaptation to altered demands occur after the body has attained full development, but they are characterized as functional hyperplasia or hypertrophy, not as tumor-formation, and are arrested when the needs giving rise to them are met. This limitation of growth does not hold in the case of tumors.

Our knowledge of the normal forces guiding and restricting the development of the tissues being so deficient, how can we expect to understand the causes underlying the development of tumors? The marvel is not that certain cells should occasionally continue to multiply and exercise their formative powers without reference to the needs of the whole body. The fact that such occurrences are so rare awaits explanation. Familiarity with what is usual is apt to blind us to the fact that it is not explained, and when our attention is directed to what is unusual we ask an explanation of the exception. A knowledge of the etiology of tumors appears to await the acquisition of a deeper insight into the nature of hereditary transmission and of the conditions which that transmission ordinarily imposes upon the tissues throughout the life of the individual.

Tumors arise from the cells of pre-existent tissues. The fact that those cells in producing a tumor form a tissue which is functionally useless is evidence that the usual guiding influences mentioned above no longer completely control their activities. The degree in which that control is lost is, however, by no means the same in all cases of tumor-production. Sometimes the tissues of the tumor attain nearly if not quite the complete structural differentiation possessed by the tissue in which it found origin. In such cases only that degree of normal control which has reference to function appears to be abolished, the cells retaining their special formative activities in nearly full measure and producing a tissue resembling the parent tissue. Such tumors may be regarded as an expression of only a moderate relaxation of the influences normally controlling growth. They are clinically benign.

While such tumors closely simulating normal tissues are of occasional occurrence, in the majority of tumors the formative powers of the cells from which they develop display certain departures from the normal types of the classes to which they belong, and the structure of the tumor becomes different from that of the tissue in which it arose. This departure from the normal formative activity is usually a reversion to a more primitive type of tissue-formation, the controlling influences normally guiding the cells being weakened to such a degree that the tissues produced fail to acquire the structural differentiation of the parent-tissue. This failure in structural differentiation may be so great that the resulting tumor resembles embryonic tissue. Such tumors are clinically malignant, and, in general, it may be said that the degree of malignancy is approximately proportional to the lack of specialization exhibited by the formative activities of the cells. Up to this point we have considered two possibilities in the production of tumors: 1. The production of a tumor by cells which no longer respond to the needs of the organism in performing work for the general good, but which remain subject to the influences controlling the structural differentiation of the parent-tissue. 2. The formation of a tumor by cells which are less restrained by normal influences and which exercise their formative powers without conforming to the special differentiation exhibited in the parent-tissue. This we may regard as a reversion of the cells to a less specialized state, in which they exercise their formative powers in elaborating tissues corresponding to those normally present at some earlier stage in the development of the individual.

There is a third possibility. The reversion just described may be conceived as affecting the cells involved in tumor-production, but those cells, instead of forming a tissue corresponding to the degree of reversion they have suffered, may become specialized along some divergent line of development and produce a tissue more or less akin to that of the parent-tissue. Thus a tumor composed of bone may be produced within some other form of connective tissue, such as cartilage or fibrous tissue. The dissimilarity between the tissues of a tumor and those of the part in which it grows would seem, from this point of view, to depend upon the degree of reversion that had taken place. Even after a tumor has once been formed, portions of it may acquire a different structure, due to reversion on the part of some of its cells or a modification of their formative activities. There appears to be a limit to the extent of these reversions. It is found in the early differentiation of the three embryonic layers. Cells derived from the mesoderm, for example, do not seem to revert to such an undifferentiated condition that they can develop tissues like those normally springing from the epiderm or hypoderm.

A still further complexity of structure may arise from the formative tendencies of different cells within the same growth developing along different lines of specialization. This occasions the production of "mixed" tumors, composed of various tissues arranged in a manner usually quite unlike that of any normal organ.

In consequence of the numerous variations in tissue-production which may participate in their development it follows that tumors have a marked individuality, and that only certain types of more frequent occurrence can be described. Departures from those types will be met with in practice, and they must each be interpreted in accordance with the insight which the observer can gain as to their nature and tendencies. The more atypical the structure of a growth —i. e., the more it departs from the structure of normal adult tissue —the less likely is it to prove benign; the more highly cellular it is, the more likely it is either to grow rapidly or to act injuriously upon the whole organism : for its cells derive their nourishment from the general system and throw upon it the task of eliminating their waste-products.

Tumors are subject to morbid changes comparable with those affecting normal tissues. They may be the seat of inflammation, infiltrations, and degenerations. In fact, the more cellular forms are exceedingly prone to degenerative changes, due probably to a relative insufficiency of nourishment consequent upon their rapid growth and active metabolism. It is quite likely that the products of those degenerations, when absorbed into the system, act injuriously upon the general health.

The effects upon the nutrition of the body occasioned by the presence of a tumor constitute that part of the clinical picture which is known as "cachexia," and is most marked when the tumor is malignant But cachexia is not necessarily a sign of malignancy, and is not always present, even when the patient has a very malignant form of tumor. The degree of malignancy is measured by the rapidity of growth, the tendency to infiltrate surrounding tissues, and the liability to metastasis, and these depend upon the reproductive activity of the cells and the extent to which their formative activity is displayed in the elaboration of firm intercellular substances. Metastasis takes place when cells become detached from a tumor and are conveyed to some other part of the body, where they find conditions favorable for their continued multiplication. They then produce secondary tumors, which usually closely resemble the primary growth to which they owe their parent-cells.

It is evident that a microscopical study of a tumor may be made the basis of pretty accurate estimates of its nature and tendencies. The general character of the tissue composing it can be determined; an approximate idea of the reproductive activity of the cells formed; the tendency to invade or infiltrate the surrounding tissues, and therefore the probability of the occurrence of metastases, estimated; and the presence of degenerative or other changes observed. The knowledge so gained will throw light upon the clinical significance of the tumor. It is evident, however, that all the knowledge required cannot, in every case, be learned from the examination of a single piece of the tumor. Some of the necessary facts are best observed at the periphery of the growth, others in the central portions, and in mixed tumors the various parts of the growth may possess quite different characters. Every tumor must be made the object of a special study, if all the information it is capable of yielding is to be acquired.

Before passing to a description of the more common types of tumors we must turn our attention for a moment to their classification and nomenclature.

Tumors are sometimes grouped in two great divisions: 1, the "malignant tumors," which threaten life because of the rapidity of their growth, their infiltration of surrounding structures, and their liability to metastasis; and, 2, "benign tumors," which are essentially harmless unless they develop in a situation where they interfere with the function of some vital organ, or unless they appropriate so much of the nutritive material of the body that the general health suffers. This classification is a purely clinical one, and deserves mention only because of its medical importance. There are many degrees of malignancy, and these can be estimated in individual cases only with the aid of deductions from the structural peculiarities of the particular growths. A classification based upon the structure of tumors is, therefore, of greater value than one based merely upon their clinical aspects, for it includes that and much more besides.

If we bear in mind the fact that any form of cell capable of multiplying may give rise to a tumor, it will become evident that those tumors composed of a single variety of tissue may be classified in a manner similar to that in which the normal tissues are classified. Such tumors are grouped under the term "histioid," to distinguish them from tumors of more complex structure not analogous to simple elementary tissues, which are collectively referred to as "organoid." The histioid tumors are designated by names formed from the word indicating the normal tissue they most closely resemble and the suffix "oma." Thus, a fibroma is a tumor consisting essentially of fibrous tissue—*i. e.*, connective-tissue cells with a fibrous intercellular substance—even if the arrangement of the tissue-elements is not quite like that of normal fibrous tissue. A myoma is a tumor composed of muscular tissue, with only so much admixture of fibrous tissue as would be comparable with that found in masses of normal muscle. But as there are smooth and striated muscular tissues, so there are leiomyomata and rhabdomyomata. When a tumor contains two varieties of elementary tissue in such proportions that neither can be considered as subsidiary to the other, it receives a compound name, in which the most prominent or important constituent tissue is placed last, being qualified by the name of the less important tissue. Thus there are myofibromata, in which the fibrous tissue is more prominent than the muscular tissue; and fibromyomata, in which the muscular tissue predominates. In like manner three or more tissues may be designated as forming a tumor by such names as osteochondrofibroma, myxochondrofibroma, etc., implying that the growths are composed of fibrous tissue with an admixture of cartilage and bone, or cartilage and mucous tissue, etc.

The problem of classification is not so simple when we take up the consideration of tumors less closely resembling the normal tissues that are found in the adult body. Those tumors which are akin to embryonic tissues still retain names that have come down from earlier times, and which were conferred on them because of some characteristic visible to the unaided eye. Those of connective-tissue origin are called sarcomata (singular, sarcoma), which means tumors of fleshy nature; and those containing tissues derived from epithelium are called carcinomata, or cancers, because by virtue of their infiltration of the surrounding tissues they possess a fanciful resemblance to a crab. The terms sarcoma and carcinoma have, in the course of time, become more defined, and are now restricted to certain well-marked types of structure. The carcinomata are composed of fibrous tissue and epithelium, the one derived originally from the mesoderm, the other from either the epiderm or hypoderm. In this dual origin they resemble the viscera of the body, and may, therefore, be regarded as among the simpler members of the group of organoid tumors. The most complex members of that group are the "teratomata," which contain structures simulating hair, teeth, bones, etc., arranged without definite order, and often present in great numbers. They spring from the reproductive organs of the body, and appear to be erratic attempts at the production of new individuals.

A new formation of bloodvessels accompanies the development of tumors, and these vessels are associated with a supporting connective tissue which may be conceived as a part of this addition to the vascular system of the body, rather than as an integral part of the tumor itself. This development of new bloodvessels is analogous to that which takes place in the course of some of the inflammatory processes, and appears to be brought about in the same manner.

## I. THE CONNECTIVE-TISSUE TUMORS.

1. **Fibroma.**—The structure of a fibroma is apt to resemble that of the particular fibrous tissue in which it develops. Very soft varieties frequently spring from the submucous tissues of the nose, pharynx,

Fig. 308.

Section of a nodular fibroma. (Birch-Hirschfeld.) The dense fibrous tissue is in irregular nodules, between which are bands of less dense fibrous tissue containing bloodvessels.

and rectum, forming polypoid growths projecting from the surface of the mucous membrane. They are composed of delicate bands of fibres, loosely disposed to form an open meshwork, which is filled with a fluid resembling serum. In the fluid occasional fibres of still more delicate structure may be seen, together with lymphoid cells, either isolated or in little groups like imperfectly formed lymph-follicles. The surface of the growth is formed by a layer of rather denser fibrous tissue, which is covered by a continuation of the epithelium belonging to the mucous membrane. Similar soft fibromata sometimes take origin from the subcutaneous tissues, but fibromata of the skin are usually of denser structure, the bands of fibrous tissue being coarser, more compact, and less loosely arranged. Œdema may make these tumors look very much like the first variety.

Harder varieties of fibroma take origin from such dense forms of fibrous tissue as compose the dura mater, the fasciæ, periosteum,

FIG. 309.

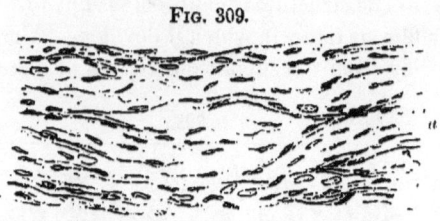

Dense form of fibroma. (Ribbert.) Section from a fibroma of the dura mater. The intercellular substance is very compact and the cells compressed. The latter are most numerous in the neighborhood of the narrow vessel, a, which, together with a branch, is cut longitudinally.

FIG. 310.

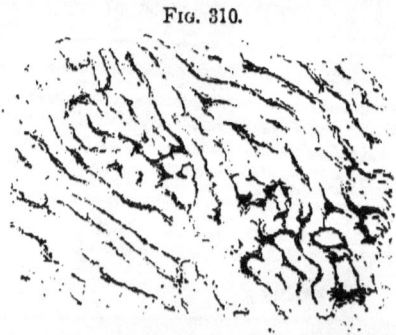

Dense form of fibroma. (Ribbert.) Section from older portion of a keloid. Dense masses of compact, apparently homogeneous intercellular substance interlace to form the chief bulk of the tissue. The cells are so few in number and so compressed that they are hardly distinguishable, and have been omitted from the figure.

etc., and those fibromata that occur in the uterus are of similar character. They are usually composed of nodular masses of dense structure, which are held together by a more areolar fibrous tissue supporting the larger bloodvessels of the tumor (Fig. 308). Among the hardest of the fibrous new-formations is the keloid, which in its oldest parts resembles old cicatricial tissue, the fibrous intercellular substance being compacted into dense, almost homogeneous masses and bands, in which the nuclei of the cells are barely discernible (Figs. 309 and 310).

Fibromata do not always have a nodular character, even when they are of dense structure. They sometimes occur in a diffuse

Fig. 311.

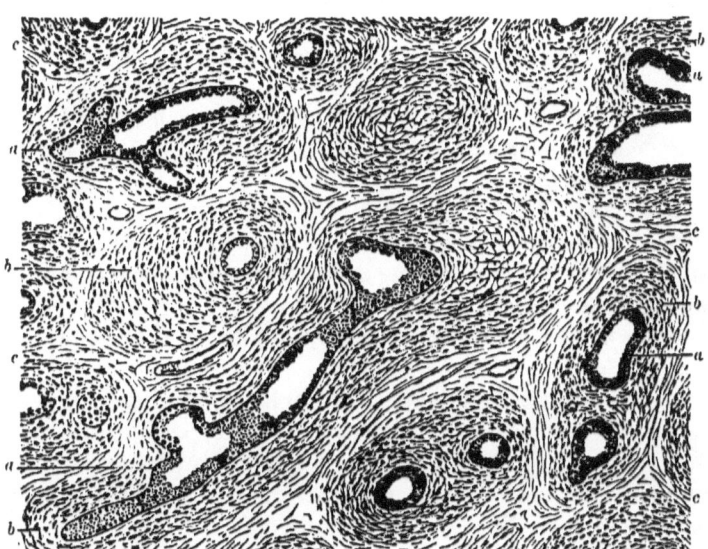

Intralobular fibroma of the breast. (Ziegler.) *a*, acini and ducts of the gland; *b*, new-formed fibrous tissue; *c*, areolar tissue of the interstitium, containing the vascular supply.

form, surrounding and enclosing the structures of the organ in which they develop. Such diffuse fibromata of the mammary gland are not uncommon, and two varieties may be distinguished: 1, those in which the fibrous tissue develops between the lobules of the gland, separating them from each other by broad bands of dense character, the *interlobular* form; and, 2, the *intralobular* form, in which the individual acini of the gland are separated and surrounded by bands of fibrous tissue (Fig. 311). These diffuse fibromata of the breast must not be mistaken for carcinomata, which they superficially resemble when the glandular epithelium has undergone atrophy due to pressure. In general appearance under the microscope these fibromata resemble the outcome of a chronic interstitial inflammation, but they do not seem to owe their origin to an inflammatory process.

Fibromata may undergo localized softening, due to fatty metamorphosis and necrosis. More frequently they are the seat of calcification, the lime-salts being deposited in granules within the intercellular substance, or in little globular masses, variously aggregated. These calcified portions are apt to acquire a diffuse blue color in sections that have been stained with hæmotoxylin.

Mixed tumors, containing fibrous tissue and some other variety of connective tissue, or smooth muscular tissue, are common. Fibrosarcomata and fibromyxomata are liable to metastasis; the other mixed tumors and pure fibromata are among the most benign of the tumors.

2. **Lipoma.**—Tumors composed of adipose tissue arise from preexistent fat, or from fibrous tissue of the areolar variety. Their structure very closely simulates and is frequently indistinguishable from that of normal fat (Fig. 312). But they reveal their independence of the general economy by not being reduced in size during emaciation of the individual. They sometimes enter into the composition of mixed tumors, such as lipomyxomata, lipofibromata, and fibrolipomata. They often grow to considerable size, may be multiple, but are not liable to metastasis and are benign.

Calcification, necrosis, and gangrene may occur in lipomata, but are usually confined to those of large size.

3. **Chondroma.**—The cartilage entering into the formation of chondromata is usually of the hyaline variety, but sometimes fibrocartilages are also present, and may, in rare instances entirely replace the hyaline form. The structure of the cartilages differs somewhat from that of the normal types. The cells are less uniform in character and in size, are more irregularly distributed through the matrix, and are frequently embedded in the latter without an intervening capsule. The tumor is rarely composed exclusively of cartilage, but is usually nodular, the cartilaginous masses being surrounded by a fibrous tissue in which the vascular supply of the growth is situated.

Chondromata generally arise from pre-existent cartilage, bone, or

Fig. 312.

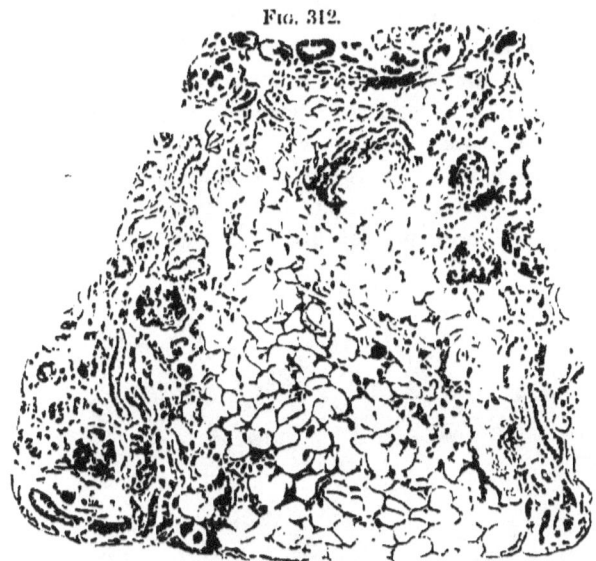

Lipoma of the kidney. (Birch-Hirschfeld.) The boundary between the adipose tissue of the tumor and the renal tissue is not sharply defined. The former occupies the middle of the section and extends to its lower edge.

fibrous tissue. When they apparently spring from bone their true origin may be from small remnants of cartilage which have escaped the normal ossification.

Fig. 313.

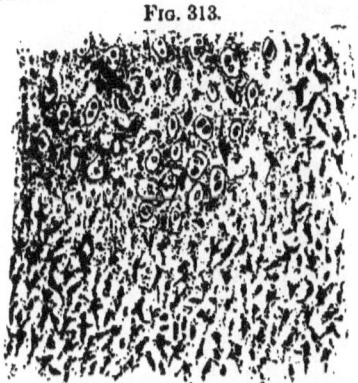

Chondrosarcoma of the rib. (Hansemann.) The lower portion of the section is exclusively sarcomatous. The upper part contains cartilaginous tissue, but there are a few spindle-shaped cells in the matrix similar to those in the sarcomatous portion of the growth.

Cartilage is a not infrequent constituent of mixed tumors, especially of the parotid gland or testis, when it is usually associated with mucous and fibrous tissue, adenomatous new formations, or sarcoma (Fig. 313).

Chondromata are subject to a number of secondary changes, the most important of which are: calcification; conversion into a species of mucoid tissue through softening of the matrix and modification of the cells, which assume a stellate form; transformation into an osteoid tissue, resembling bone devoid of earthy salts; or into a fairly well-developed calcified bone (Fig. 314). Local soften-

FIG. 314.

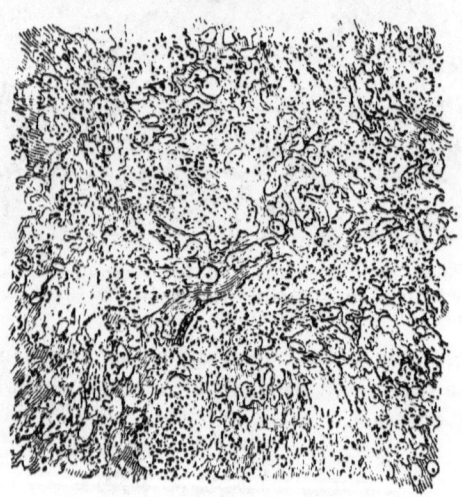

Osteoid endochondroma. Section from a metastatic nodule in the lung. The cartilage is atypical, and is arranged in a manner simulating that of cancellated bone. Between the bands and lamina of cartilage is a mixture of mucous and sarcomatous tissue, myxosarcoma, which has rendered the tumor subject to metastasis. The whole tumor may, then, be called a chondromyxosarcoma.

ing of the tumor may also take place through a liquefaction of the matrix and disintegration of the cells. The latter may also undergo a fatty degeneration in parts of a tumor which show no signs of softening of the matrix.

Chondromata are classed with the benign tumors, but occasional instances of metastasis are on record. It is difficult to understand how this could take place in the case of the harder chondromata, in which the cartilage is surrounded by a somewhat dense fibrous tissue resembling the normal perichondrium. Where there is an admixture with either sarcomatous or myxomatous tissues, these confer a malignant character upon the mixed tumor, and it is quite possible for fragments of cartilage to become detached from the primary growth and appear in the secondary tumors, should metastasis occur.

4. **Osteoma.**—The most important tumors containing bone are mixed tumors that are significant chiefly because of their other constituents. Small growths consisting of bone alone, either in its compact or its spongy form, occur in the lung, walls of the air-passages, and, rarely, in other situations (Fig. 315). Where bony

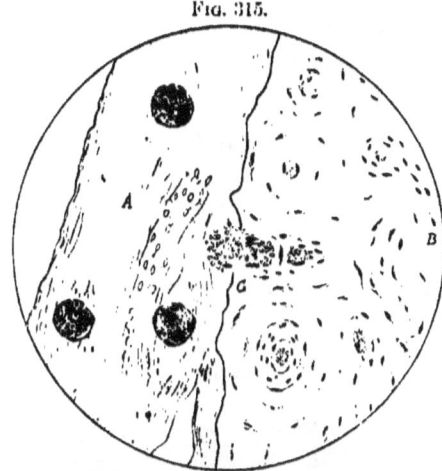

Fig. 315.

Developing osteoma of the arachnoid. (Zanda.) *A*, dura mater; *B*, as yet non-calcified osteoid tissue; *G*, bloodvessel.

new formations spring from pre-existent bone—*e. g.*, from parts of the skeleton—they are usually the result of some inflammatory process, and are not to be grouped among the tumors.

In mixed tumors bone is frequently associated with fibrous tissue, myxoma, sarcoma, and chondroma.

The structure of the bone in tumors presents slight departures from the normal type, just as that of cartilage in chondromata is somewhat atypical. The lacunæ are apt to vary in size, shape, and distribution more than in normal bone, and the system of canaliculi is less perfectly developed.

5. **Myxoma.**—The mucous tissue of myxomata has its normal prototype in the Whartonian jelly of the umbilical cord. In its purest form it consists of stellate or spindle-shaped cells, with long fibrous processes that lie in a clear, soft, gelatinous, intercellular sub-

stance containing mucin in variable quantities (Fig. 316). This tissue is closely allied to the other forms of connective tissues and tumors are rarely composed of mucous tissue alone. There is usually an admixture with fibrous tissue, bone, cartilage, fat, or sarcoma; form-

FIG. 316.

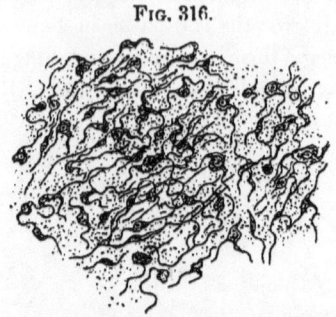

Section from a subcutaneous myxoma. (Birch-Hirschfeld.)

ing fibromyxoma, osteomyxoma, chondromyxoma, lipomyxoma, or myxosarcoma (Fig. 317). The flat endothelial cells of connective tissue also sometimes proliferate to such an extent as to form an

FIG. 317.

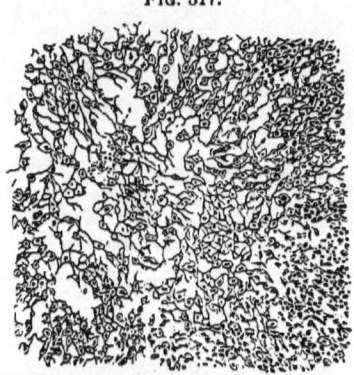

Myxosarcoma of the femur. To the left of the section the tissue is nearly pure mucous tissue. Toward the right, this tissue gradually merges into a more highly cellular structure, constituting the sarcomatous element in the growth. It is this admixture with sarcoma that gives the tumor a malignant character.

appreciable constituent of the tumor, the cells being large, rather rich in protoplasm and frequently multinucleated. When this development is pronounced the tumor may be designated a myxendothelioma, and approaches the myxosarcomata in character.

Mucous tissue is best studied in the fresh condition by pressing small bits flat between a cover-glass and slide. The processes of the cells may then be seen in their continuity; while, if sections are prepared after hardening, many of those processes will be cut in such a way that their connections with the cells in the contiguous sections are destroyed, and they appear as fibres lying free in the intercellular substance.

Mucous tissue must be carefully distinguished from œdematous fibrous tissue. Such œdematous tissue possesses cells of a spindle or flat shape, like those usually met with in fibrous tissue; but the usual fibrous intercellular substance has a loosened texture, due to the presence of fluid between the fibres, which gives the tissue a soft, transparent character not unlike that of mucous tissue. It must also be borne in mind that fibrous and adipose tissues are liable to undergo a mucous degeneration in which the cells assume a more stellate form than is usual with those tissues, and the intercellular substances lose their fibrous character and become more homogeneous. Such degenerations are distinguished with difficulty from the tissue which originally develops as mucous tissue, but they have nothing in common with tumors.

Myxomata usually develop in fibrous tissue, adipose tissue, or the medulla of bone. In association with cartilage they are not uncommon in the parotid gland. When pure they are benign, but their association with sarcoma often gives them a malignant character, the degree of malignancy depending upon that of the sarcomatous tissue present.

6. **Endothelioma.**—The endotheliomata are connective-tissue tumors which owe their origin to a proliferation of the flat endothelial cells that line the serous cavities, line or form the walls of the blood-vessels and lymphatics, and are present in some of the lymph and other spaces of the fibrous tissues. Young cells of this variety do not have the membranous bodies that characterize the fully developed older cells, but closely resemble the cells of epithelium. It follows that in this class of tumors it is not always easy to determine the origin of the cells from a mere inspection of their shapes and sizes. The situation and general structure of the tumor will often decide this point. Epithelial tumors spring from pre-existent epithelium, either in some normal site or in an unusual situation because of some anomaly of development (*e. g.*, in the neck, owing to imperfect obliteration of the branchial clefts).

Endotheliomata, on the other hand, spring from the connective tissues, often at a point remote from any epithelial structures; e. g., the dura mater.

When the endothelioma owes its origin to a proliferation of the flat cells lining the lymph-spaces or vessels it has a plexiform structure, the young cells occupying pre-existent interstices in the tissues or following the arrangement of the vessels (Figs. 318 and 319). As the cells grow older they may become flattened, and are then

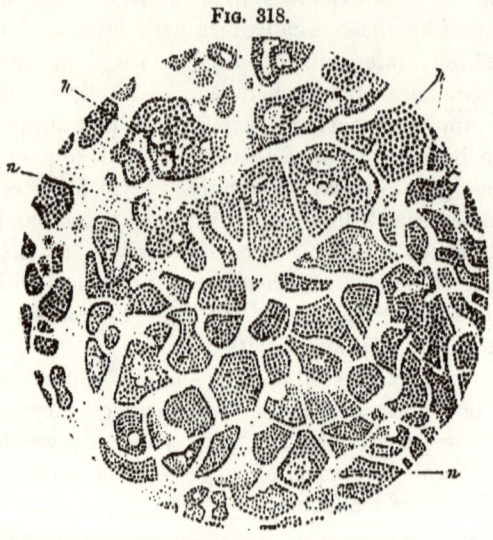

Fig. 318.

Endothelioma from the floor of the mouth. (Barth.) Older portion of the growth. This has a general alveolar structure, the alveoli being separated by a vascularized areolar tissue. *n, n*, necrosed groups of endothelial cells; *h, h*, similar necrosed masses that have undergone hyaline degeneration.

often imbricated, forming little, pearl-like bodies. These may subsequently undergo degenerative changes, such as hyaline degeneration, which convert them into homogeneous masses or bands. Where this takes place the tumor has received the name, "cylindroma." Or the degenerated cells may be the seat of calcareous infiltration. This is the origin of the psammomata or "sand-tumors" of the cerebral membranes (Fig. 320). In other cases the cells may not acquire the membranous character of adult endothelium, but continue to multiply without such specialization. Then the tumor partakes of the sarcomatous nature of the other connective-tissue tumors of highly cellular structure and devoid of any marked

Fig. 319.

Endothelioma from the floor of the mouth. (Barth.) Section showing the advance of the growth into the lymph-spaces: *a*, karyokinetic figure in an endothelial cell. Other less well-preserved figures are seen in other portions of the section.

intercellular substance. This is more particularly the case when the endothelial cells in the adventitia of the bloodvessels mul-

Fig. 320.

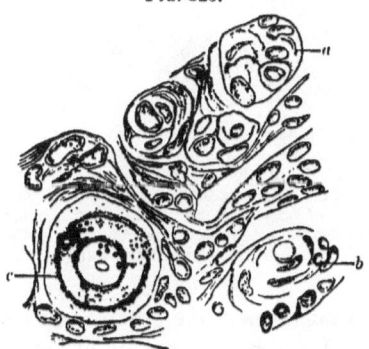

Early stages in the formation of a psammoma. (Ernst.) *a*, collection of endothelial cells; *b*, similar group showing imbrication of the cells and beginning hyaline degeneration; *c*, hyaline mass containing a slight deposit of infiltrated calcareous matter, appearing as granules.

tiply to form the growth. The cells of the growth are then in intimate relation with the walls of the vessels, and the tumor is designated as an angiosarcoma or alveolar sarcoma, according as the cells show a grouping around the vessels or form collections occupying the meshes between them (Figs. 321, 322, and 323).

This brief outline of a complicated group of tumors will serve to show that some members of that group closely simulate epitheliomata in their structure, though they are quite different in their

Fig. 321.

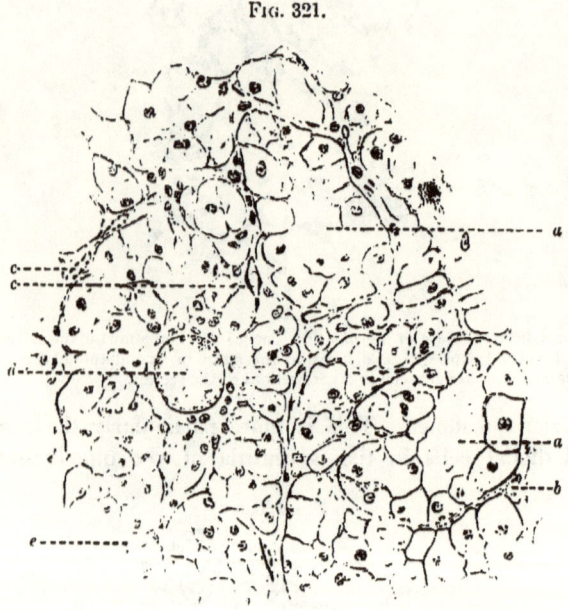

Endothelioma of the ulna. (Driessen.) *a, a*, alveoli lined with endothelial cells and occupied by blood; *b*, areolar tissue between the alveoli, containing capillary vessels, *c; d*, large vessel closely surrounded by proliferated endothelium. The structure of this tumor is difficult of interpretation. It appears most probable that its origin lay in the proliferation of the endothelium of lymphatics, and that the blood in *a*, *a* is due to communications established between the bloodvessels and elongated and anastomosing alveoli of the tumor. The cells of this growth contained glycogen (see Fig. 249).

origin; while other members of the group are essentially sarcomata, owing their origin to a particular variety of connective-tissue cells and having peculiarities of structure due to the situations in which those cells normally occur. The significance of the tumor will depend in each case upon its tendency to grow rapidly and to infiltrate the surrounding tissues, and its liability to metastasis. These qualities must be estimated by a consideration of the history of the case and the structure and evidences of proliferation presented by the tumor itself.

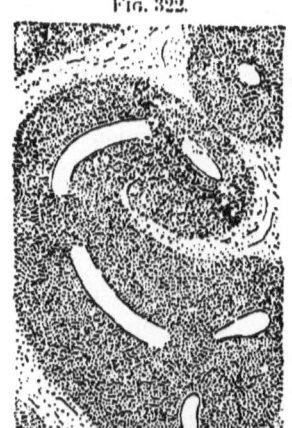

Fig. 322.

Angiosarcoma of bone. (Kaufmann.) The lumina of bloodvessels are seen in longitudinal and in cross-section. They are surrounded by a highly cellular tissue, derived from the proliferation of the endothelium forming the perivascular lymphatics. Such tumors are also called "peritheliomata."

7. **Sarcoma.**—This term includes a variety of tumors differing in the details of their structure and in their clinical significance, but

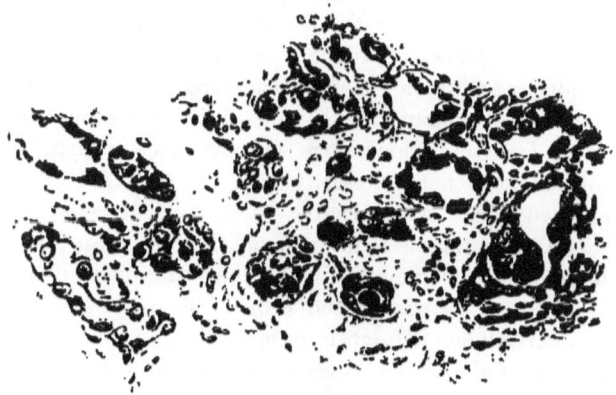

Fig. 323.

Endothelioma of the thyroid. (Limacher.) In this example the endothelial cells of the tumor spring from the endothelium of the capillary bloodvessels. Various stages in the proliferation of that tissue are represented in the section.

having in common a general resemblance to imperfectly developed or embryonic connective tissue. Such tissues are not infrequently associated with other neoplasmic tissues of higher differentiation, forming mixed tumors; but in such cases the tissues of higher type are not the result of a progressive development on the part of the sarcomatous tissue, for the essential feature of the latter is that it remains in a primitive condition, the formative powers of its cells being chiefly confined to a reproduction of fresh cells, and not to the elaboration of intercellular substances which would convert the tissue into some variety of adult connective tissue. In this respect, as well as in the absence of any natural limitation of growth, the sarcomata differ from the tissues of somewhat similar structure which result from the rejuvenescence of connective tissue in the productive stages of inflammation leading to repair. Some forms of sarcoma closely resemble granulation-tissue, for both have the same origin from the cells of the connective tissues; but the two must be sharply distinguished from each other, for their tendencies and usefulness are extremely different. The formation of granulation-tissue has a definite cause, and it undergoes a progressive differentiation into a dense fibrous tissue, which terminates the process (with the possible, but notable, exception of the development of keloid; which is, however, not sarcoma). Sarcoma, on the other hand, arises without a well-defined cause, shows no tendency to higher differentiation, and continues to grow without any assignable limitations. A further difference that may aid in the decision of whether an undifferentiated tissue of connective-tissue type is sarcoma or due to inflammatory processes lies in the fact that sarcoma has a tendency to infiltrate the surrounding tissues, while the young connective tissue that results from an inflammatory rejuvenescence has not.

Sarcomata need not necessarily have the structure of the least differentiated forms of connective tissue. Their cells may show a greater differentiation than is found in that tissue, and there may be a certain amount of intercellular substance of a fibrous or other nature separating the cells. The presence of such a fibrous intercellular substance is an evidence that the formative activity of the cells is not wholly concentrated in the production of new cells, but is partly diverted to the formation of intercellular material. It is therefore a sign of less active growth than would be the case were there no such diversity of activity. The intercellular substances also tend to confine the cells to the growth itself, impeding their penetration into the interstices of the surrounding tissues (infiltration) and reducing the probability that some of the cells will be carried to distant parts by the currents of the fluids circulating in the tissues (metastasis). It follows that the presence of intercellular substances having these effects must reduce the degree of malignancy of the whole growth if they are present throughout its substance. This argument is borne out by the results of experience. The sarcomata might be arranged in a series according to their degrees of malignancy, beginning with those that are most malignant, and have little intercellular substance, and cells which are only slightly, if at all, differentiated, and ending with those that can hardly be considered malignant, and which have such an abundant fibrous intercellular substance that their structure closely agrees with that of fibroma. In fact, no sharp line between these sarcomata and the fibromata can be drawn. The two classes of tumor merge into one another: they have the same origin, and differ only in the behavior of their cells in the exercise of their formative activities. Those differences are, however, of the utmost clinical importance.

The sarcomata are classified, according to the characters of their component cells, into the round-cell, spindle-cell, giant-cell, melanotic, etc., varieties. They are also subdivided according to the way in which those cells are arranged. The alveolar sarcomata, for example, consist of groups of cells enclosed in the meshes of a fibrous network. These names are, however, more descriptive than indicative of essentially distinct kinds of tumor, and the demarcation between the different varieties is not a sharp one. Many sarcomata consist of cells of various shapes, either in different parts or intermingled throughout the growth. This necessitates the insertion of mixed varieties between the above groups of distinct and relatively pure types. Furthermore, the cells not only differ in shape, but also in size, so that a distinction may be made between the small round-cell sarcomata and the large round-cell variety; but notwithstanding the fact that this grouping is somewhat artificial, it has a certain clinical value, because it indicates in a rough way the degree of differentiation attained by the tumor, and for this reason it will be well to adhere to this classification and to consider the purer types separately, bearing in mind that the mixed forms of sarcoma possess characters intermediate between those of the simpler forms upon which the classification is primarily based.

*a.* SMALL ROUND-CELL SARCOMA.—This variety presents the least degree of structural differentiation. The substance of the tumor is composed of small, round cells with single vesicular nuclei enclosed in very little cytoplasm. They are so closely aggregated that they appear to be in contact; but careful examination will often reveal a small amount of a nearly homogeneous, finely granular, or slightly fibrillated intercellular substance (Figs. 324 and 325). The tumor is supplied with blood-vessels having very thin walls,

FIG. 324.

FIG. 325.

Small round-cell sarcoma of the neck.

Fig. 324.—Section only moderately magnified, showing the extremely cellular character of the growth; the great friability of the tissue is owing to the minimal amount of intercellular substance it contains and the intimate relations between the tissue of the tumor and the walls of relatively large, thin-walled bloodvessels.

Fig. 325.—Sketch of a fragment of the tumor, more highly magnified. The cytoplasm around the nuclei is hardly distinguishable, and the cells are separated by only a small amount of an indefinite intercellular substance.

formed of a single layer of cells, which are usually more protoplasmic than those of fully developed endothelium. These vessels may be very abundant, but, especially if the tumor has been removed by operation, they are likely to be empty and their walls so collapsed that they are not easy of recognition. When seen in longitudinal section these emptied vessels appear as a double line of elongated, somewhat fusiform cells, lying in close contact with the cells of the rest of the tumor. In cross-section they are still more difficult of detection, since the swollen endothelial cells then look very much like the contiguous cells of the growth itself.

Where the sarcoma is infiltrating the surrounding tissues groups of the round cells, distinguished from the leucocytes which may be present by the character of their nuclei, appear in the interstices of the tissue, the formed elements of which undergo atrophy, either because subjected to increased pressure or because their nutrition is interfered with (Fig. 326). In this way the tumor increases the territory which it occupies, but the more central portions also grow. After a certain stage of growth has been attained the older portions of the tumor are liable to undergo degenerations or necrosis.

It is evident, from the structure of this variety of sarcoma, that it must be very prone to suffer metastasis. This may take place through the lymphatics of the surrounding tissues, favored by the infiltrating qualities of the growth; or it may take place through the bloodvessels, some of the cells finding their way through the thin walls of the vessels in the tumor itself, or into the lumina of larger vessels through an infiltration of their walls. In either

Fig. 326.

Small round-cell sarcoma of the pelvis, infiltrating dense fibrous tissue.

of these ways a generalization of the growth may take place, secondary nodules appearing in many parts of the body.

Round-cell sarcomata of this type are liable to arise in the connective fibrous tissue between the muscles, in the fasciæ, etc. They also find their origin in the skin, testis, and ovary. They are among the most malignant of the sarcomata, growing rapidly, infiltrating their surroundings, and undergoing metastasis.

*b*. Lymphosarcoma.—This variety of sarcoma differs only slightly in structure from the small round-cell form in possessing a somewhat more elaborate stroma, a term which could hardly be applied to the small amount of intercellular substance found in the latter. In the lymphosarcomata the cells closely resemble those of the small round-cell variety of sarcoma, but they lie loosely aggregated in the meshes of a reticulum of fibres, many of which constitute the processes of stellate cells penetrating the substance of the growth and possibly joining each other This reticulum is somewhat more pronounced around the bloodvessels which it supports. The cells may be shaken out of this reticulum, if unembedded sections are agitated with water (Fig. 327).

FIG. 327.

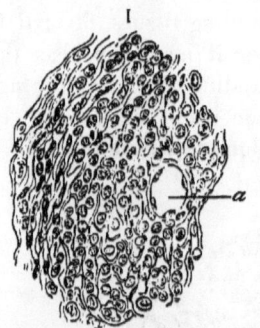
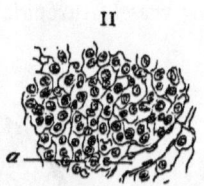

Sections from lymphosarcomata. (Kaufmann.) I, firmer variety, with a pronounced stroma; from a mediastinal tumor. II, softer variety, with a more delicate stroma; from a tumor of the small intestine. *a*, capillary bloodvessel.

This structure closely resembles that of the lymphadenoid tissue found in the normal lymph-nodes, and there is danger of confounding the growth with a simple or inflammatory hyperplasia of those organs. This danger is enhanced by the fact that these sarcomata frequently find their origin in a lymph-gland or the lymphadenoid tissue in the mucous membranes. When the enlargement of the gland is the result of hyperplasia the superabundant tissue is confined within the capsule of the gland, which enlarges as its contents increase in amount. There is also a history of some inflammatory process within the lymphatic province to which the gland belongs. Such is not the case when the increase of tissue is due to the development of a tumor. The growth usually pierces the capsule of the gland, and cannot be traced to inflammatory causes. This penetration of the capsule is an evidence of the infiltrating power of the growth. Like the small round-cell sarcoma, this variety is liable to early and extensive metastases, and is hardly less malignant than that form.

*c*. LARGE ROUND-CELL SARCOMA.—As the title implies, this tumor is composed of larger cells than those found in the small round-cell sarcomata. The greater size is due to a larger amount of cytoplasm, in which are rather large round or oval vesicular nuclei, usually one in each cell, but not infrequently cells with two or even more nuclei are observed. The intercellular substance is more abundant and more distinctly fibrillated than is the case in the small round-cell sarcomata, but it is not uniformly distributed between the individual cells. These are usually aggregated in groups, which are surrounded by the denser bands of fibrous tissue. From these, little fibrous twigs may sometimes be seen penetrating between the individual cells of the group. This arrangement gives sections of the growth an alveolar appearance (Fig. 328). The

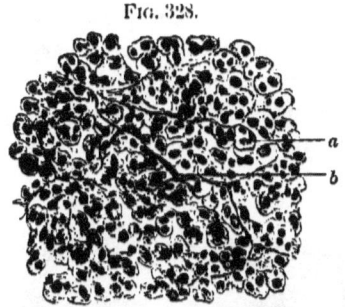

Fig. 328.

Large round-cell sarcoma of the tongue: *a*, large round cell containing three nuclei; *b*, delicate fibrous stroma supporting the cells of the growth. At the point *b* this stroma contains a collapsed capillary bloodvessel. The large round cells are probably of endothelial origin. The growth occurred in a man aged sixty-one years, and in the course of eight months had attained the size of a hickory-nut.

fibrous tissue itself may be highly developed, resembling the adult form; or it may be more highly cellular and contain large spindle-shaped cells. When this is the case the tumor becomes a mixed-cell sarcoma composed of large cells, partly round, partly fusiform.

The large round-cell sarcomata spring from the same tissues that give rise to the small round-cell sarcomata, but it is probable that they owe their production in large measure to a proliferation of the endothelial cells of those tissues, and are, therefore, etiologically related to the endotheliomata. They grow less rapidly than the small round-cell and lympho-sarcomata, and, as would be expected from a study of their structure, they are less prone to infiltrate their surroundings or to be subject to metastasis. They are, to a corresponding degree, less malignant in their clinical manifestations.

*d*. SPINDLE-CELL SARCOMATA.—The shape of the cells of this group of tumors betokens a higher state of differentiation than is found in the small round-cell sarcomata, the cells having more nearly approached the character of those found in the adult fibrous tissues; but although in this respect all the tumors of this group are more nearly like the normal tissues, they differ greatly among themselves in regard to the extent to which the formative activities of their cells are displayed in the production of intercellular substances. Some possess hardly more intercellular substance than the small round-cell varieties, while others have the appearance of a rather highly cellular fibrous tissue, the intercellular substances being abundant.

The fusiform cells of the tumor possess oval vesicular nuclei, around which is an amount of cytoplasm varying in the different individual growths. Sometimes the cytoplasm is abundant, and the tumor appears composed of large spindle-shaped cells, tapering at their ends to form processes of various lengths (Fig. 329). In

FIG. 329.

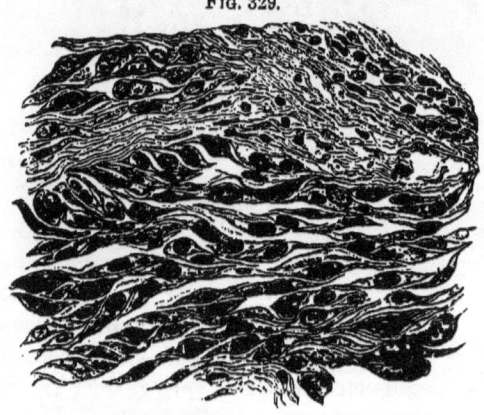

Large spindle-cell sarcoma. (Birch-Hirschfeld.)

other cases the cells are small and the cytoplasm is reduced to a thin investment of the nucleus, at the ends of which it rapidly dwindles to a very thin fibrous process. The spindle-cell sarcomata may, therefore, be divided into large- and small-cell varieties.

The cells are usually arranged with their long axes parallel to each other, forming bundles or broad bands of tissue, in which the cells all have the same general position. This direction is generally the same as that taken by the bloodvessels (Fig. 330). These have very thin walls, as in the preceding varieties of sarcoma, and the cells of the tumor appear to be in direct contact with the outside of the vessels. The cellular bundles may not all lie parallel to each other, but frequently are interwoven, so that a given section will contain longitudinal, cross, and oblique sections of the individual cells. Such appearances must not be mistaken for the somewhat similar aspect of sections of mixed-cell sarcomata.

The spindle-cell sarcomata are among the most common of tumors. They may arise from any of the connective tissues. When they spring from the periosteum they are apt to have an imperfectly formed bony tissue associated with the structure of the sar-

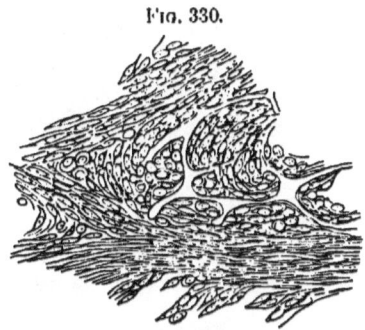

FIG. 330.

Spindle-cell sarcoma. (Rindfleisch.) Where the cells of the tumor lie parallel to the plane of the section their spindle shape is manifest; where they are perpendicular to the plane of the section their cross-sections appear round. The bloodvessels appear to have no proper walls, but to be bounded by the tissue of the neoplasm.

coma. They then form osteosarcomata or osteoid sarcomata, according to the perfection with which the structure of normal bone is reproduced.

In judging of the probable malignancy of a given specimen of spindle-cell sarcoma, the rapidity of its growth, as evidenced by the number of mitotic figures seen in the cells, and the abundance of fibrous intercellular substance, must be taken into consideration. As a group, the spindle-cell sarcomata are less malignant than the small round-cell sarcomata; but this is because the majority of spindle-cell sarcomata have a well-marked intercellular substance of fibrous character. Those forms which are almost destitute of this are hardly less malignant than the small round-cell variety (Figs. 331 and 332).

*c.* GIANT-CELL SARCOMA.—This form of sarcoma is characterized by the presence of large, multinucleated cells lying among the other cells of the growth. These giant-cells may be scattered

368  HISTOLOGY OF THE MORBID PROCESSES.

Fig. 331.

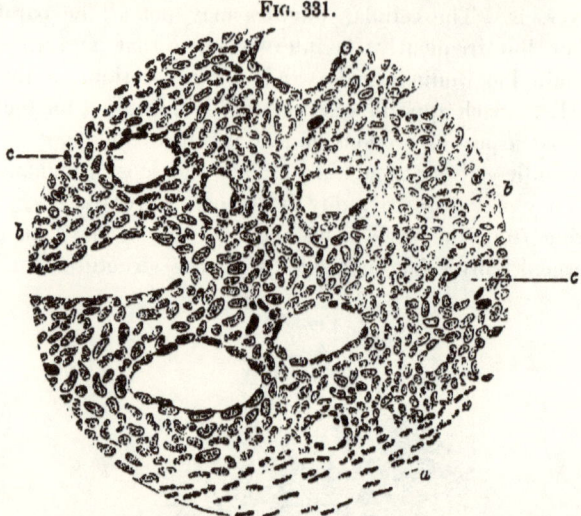

Example of a highly malignant variety of spindle-cell sarcoma. Sarcoma of the uterus with oval nuclei, indicating somewhat spindle-shaped cells. In other respects the character of the tumor resembles that of a small round-cell sarcoma. *a*, contiguous fibrous tissue of the uterus; *b*, sarcomatous tissue; *c*, bloodvessels. (v. Kahlden.)

Fig. 332.

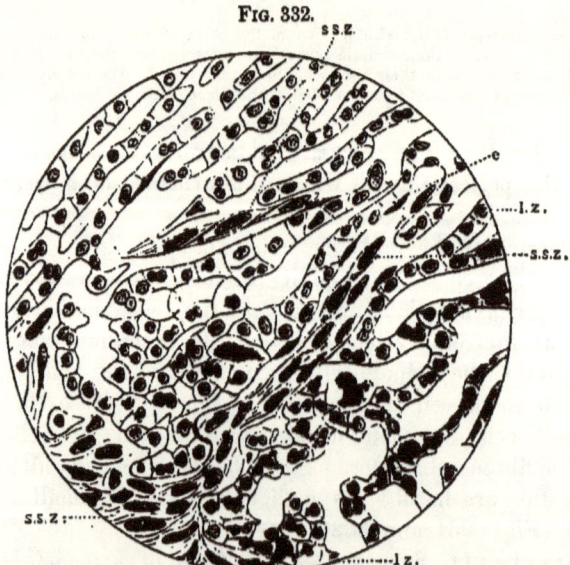

Example of a highly malignant spindle-cell sarcoma. Spindle-cell sarcoma infiltrating the liver. *l z*, liver-cells; *s s z*, spindle-cells of the sarcoma; *e*, endothelium of the intralobular capillaries. (Heukelom.)

pretty uniformly throughout the growth, or they may be much more abundant in some places than in others. The cells with single nuclei, among which the giant-cells are found, may be of the spindle-shaped variety, or they may be polymorphic, in which case cells of various shapes and sizes are met with.

The giant-cell sarcomata are usually derived from the medulla of bone. They constitute the most common form of epulis (Fig. 333), and frequently attain very large dimensions when they take

Fig. 333.

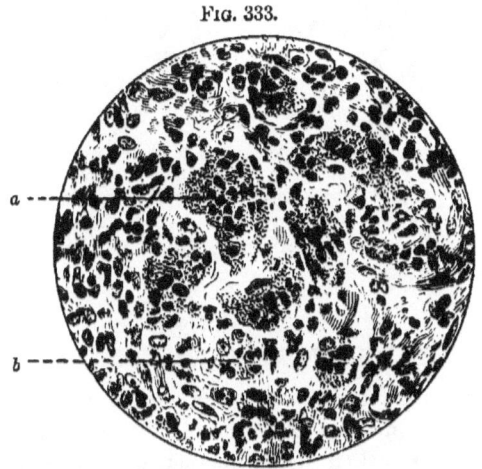

Giant-cell sarcoma of the superior maxilla: epulis: *a*, large giant-cell, with numerous nuclei; *b*, tangential section of a similar cell. Aside from the giant-cells, the growth is composed of spindle-cells and a moderate amount of a fibrous intercellular substance. The tumor was removed from a man forty-one years of age, and was of slow growth, having attained the size of a filbert in two and a half years.

their origin in the marrow of the larger bones, such as the femur or tibia. They are not, however, confined to bone, but may occur in other situations; *e. g.*, the breast.

The malignancy of giant-cell sarcomata must be estimated in individual cases according to the principles already elucidated.

*f.* MELANOSARCOMA.—Sarcomata which spring from pigmented tissues, such as the choroid of the eye, pigmented moles, etc., frequently show a pigmentation of their constituent cells, the pigment appearing as brown granules of various size within the cytoplasm of the cells. The cells are not all equally affected, and many may be seen without any sign of pigmentation. The tumors are apt to be of the spindle-cell or large round-cell varieties, and are

FIG. 334.

Melanosarcoma of the skin. (Ribbert.) The growth is an alveolar large round-cell sarcoma, containing cells that have undergone a pigmentary degeneration. Some of these cells contain so much pigment that the cellular constituents are invisible.

considered as rather more malignant than the non-pigmented forms of those tumors (Fig. 334).

## II. THE MUSCULAR TUMORS.

Muscular fibres of either the smooth involuntary or the striated variety may enter into the formation of tumors. Tumors made up of the former are called leiomyomata; those containing striated muscle, rhabdomyomata.

1. **Leiomyoma.**—The cells of the tissue forming leiomyomata very closely resemble those of normal smooth muscular tissue, but they may show a greater variation in size. They are arranged in bundles, their long axes parallel to each other; and these bundles are interwoven in such a way that sections of the tumor contain longitudinal, oblique, and cross-sections of the individual fibres (Fig. 335). Between the bundles there is a variable amount of fibrous tissue, giving support to the bloodvessels of the tumor. This fibrous tissue may be so abundant as to form a large element in the structure of the tumor, which is then denominated a fibromyoma. It may, also, occasionally be imperfectly developed, converting the growth into a leiomyosarcoma. The muscular tissue may undergo a hyaline degeneration and become the seat of calcareous infiltration, or the cells may be the seat of fatty degeneration with subsequent softening.

Leiomyomata arise in parts which normally contain smooth mus-

# TUMORS. 371

Fig. 335.

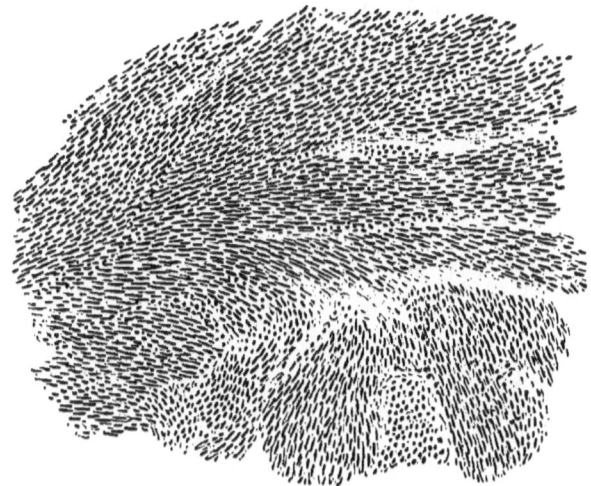

Leiomyoma of the uterus.  (Birch-Hirschfeld.)

cular tissue. They are common in the uterus, but may occur in the

Fig. 336.

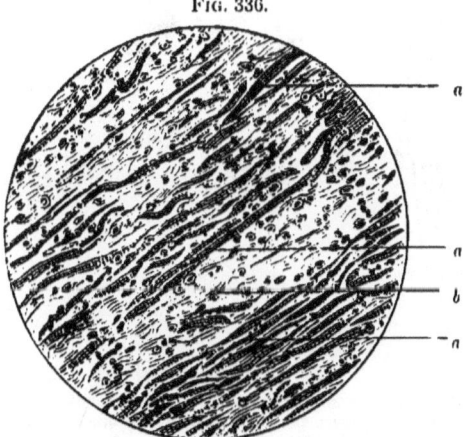

Rhabdomyosarcoma of the kidney: *a, a, a*, imperfectly developed striated muscle-fibres; *b*, tissue composed of small round and spindle-shaped cells, separated by considerable delicate fibrous intercellular substance. In other parts of the growth, which was the size of the fist, this tissue was more distinctly sarcomatous and the amount of muscular tissue smaller. The child from which this tumor was removed was about two years old.

intestinal walls, the urinary tract, or the skin. When pure, or when associated with fibrous tissue alone, they are benign.

## 372   HISTOLOGY OF THE MORBID PROCESSES.

**2. Rhabdomyoma.**—The striated muscle-fibres of rhabdomyomata are often so imperfectly developed that they are difficult of recognition. They are much more attenuated than the normal fibres, and may be reduced to very narrow and tapering structures that possess only traces of striation. Staining with eosin will often aid their detection among

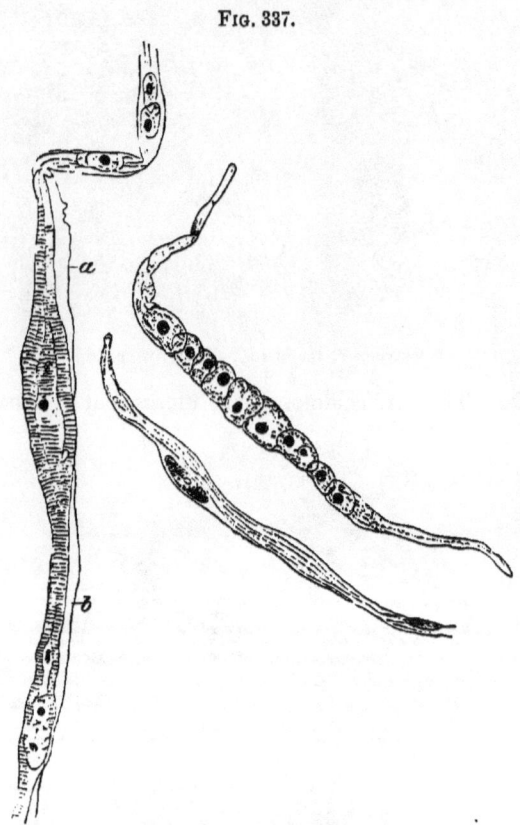

FIG. 337.

Isolated muscle-fibres from a rhabdomyoma of the œsophagus. (Wolfensberger.) *a, b,* appearances simulating a sarcolemma, probably due to adherent fragments of the intercellular substance.

the fibres of the connective tissue surrounding them, as it stains the contractile substance a coppery-red. The nuclei of the muscle-cells are frequently numerous, and may occupy the centre of the fibre, the imperfectly formed contractile substance lying at the periphery. In some rare cases the tumor is composed almost exclusively of striated muscle-fibres, arranged in irregular, interwoven bands, with a little vascular fibrous tissue among them. In other cases the muscular fibres are sparsely distributed through the growth, and can often be found only after a prolonged search. In these cases the tissue in which the muscle is situated is usually some variety of sarcoma, when the whole tumor is known as a rhabdomyosarcoma (Figs. 336, 337, and 338). Such mixed tumors are most frequently found in the genito-urinary tract, especially in the kidney, and may attain very large size. They are apt to occur in the early years of

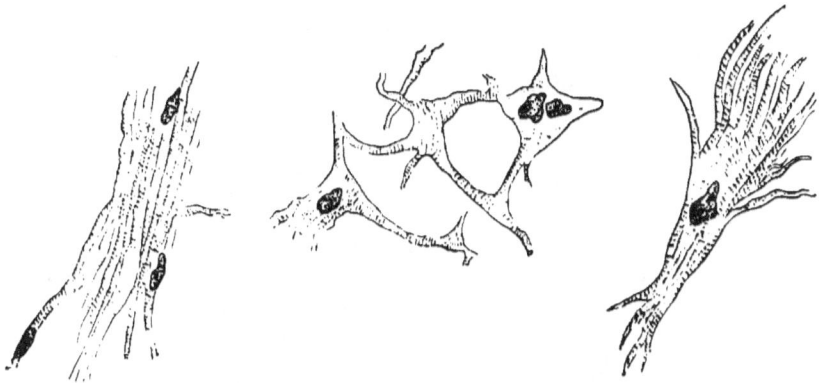

Isolated cells from a rhabdomyoma of the heart. (Cesaris-Demel.)

life, and are probably due to developmental anomalies. The sarcomatous element, which is usually predominant, gives them a highly malignant character.

## III. THE ANGIOMATOUS TUMORS.

Reference has already been made to the manner in which the bloodvessels of a part may proliferate under the influence of the inflammatory process, and also to the fact that when tumors develop the bloodvessels proliferate in a similar way to form new vascular areas within the tumor, from which the latter derives its nourishment. These instances of proliferation may be regarded as the natural response on the part of the vascular system to the demand thrown upon it by the formation of new tissues. In a general way, they are limited to the needs of the tissues which they supply. A vascular proliferation may, however, take place irrespective of any such demand, and continue without any such limitation. In this way the vascular tumors, or angiomata, are produced. We may regard them as springing, not from a single tissue or an adventitious combination of tissues, but from one of those anatomical "systems" in which several tissues are normally associated in a definite arrangement, and, under normal conditions, develop together to form well-defined structures distributed throughout the body. There are three such systems of associated tissues: the bloodvessels, the lymphatic system, and the nervous system. Each of these may enter into the formation of an apparently purposeless neoplasm, forming the hæmangiomata, lymphangiomata, and neuromata. Of these, the first two are of vascular character and mesodermic origin, and their consideration naturally follows that of the other tumors arising in tissues of similar embryonic origin.

1. **Hæmangioma.**—The bloodvessels entering into the formation of hæmangiomata are usually relatively deficient in the development of their muscular coats. They resemble large capillaries which have been reinforced by a covering of fibrous tissue. The vessels may lie with their walls almost in contact with each other, or there may be a considerable amount of interstitial tissue between them. It is not always possible to decide in a given case whether the vessels are strictly of new formation or not. Masses consisting essentially of bloodvessels may arise through dilatation of pre-existent vessels, with atrophy of the tissues that normally lie between them. This is the origin of the angiomata of the liver, and many of the angiomata of the skin (nævi) are explicable in the same manner. In the liver the capillaries of the lobules become dilated and their walls thickened, the parenchymatous cells between them disappearing by atrophy, and, as the capillary walls come in contact and exert mutual pressure, they may undergo atrophy, permitting a communication between their lumina, so that a spongy mass of tissue results, with large cavities filled with blood (Fig. 339). Such "cavernous angiomata" hardly constitute tumors in the restricted sense in which that term has been used hitherto. They are rather ectatic states of the vessels normally present in the parts where they are found.

Somewhat more akin to the true tumors are the masses which arise through elongation and widening of the vessels of a part (aneurisma racemosa), for in this case there is a real reproduction or growth of the vessels.

Angiosarcomata are tumors in which a new formation of blood-vessels with a sarcomatous adventitia springs from connective tissue either in the general fibrous structures of the body or the interstitial tissue of the viscera. Sections of these tumors sometimes reveal thin-walled vessels with a distinct, broad zone of sarcomatous tissue around them, resembling an enormously thickened adventitia of embryonic tissue (Fig. 322). In other cases the embryonic tissue that represents the adventitia of the separate vessels is fused into a mass of sarcomatous tissue lying between the vessels. The tumor

Fig. 339.

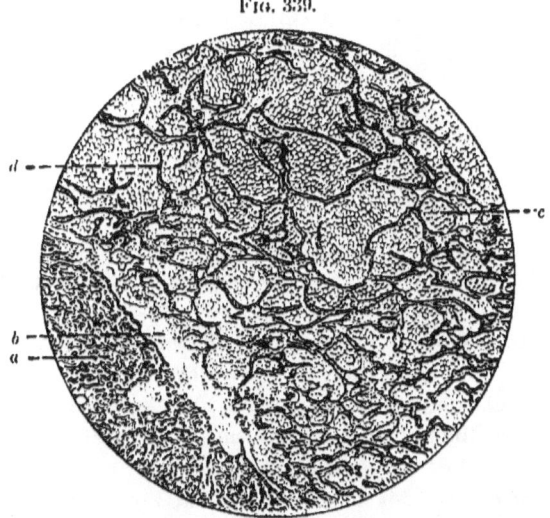

Cavernous haemangioma of the liver: *a*, substance of the liver; *b*, fibrous capsule formed at the margin of the angioma, probably the result of a chronic productive inflammation; *c*, space filled with blood; *d*, atrophic wall between two of the spaces of the angioma.

is then similar in structure to an ordinary sarcoma, in which the vessels are more abundant, perhaps, than is usual.

When the angiomata have been removed by operation the vessels are usually emptied by the pressure that has been exerted upon their tissues by the operative manipulations. This condition often gives rise to puzzling appearances, when the endothelial cells of the vascular walls are swollen or richer in cytoplasm than normal adult endothelium. Sections of the tumor then look like sections through a gland. The true nature of the tubules can generally be determined by the appearance of the lumina, which in the collapsed vessels is not circular, while in the glands it is nearly so if the section be transverse to the direction of the tube. In glandular tubules the epithelial cells are usually well-defined and clearly distinguishable from each other. This is not apt to be the case in immature endothelium.

2. **Lymphangioma.**—What has already been said with respect to the hæmangiomata applies to the lymphangiomata. Many of these tumors appear to be the result of a dilatation of the lymphatic vessels normally present in the tissues; but cases may arise in which there is a real reproduction of those vessels. The spaces in the tumor are either empty and collapsed, or they contain lymph and not blood. The walls of the vessels are frequently thickened by the production of fibrous tissue around them.

## IV. THE EPITHELIAL TUMORS.

The epithelium, which by its proliferation gives rise to tumors, may be situated either within a glandular structure of the body or upon one of its free surfaces, such as the skin or a mucous membrane. The tumors which result are not wholly composed of epithelium. There is always a development of the connective tissue of the part, furnishing a vascularized nutrient substratum for the epithelium. The epithelium of glandular organs may give rise to two sorts of tumors, the adenomata and the carcinomata. The stratified epithelium of the skin and some of the mucous membranes proliferate to form the epitheliomata.

1. **Adenoma.**—In this form of epithelial tumor there is a more or less perfect adherence to the structure of a normal gland. When adenomata spring from the epithelium of tubular or acinous glands the lobules of the tumor are composed of tubes or acini with a distinct lining of epithelium enclosing their lumina (Fig. 340). But there is almost always some departure from the typical structure of a gland; the lobules may be of unequal size in a more marked degree than is usual, the character of the epithelial lining may be abnormal, or the distribution and arrangement of the lobules may betray an abnormal tendency on the part of the growth. The latter feature is exemplified in the adenomata of the rectum, in which the new-formed glandular structure is apt to penetrate the muscularis mucosæ and develop abundantly in the submucous coat or even in the deeper, muscular tissues of that part of the intestine.

The adenomata of the breast deserve a rather close study. A perfectly simple adenoma of this gland appears to be a very rare growth. There is nearly always an association with diffuse fibroma, forming an adenofibroma. These are often cystic, an accumulation of a serous fluid in the acini causing their dilatation (cystic adenofibroma) (Fig. 341). In other cases the fibromatous tissue grows

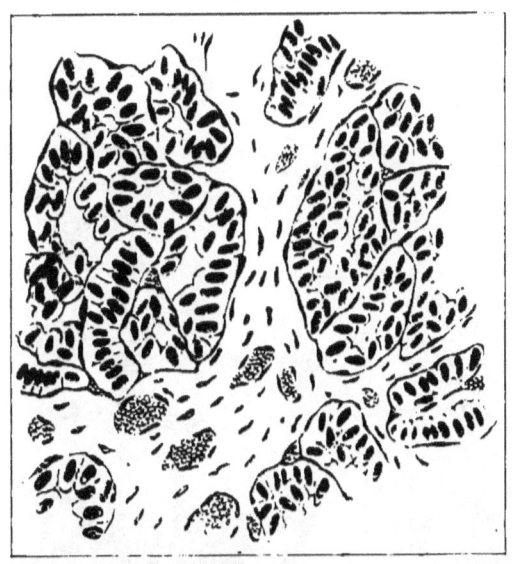

Fig. 340.

Adenoma of the pancreas. (Cesaris-Demel.) The atypical nature of the growth is revealed by the character of the epithelial cells, their arrangement within the alveoli, and the disposition of the latter with respect to each other and the interstitial tissue.

into the acini, which are enlarged to receive these ingrowths from their walls. The ingrowing masses of fibrous tissue are covered with epithelium like that lining the rest of the acinus, a fact which would be expected when we reflect that the ingrowth is a sort of intrusion of the wall of the acinus itself. Sometimes these ingrowths have a papillomatous character, but more frequently they have a globular form and give off globular branches within the acinus. Sections of such growths often have a complicated appearance. Irregular and branching bands of epithelium are seen coursing through a mass of fibrous tissue. They are the epithelial linings of the acini which have been brought into contact by the ingrowths of fibrous tissue, obliterating the lumina of the acini.

Part of this epithelium is, therefore, that which may be said to line the dilated acini; the rest, that which covers the fibrous tissue which has grown into the acini and caused contact of the epithelial layers with obliteration of the lumina. Where the pedicles of these ingrowths are small, sections may contain rings of epithelium surrounding an isolated mass of fibrous tissue if the section does not include the pedicle of that particular ingrowth (Fig. 342).

FIG. 341.

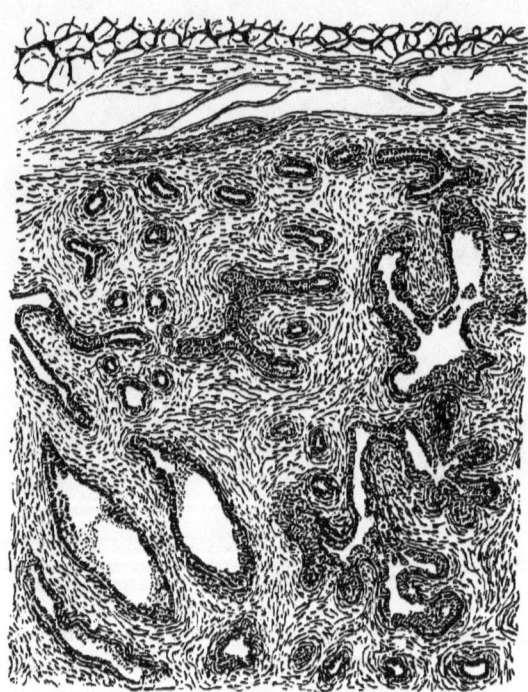

Adenofibroma of the breast. (Birch-Hirschfeld.) The section shows a tendency toward cystic dilatation of the glandular acini.

If the tumor is examined macroscopically, the ingrowths may often be lifted from the acini in which they lie. These tumors have received the name "intracanalicular adenofibroma." They must be carefully distinguished from the scirrhous carcinomata of the breast, which, upon superficial examination, they somewhat resemble.

In examining sections of the breast with a view to determining

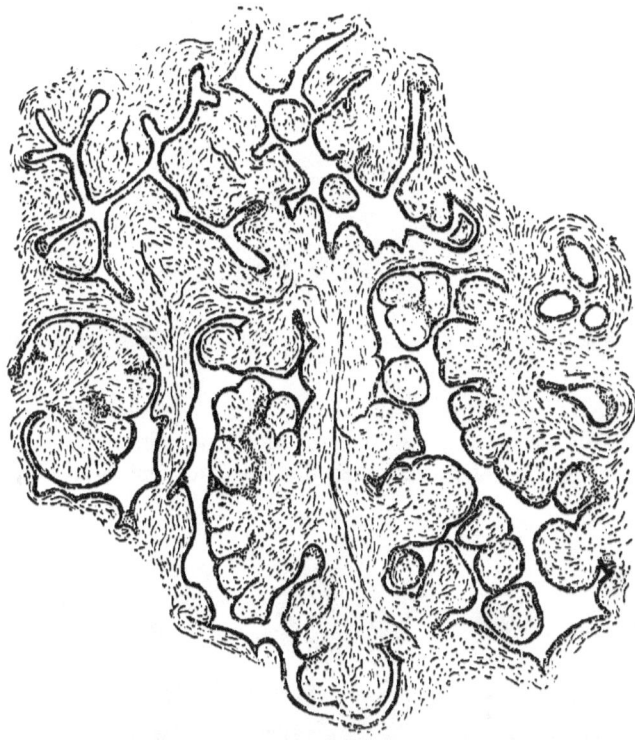

Intracanalicular adenofibroma of the breast. (Kaufmann.) In this example the lumina of the acini have not been obliterated, and a correct interpretation of the appearances presents no difficulty.

Section from the mammary gland of a nullipara, aged eighteen; moderately magnified. (Altmann.)

the question of the existence of a tumor the normal variations in that organ must be carefully considered. In the description of the normal mammary gland it was stated that the microscopical structure differed greatly according to the functional activity of the

FIG. 344.

Section from the mammary gland of a nullipara, aged eighteen; more highly magnified. (Altmann.)

organ. It is proper to recur to those differences in this connection because of the importance of many of the mammary tumors, that gland being one of the common sites of carcinoma and adenoma.

FIG. 345.

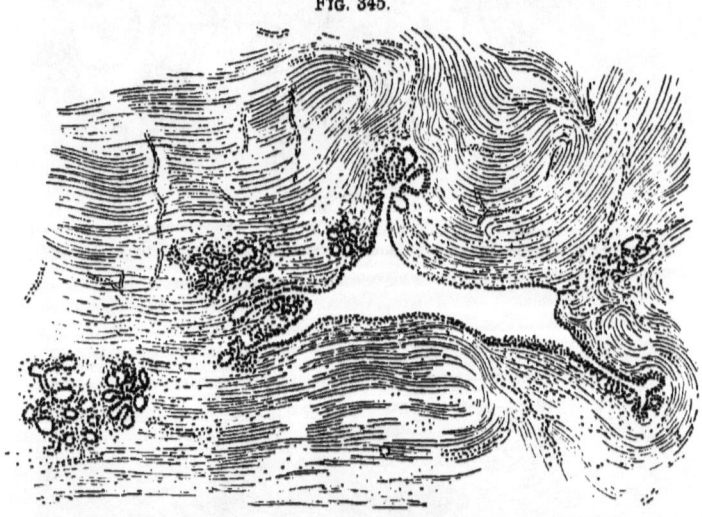

Section from the mammary gland of a nullipara, aged twenty-two; slightly magnified. (Altmann.)

In Figs. 343 to 350 sections of the gland in various stages of development and involution are represented. Figs. 343 to 346 represent sections from the breasts of nulliparæ, aged respectively eighteen and twenty-two years. The parenchyma of the gland has a general tubular structure, the acini being in an undeveloped state.

Figs. 347 and 348 show sections of the mammary gland of a

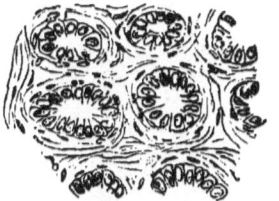

Section from the mammary gland of a nullipara, aged twenty-two; more highly magnified. (Altmann.)

woman, aged thirty-eight, who had born five children. The sections were taken at the beginning of functional activity of the gland.

Figs. 349 and 350 represent involuted mammary glands, respec-

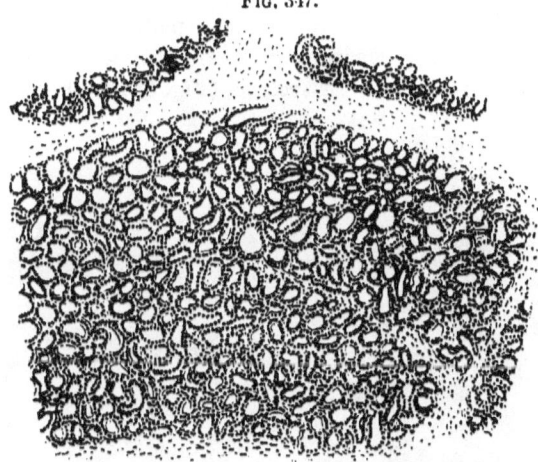

Section of the mammary gland at the beginning of lactation; moderately magnified. (Altmann.)

tively nine and fourteen months after functional activity had been arrested.

Adenomata are usually of benign character; but, as is the case with all neoplasms, it will not do to conclude that a growth is harmless merely because it can be included in a group of tumors that are usually benign. The evidence as to its tendencies revealed by the structure of each individual tumor must be carefully weighed before a conclusion as to its benignancy or malignancy is reached. Adenomata are benign in proportion as they adhere to the structure of a normal gland of the type which they simulate. They approach

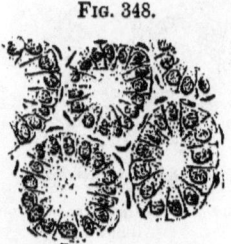

FIG. 348.

Section of the mammary gland at the beginning of lactation; more highly magnified. (Altmann.)

malignancy when they become atypical and show a tendency to infiltrate their surroundings. The adenomata of the rectum, already referred to, are likely to prove malignant, and in their structure they show a departure from the simple type of tubular gland normally present in the rectum (Fig. 351). They also dis-

FIG. 349.

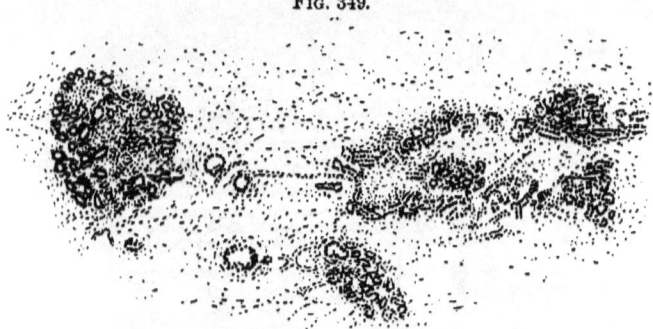

Section of the mammary gland in a state of involution. (Altmann.) From a woman, aged twenty-five, nine months after the cessation of functional activity.

play a marked tendency to infiltrate their surroundings. While they belong to a group of generally benign tumors, they possess an atypical structure and a power of infiltration that reveal their malignant character.

2. **Carcinoma.**—The epithelium of developing secreting glands

first appears as little solid columns of epithelial cells, which spring from the epithelium covering the part and penetrate the underlying

Fig. 350.

Section of mammary gland in a state of involution. (Altmann.) From a woman, aged thirty-two, fourteen months after functional activity had ceased.

tissues (see Fig. 181). These columns subsequently become hollowed to form tubes or sacs lined with secreting epithelium. In carci-

Fig. 351.

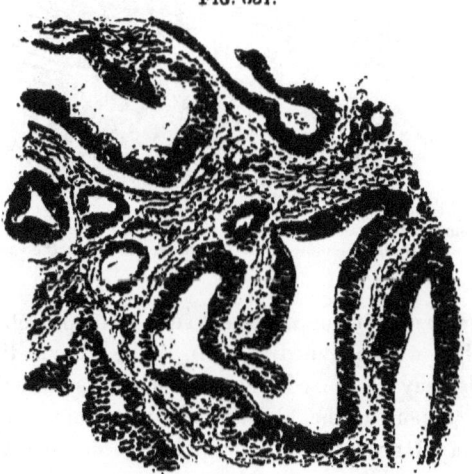

Infiltrating adenoma of the rectum. (Hansemann.) The figure represents alveoli of atypical character, differing greatly from the normal glandular structures of that part of the body. The section does not include the infiltrating portion of the growth.

nomata the embryonic state of gland-formation is simulated by the growth, so that a carcinoma may be considered as formed upon the type of a developing gland in the same sense as a sarcoma is analogous to developing connective tissue.

As a result of this structure, sections of carcinomata appear to be composed of alveoli, which are filled with epithelial cells and have walls of fibrous tissue. The character of the epithelium depends chiefly upon the variety from which the tumor sprang. The sizes of the alveoli and the amount of fibrous tissue that separates them from each other vary in different tumors, and the carcinomata are divided into rather ill-defined groups, according to the relative abundance of the epithelium they contain as compared with the amount of fibrous tissue They are also subdivided according to the character of the epithelium.

*a*. MEDULLARY CARCINOMATA (Fig. 352) are those in which

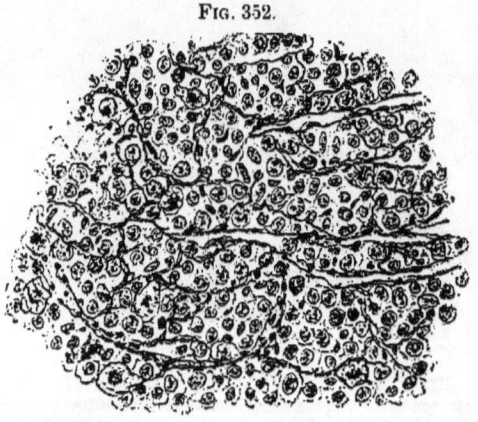

FIG. 352.

Medullary carcinoma of the mammary gland. (Hansemann.) The stroma of the tumor is here reduced to a minimal amount of areolar tissue containing the vascular supply of the growth.

there is the least amount of fibrous tissue. The alveoli are usually large and filled with polyhedral cells. The fibrous tissue of the alveolar walls may be so reduced in amount as virtually to serve merely as a support to the bloodvessels it contains. Such tumors are soft, of rapid growth, and very prone to degenerative changes and metastasis.

*b*. SIMPLE CARCINOMATA contain about an equal amount of epithelial and fibrous tissues (Fig. 353).

*c*. SCIRRHOUS CARCINOMATA (Fig. 354) are characterized by small alveoli separated by large quantities of dense fibrous tissue.

The latter may so greatly preponderate over the epithelium that there is a possibility of mistaking the tumor for a simple fibroma.

Fig. 353.

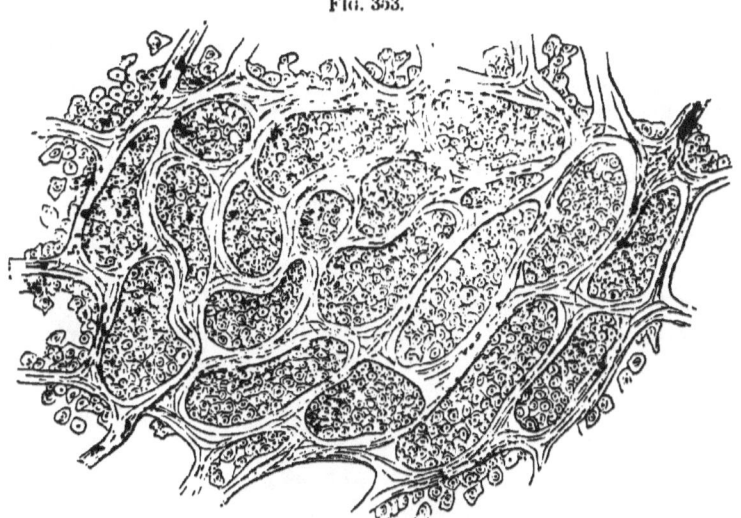

Carcinoma simplex mammæ. (Kaufmann.) In this growth the stroma is well developed and divides the tumor into a number of intercommunicating alveoli, filled with epithelial cells.

Care must be taken not to confound these carcinomata with the intracanalicular adenofibromata already described. In the carcinoma

Fig. 354.

Scirrhous carcinoma of the breast. (Ribbert.) The bulk of the section is composed of dense fibrous tissue, in which there are a few rows of epithelial cells, a.

there is no ingrowth of fibrous tissue into the alveoli, as in the case of the adenofibroma. The development of the fibrous tissue in

25

these cancers is probably induced by the proliferation of the epithelium, but it sometimes happens that the fibrous tissue forming the stroma of the tumor compresses the epithelium after the growth has attained a certain stage of maturity, and causes an atrophy of its cells (atrophying carcinoma). As a result the tumor may suffer a diminution in size, but this shrinkage occurs only in the older parts of the tumor; the peripheral portions continue to grow. It is no indication of a spontaneous cure.

Carcinomata are malignant, but differ in the rapidity of their clinical course. Those which are softer—*i. e.*, contain a larger proportion of epithelium—are of more rapid growth than the harder varieties; but they all tend to infiltrate their surroundings and are liable to metastasis. The usual mode of infiltration is for the proliferating epithelium to penetrate the lymph-spaces or lymphatic vessels of the neighboring tissues. The cells may advance as solid

Fig. 355.

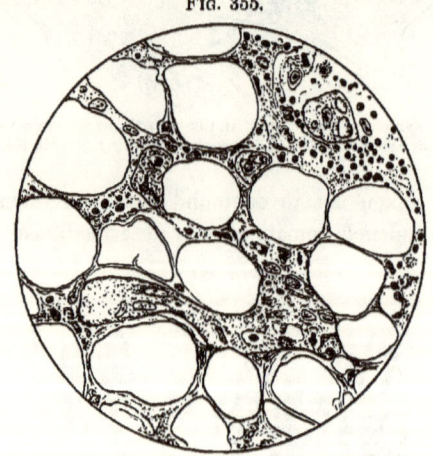

Carcinoma invading adipose tissue. The figure represents a section of the fat surrounding the breast in a case of mammary carcinoma. Masses of epithelium are present in the lymphatic spaces of the areolar tissue between the fat-cells. The nuclei of some of the epithelial cells show imperfectly preserved karyokinetic figures. To the right, above, is a group of four epithelial cells surrounded by a round-cell (inflammatory) infiltration.

columns pushed out from the growth along these lymph-channels, or cells may become detached from the main growth and be carried by the lymph-current for a greater or less distance from the original tumor, to find lodgement in some situation in which the conditions may be favorable for their continued multiplication

(Fig. 355). The connective tissue of the new site is then induced to proliferate and form the cancerous stroma. If this transfer of cells is only for a short distance, the process is called infiltration; if the distance is greater, metastasis. It appears, then, that metastasis usually occurs through the lymphatics, as it is through them that the natural extension of the carcinoma takes place. The cells that gain entrance to the lymphatic vessels are most likely to be arrested in the nearest lymph-node, giving rise, if they multiply,

Fig. 356.

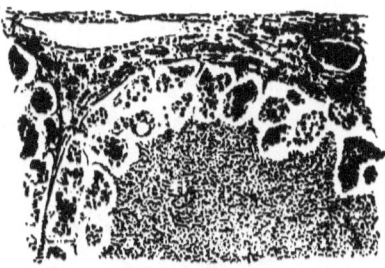

Secondary carcinoma of a lymph-gland. (Ribbert.) Epithelial cells from the primary carcinoma have been carried by the lymph-current to the node, where they have been arrested in the lymph-sinus. Here they have continued to proliferate, giving origin to a secondary, or metastatic, nodule of carcinoma.

to a secondary tumor within it (Fig. 356). These secondary tumors in the lymph-nodes may, after a period of development, furnish cells for a still wider metastasis.

Metastasis through the lymphatics is not the only means by which carcinomata may become generalized. They may infiltrate the walls of bloodvessels, usually veins, and finally discharge cells into the blood, giving rise to cancerous embolism with a general diffusion of secondary nodules in the first capillary district through which the blood is distributed. In this way multiple carcinomata of the liver or lung are produced. The secondary carcinomatous nodules usually resemble the primary tumor, especially as regards the character of the epithelium; but the relative amount of stroma is very frequently considerably less. A scirrhous carcinoma may give rise to secondary nodules of medullary carcinoma. The distinction between the different varieties is, therefore, more descriptive than essential.

Carcinoma is apt to occasion the development of a cachexia in the patient. The reason for this is probably to be sought in the absorption of the products of metabolism from the tumor, rather than in the abstraction of nourishment from the organism. Epithelium, especially of the glandular form, is a tissue of great chemical activity, and in carcinomata there is no special outlet for the products of that activity, such as is furnished by the ducts of normal glands. It may, therefore, be reasonable to infer that the products resulting from the chemical activities of the epithelial cells must be absorbed into the system, and that they may injuriously affect the nutrition and the functions of distant organs. Carcinomata are also liable to undergo degenerations, the products of which may be deleterious to the organism.

A form of carcinoma which differs somewhat in appearance from those that have been mentioned, though it is of essentially the same nature, is the "colloid carcinoma" (Fig. 357). This variety springs

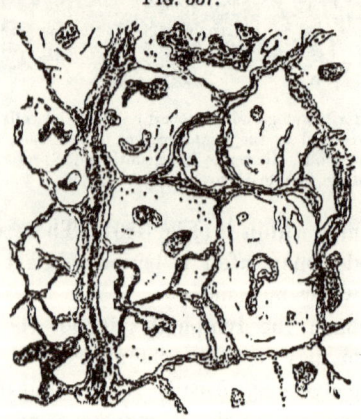

FIG. 357.

Colloid carcinoma. (Ribbert.) The section represents a delicate stroma of areolar tissue separating alveoli, which are not filled with cells, but contain the products of their mucous degeneration and a few cells which have not yet undergone complete destruction.

from epithelium that under normal conditions secretes mucus. This function renders the cells of the cancer particularly liable to mucoid degeneration, and this may be so extensive as to destroy all or nearly all of the cells in some of the alveoli of the tumor, converting them into a soft mucous mass that usually does not appear quite uniform under the microscope. The epithelial cells are generally of columnar form, arranged at the periphery of the alveoli, with their ends in contact with the alveolar wall. This arrange-

Fig. 358.

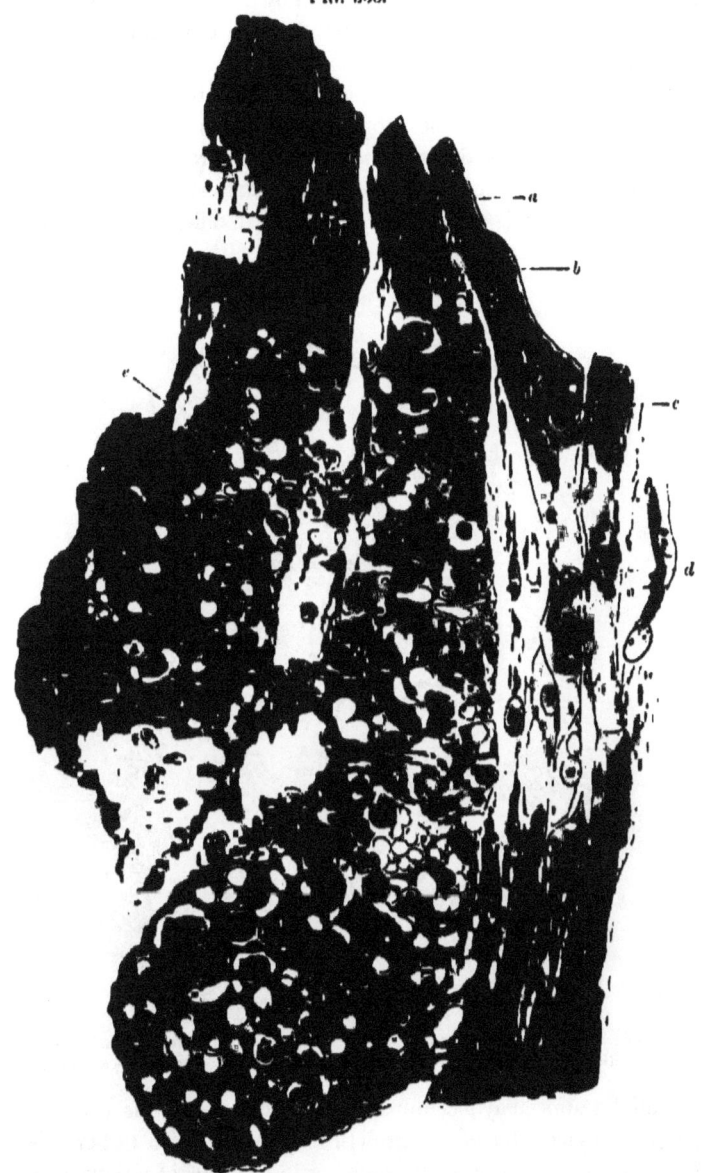

Adenocarcinoma of the liver. (v. Henkelon.) *a*, normal liver-cell; *b*, modified epithelial cell entering into the formation of the neoplasm; *c*, normal nucleus; *d*, nucleus abnormally rich in chromatin preparatory to cell-division; *e*, fat-globule in the epithelium of the tumor, showing a tendency to fatty degeneration.

ment of the cells is often strikingly shown in secondary tumors of the lung, in which the cells have appropriated the pulmonary alveoli for their stroma.

It occasionally happens that the connective tissue that forms the stroma of a carcinoma does not progress in its development to the formation of fibrous tissue, but assumes a sarcomatous character. Such tumors are called "carcinoma sarcomatosum." A more frequent association is one of carcinoma with adenoma, "adenocarcinoma" (Fig. 358). In these neoplasms, either the two forms of

FIG. 359.

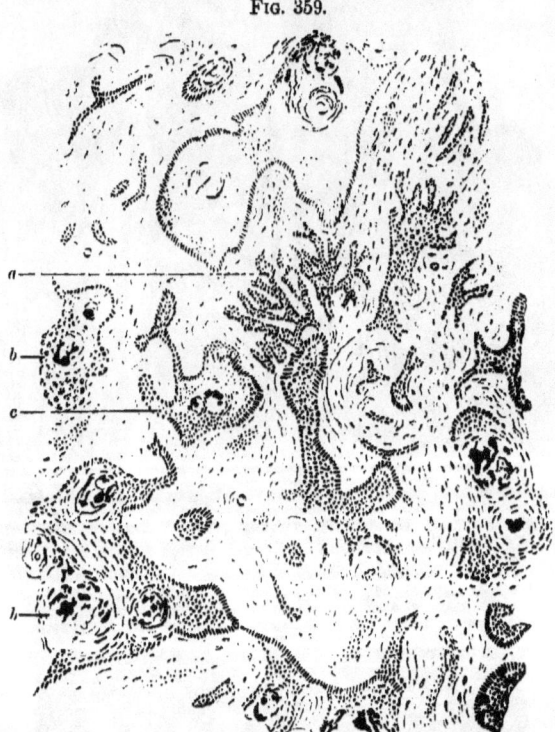

Epithelioma of the cheek. (Ernst.) *a*, delicate tongues of epithelium extending into the lymphatics of the part; *b, c*, larger masses of epithelium containing pearl-bodies.

epithelial tumor may occupy different portions of the growth, or the general character of the growth may be that of a rather typical carcinoma—*i. e.*, a carcinoma showing indications of a development beyond the undifferentiated state analogous to an embryonic gland —or that of a rather atypical adenoma.

3. **Epithelioma.**—This tumor is essentially a carcinoma springing from stratified epithelium. Under normal circumstances the cells of this variety of epithelium multiply in its deeper layers and are gradually pushed toward the surface while they mature. Epitheliomata are produced when the proliferating cells penetrate the underlying tissues in columns, which ramify through those tissues and ultimately appear as the contents of well-defined alveoli surrounded by a fibrous-tissue stroma similar to that present in carcinomata (Fig. 359). The epithelium retains its general characters: the cells at the periphery of the alveoli multiply, and either further

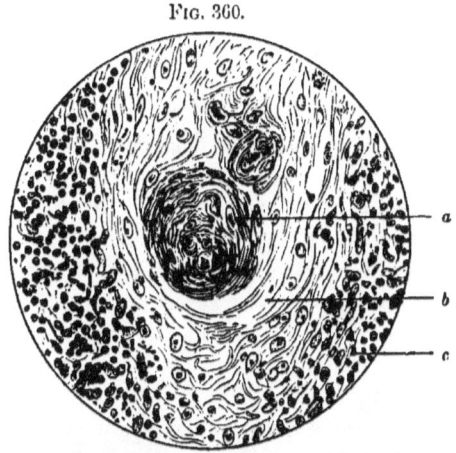

Fig. 360.

Epithelial pearl-body from an epithelioma of the lip: *a*, pearl-body; *b*, surrounding epithelium, forming one of the epitheliomatous tongues or columns; *c*, round-cell infiltration of the contiguous fibrous tissue.

infiltrate the surrounding tissues or crowd each other toward the centres of the alveoli as they increase in number and size. Here they eventually undergo keratoid transformation, just as they would upon the surface of the normal epithelium; only here they are crowded toward the centres of the alveoli, where the horny scales become imbricated to form globular masses, called epithelial "pearl-bodies" (Fig. 360). The epitheliomata may penetrate into the lymphatics and be subject to metastasis in a manner entirely comparable to that already described above. They are, therefore, malignant, though of slower growth than the medullary or simple carcinomata, at least during the early stages of their development.

It should always be borne in mind, when considering the prognosis in a case of carcinoma or epithelioma, that metastasis may take place while the primary growth is still of very small size, even before attention has been called to the existence of a tumor. An examination of the peripheral portion of the growth will often throw considerable light upon the probability that this has occurred, by revealing an extension of epithelial cells into the lymphatics of the surrounding tissues. Cases of speedy recurrence of such a growth after operation are really cases in which tissues that have thus been infiltrated have not been completely removed.

Much has been written within late years advocating the theory of a parasitic causation of carcinomata and epitheliomata. The appearances which have led to this belief are probably due to degenerative or morbid processes within the epithelial cells of the tumor, and not to the presence of parasites; but further study of this subject may show that parasites have the power of causing rejuvenescence of cells and an emancipation from the ordinary restraints that regulate their development.

4. **Cystoma.**—Attention has been called to the cystic adenomata of the mamma. Similar cysts may occur in other regions through dilatation of cavities normally present in the tissues by some fluid, usually of a serous character. It is best to exclude cystic growths in which the cystic character is evidently a secondary feature of the tumor, or where a cyst arises from the retention of a secretion or is due to the accumulation of a fluid in a normal cavity, from the group of tumors that are essentially cystic. Thus, for example, simple hydrops folliculorum of the ovary should not be classed with the cystic tumors of that organ.

The ovary is the favorite site for cystic tumors of new formation, which may contain only a single cavity (unilocular) or several cavities (multilocular). Histologically, they may be grouped in three divisions: 1, simple, in which the walls of the cyst are smooth and covered with epithelium; 2, papillary, in which there are ingrowths from the walls of the cysts into their cavities, either simple or branching (Fig. 361); and, 3, dermoid, which contain structures simulating the normal skin: hair, imperfectly developed teeth, or other highly differentiated tissues, such as bone, etc. In the first two forms the fluid in the cystic cavities may be serous, mucoid, or colloid; frequently it is different in the various cavities of the tumor. In dermoid cysts there is often a greasy substance, similar to the sebum of the skin, derived from sebaceous glands in the cutaneous struct-

FIG. 361.

Section from a papillary cystoma of the ovary. (Birch-Hirschfeld.) Part of the wall separating two cystic cavities is represented. From this wall, papillary ingrowths arise, which project into the cavity of the cyst. They are composed of a delicate areolar tissue covered with columnar epithelium similar to that lining the cysts.

ures of the growth. Similar dermoid cysts occasionally develop from the skin, but are usually lined with merely an epidermis, the

FIG. 362.

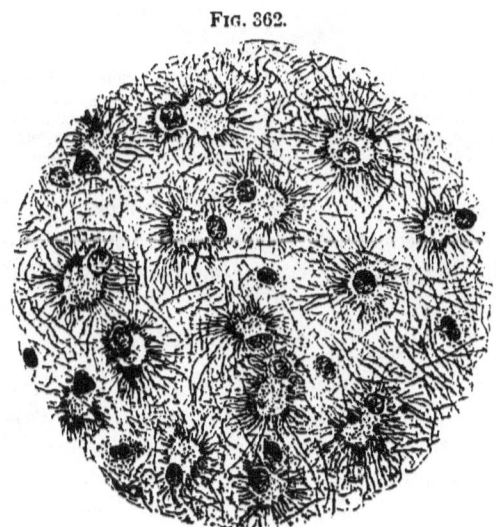

Gliomata of the brain. (Stroebe.) Composed of glia-cells of small type, with fine processes.

scales from which accumulate in the cavity of the tumor, where they may be mixed with sebum (wens).

5. **Glioma.**—The neuroglia, originally of epithelial origin from

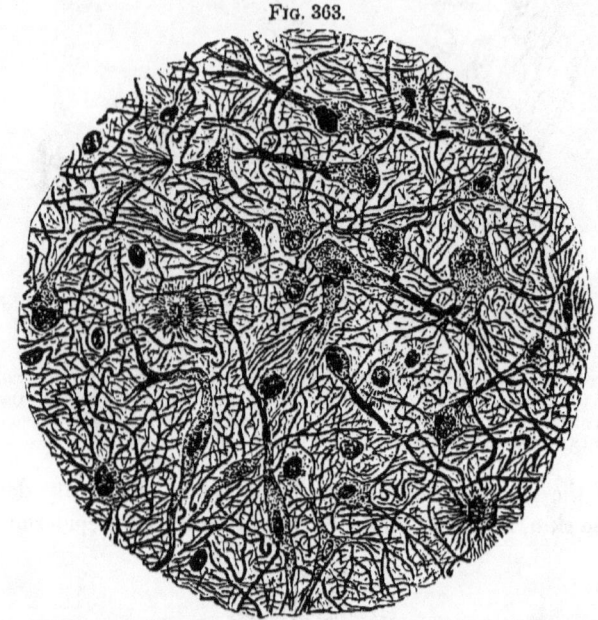

Fig. 363.

Gliomata of the brain. (Stroebe.) Mixed type, containing cells like those in Fig. 362, but also large branching cells simulating ganglion-cells, "glioma gangliocellulare." In sections of gliomata stained by the methods in more general use the delicate processes are often not visible, but the nuclei are prominent. The tumor, therefore, appears highly cellular with a finely granular material (the unstained processes) between the cells.

the ectoderm, may proliferate to form tumors, called gliomata. These differ in their structure according to the variations in type presented by the glia-cells composing them (Figs. 362 and 363).

## V. PAPILLOMATA.

Before leaving the subject of tumors it will be necessary to devote a few words to the consideration of growths that cannot be considered as primarily arising from either epithelium or connective tissues. The papillomata are examples of such growths. These are over-developments of papillary structures normally present, or spring from mucous surfaces where such structures are normally either not present or are but poorly developed.

A papilloma consists of vascularized fibrous or areolar tissue springing from a surface which is covered with epithelium. The denser forms which occur—*e. g.*, upon the skin—constitute "warts"; but much more delicate papillomata may spring from mucous membranes, such as that of the bladder, and are then known as villous tumors or villous papillomata. In many cases the denser forms of papilloma appear to be hypertrophies due to irritation. But papillomata which seem to be true neoplasms or tumors in the restricted sense of that term hitherto employed appear to be among the possibilities of morbid development.

# PART III.
## HISTOLOGICAL TECHNIQUE.

### CHAPTER XXVI.

PRACTICAL SUGGESTIONS FOR THE CARE AND USE OF THE MICROSCOPE.—MICROSCOPICAL TECHNIQUE.

IN selecting a microscope the following considerations are of importance:

The stand should be supported on three points and rest firmly on the table; have a rack-and-pinion coarse adjustment, and a fine adjustment free from all loss of motion. It is rarely used in an inclined position, and a jointed stand is unnecessary. A triple nose-piece, or revolver, is a great convenience, and an Abbe condenser with iris-diaphragm is almost indispensable.

Three objectives are needed: a Leitz No. 3 or No. 4, No. 7, and $\frac{1}{10}$th or $\frac{1}{12}$th oil immersion, or their equivalents of other manufacture, are suitable powers for general use. Two oculars, No. 2 and No. 4, will answer.

The microscope should be protected from direct sunlight and acid fumes, and be kept in a dry, moderately cool place. When not in use it should be covered or placed in its case, to protect it from dust. If the lenses become dirty, they may be wiped with a soft, clean cloth or Japanese paper, either dry or moistened with water, and followed by a dry cloth or paper. Balsam or cedar oil may be removed with a cloth or soft paper moistened with xylol, after which the parts should be carefully wiped dry.

In making microchemical tests special care should be taken not to let the reagents come in contact with the objectives.

Objects should always be examined in a liquid, unless there is some special reason for examining them in a dry state; and should be covered with a cover-glass, unless a cursory inspection with a very low power is all that is required.

In studying a specimen always use the lowest power that will reveal the structures it is desired to see; and, in any event, use a low power first, to get a general idea of the topography of the specimen. In this way the portions for more minute study can be readily selected, with a great saving of time.

The proper illumination of the specimen is just as important as careful focussing. If the Abbe condenser is in use, employ the plane surface of the mirror during the day; either the plane or the concave surface when artificial light is used, selecting the surface which causes less glare. The iris-diaphragm should be kept adjusted so as to give the best definition of the specimen under examination when the latter is in focus. It will be found that when colorless objects are examined a small opening gives the clearest picture, while with colored objects a larger opening is preferable. A small diaphragm serves to bring out the "structure-picture"; a large diaphragm, the "color-picture" (see p. 402).

A bottle of oil of cedar-wood, having approximately the same refractive index as the glass from which the cover-glasses are made, is furnished with the immersion-objectives. When these are used a drop of this oil is placed on the cover, and the end of the objective immersed in this drop. This arrangement permits the light to pass from the object to the bottom lens of the objective without sensible refraction, increasing the amount of light entering the objective, the sharpness of definition, and the purity of the color-picture. When the lens has been used the oil should be removed with a soft cloth or Japanese paper. The oil on the cover may be wiped off at once, or it may be allowed to dry and then removed with a cloth moistened with xylol.

**Microscopical Measurements.**—These may be made, with a fair degree of accuracy, by means of an eye-piece micrometer-scale. This is a ruled disc of glass that can be placed upon the diaphragm within the ocular, where its scale should be well defined when seen through the upper lens of the eye-piece. Special micrometer oculars are made which permit of focussing the scale, but these are unnecessary if the diaphragms of the ordinary oculars are in the right places within the eye-piece tubes. The value of the divisions of the eye-piece micrometer-scale must be determined by comparing it with the scale of a micrometer-slide which is placed upon the stage of the microscope. These scales usually consist of 1 mm. divided into hundredths, and the eye-piece scale will have different values for each combination of lenses used and for every variation in the length of the microscope-tube. The unit for microscopical measurements is one-thousandth of a millimeter, or one-millionth of a meter; it is called a "micrometer," and is designated by the Greek letter $\mu$. One division of the micrometer-slide mentioned above would, therefore, equal 10 $\mu$. From these data it is possible to calculate the value of each division of the eye-piece micrometer-scale in terms of $\mu$ for each combination of lenses, the length of the microscope-tube being fixed. (Most Continental stands and many American stands have graduated tubes, and the objectives are constructed for a standard tube-length of 160 millimeters.)

It is well for the student to get into the habit of estimating the sizes of the objects he examines. A good standard for mental comparison is the diameter of the unaltered red blood-corpuscle, which is about 7.5 $\mu$.

## MICROSCOPICAL TECHNIQUE.

Useful preparations for study under the microscope may be prepared from tissues in one of three ways: 1, simple scrapings of the tissues may be mounted on a slide in the fluids derived from the tissues themselves, or in a neutral solution—*e. g.*, 0.75 per cent. salt solution; 2, the tissue-elements, cells, and intercellular fibres, etc., may be separated from each other by treatment with some macerating-fluid—*e. g.*, very weak chromic acid (1 : 10,000), 36 per cent. caustic potash, ⅓ alcohol; 3, sections of the tissue may be prepared either while they are fresh, with a razor or a freezing-microtome, or after hardening.

The first method has a limited application. It is serviceable only when the tissue-elements are so loosely held together that they readily separate from each other and can be examined in an isolated condition. This is the case with a considerable number of tumors, the superficial tissues of mucous membranes, the spleen, etc. If the inside of the cheek be scraped with the finger-nail, and the material thus removed be diluted with saliva, placed upon a slide, and covered with a cover-glass, the squamous epithelial cells lining the cavity of the mouth will be readily seen in an isolated state. An appropriate dye may now be introduced under the cover, and by its means the nuclei of the cells stained, thus bringing them into clearer view.

When a simple scraping of the natural or freshly cut surface does not yield useful preparations, showing isolated tissue-elements, some process of maceration may be employed. Bits of the tissue are soaked for a time in some solution that serves to soften the cement-substances lying between the elements of the tissues, so that they may be easily separated with needles (teasing). Such specimens are usually best examined when mounted on a slide in some of the macerating-fluid. Many of the macerating-solutions not only favor the separation of the constituents of tissues, but also preserve them, so that a fair idea of their natural size and shape may be obtained from such preparations. It is evident, however, that with this method very little can be learned of their arrangement in the tissues before they were separated, and a knowledge of that arrangement is often of greater importance in the determination of the character of the tissue than a knowledge of the exact shape and size of the tissue-elements.

The third method, that of preparing sections of the tissues, is the one most commonly employed, because it yields the most useful results. The structural elements composing the tissues are seen in their natural relative positions, and can be distinguished from each other and identified by the use of dyes and other reagents that affect them in some characteristic manner. But in order to obtain useful sections the tissues must almost always undergo some preliminary treatment with reagents, to give them a proper consistency for cutting and to hold the tissue-elements together so that the sections shall not fall apart after they have been cut. This may be accomplished by freezing the tissue before cutting it; but more satisfactory results are obtained by causing a coagulation of the albuminous substances and subsequently extracting some or all of the water contained in the tissues. These changes in the tissues give them a firmness which favors the preparation of very thin sections; but sometimes even they are inadequate, and then the tissues are usually impregnated with some substance, like paraffin or collodion, which fills the interstices of the tissues and can then be hardened, when it serves to hold the tissue-elements together and retain them in their natural positions. The paraffin or hardened collodion is cut with the tissues and keeps the sections from disintegrating. Before mounting the section, the substance used for impregnation may be removed from the section, or it may be retained permanently, since it is usually easily recognized in the specimen and does not interfere with its study under the microscope.

The study of tissues by means of sections has the disadvantage that the elements of the tissues are cut, and the sections contain the resulting portions as well as complete elements. The incomplete portions lie near and at the surfaces of the sections, where they are in clearest view, while the uncut elements are situated in the body of the section, more or less obscured by the overlying portions that have been cut by the knife. Moreover, the tissue-elements may lie obliquely to the plane of the section, so that only a portion of them can be seen at a time, the rest being brought into clear view only when the focal plane is raised or lowered. These circumstances and the fact that the tissue-elements are frequently closely crowded together make the interpretation of sections a matter of some difficulty in many cases. These difficulties are in a measure overcome by examining sections of different thicknesses, but a more satisfactory way of studying the structure of a tissue is to examine portions after maceration as well as in section.

The processes of coagulation and dehydration, which have already been mentioned as usual preliminaries to the cutting of sections, deserve a few words in explanation of their purposes.

The coagulation of the albuminous substances in the tissues has for its chief aim the preservation of the minute structure of the tissue-elements, so that a lapse of time or the subsequent manipulations of the tissues shall not cause an alteration in the details which it is desired to study. If this precaution be omitted, the tissues undergo post-mortem changes which seriously alter the appearance of the elements of which they are composed. Coagulation brought about for this purpose is called "fixation" of the tissues. It may be induced in a variety of ways: the tissues may be subjected to heat for a few moments, thus rendering the albumins they contain both solid and insoluble; but the more usual procedure is to immerse the tissues in a solution of some substance that causes rapid death with coagulation. These solutions are called fixing-solutions, and not infrequently the substances they contain not only cause death and coagulation, but also form a union with some of the structural materials of the tissues which may facilitate their subsequent recognition.

The number of formulæ that have been devised for the preparation of fixing-solutions is very great, and some of the solutions are better for the fixation of some tissues than for others. As a rule, those solutions that most perfectly preserve the finer intracellular details of structure have very little power of penetrating masses of tissue. They can, therefore, only be employed when very small bits of tissue are to be fixed. Other fixing-solutions penetrate much better, but fail to fix the most delicate structures, which may undergo changes before they are preserved. It follows that the choice of the method of fixation must in each case depend upon the object to be attained.

The removal of water from the fixed tissues is accomplished by means of alcohol. The fixing-agents are nearly all aqueous solutions, and while they increase the consistency of the tissues to a certain extent, they do not usually render them sufficiently firm for the preparation of thin and uniform sections. If the water in the tissues be replaced by alcohol, a greater and more uniform consistency is obtained, and the tissues are also partly prepared for impregnation with an embedding-material (collodion or paraffin) should that be necessary for section-cutting.

After sections of fixed tissues have been obtained they usually require staining before they can be profitably studied. The chief reason for this will appear in the following explanation:

When a specimen is examined under the microscope differences in structure among the colorless elements of the specimen may be seen, or differences in color between the different elements may be perceptible. We may, then, distinguish between a "structure-picture," due to differences that are not those of color, and a "color-picture," due solely to such differences. The manner in which the latter is produced is, perhaps, self-evident. The structure-picture is the result mainly of differences in refraction due to the various densities of different parts of the specimen. But the processes of fixation and hardening have for their purpose the rendering of the tissues of a relatively uniform density. They must, in consequence, tend to obliterate the details of the structure-picture which the sections yield when viewed under the microscope. For this reason the sections are stained, which converts the structure-picture into a color-picture.

The substances composing the tissues have various affinities for dyes, and it is possible to take advantage of this in staining sections, so that structures of the same nature shall receive one color, while those of different composition shall be dyed of a different line or an entirely different color. The coloring-matters, when so employed, not only bring out the structure of the tissue by creating a color-picture, but they also serve as valuable reagents in revealing the nature of the substances to which they impart a color. Again, it is often necessary that a certain method of fixation or other preliminary treatment should be used before the particular dye selected can display its greatest selective power for a particular substance. These facts explain the great number of formulae for stains and the preparation of specimens that are found in the technical text-books and journals. The subject has become so expanded within recent years that it has almost created a distinct branch of learning; but it will only be necessary for the student of medicine to acquire a knowledge of a few methods that will serve to reveal the general structure of cells and the characters of the intercellular substances. The general outline of the procedures in common use for this purpose are as follows: 1, fixation; 2, hardening; 3, impregnation; 4, embedding; 5, cutting; 6, staining; 7, dehydration; 8, clearing; 9, mounting.

Some methods of preparation combine one or more of these steps in a single manipulation, thus considerably reducing the time requisite for the completion of the process. Other methods necessitate the intercalation of still other manipulations, or the subdivision of those already enumerated.

### Methods of Fixation.

1. **Müller's Fluid.**—This classic fixing- and hardening-solution consists of potassium bichromate, 2.5 per cent., and sodium sulphate, 1 per cent., dissolved in water (preferably distilled water). It is slow in action, requiring from six to eight weeks for the preservation of an average specimen, but with proper care can be made to yield excellent results when the finer details of structure are not to be studied. It is important to use large quantities of the fluid, at least ten times the volume of the tissues immersed in it, and to renew the fluid so frequently that its strength shall be constantly maintained. When fresh tissues are placed in Müller's fluid they speedily render it cloudy. This is a sign that the fluid should be renewed, even if only an hour has elapsed since the tissues were placed in it. When cloudiness no longer appears the fluid should be renewed once a day for the first two weeks: after that, two or three times a week till the process is completed.

After fixation in Müller's fluid specimens should be washed in running water over night, or for twenty-four hours, and then hardened in alcohols of progressively greater strengths. While in the weaker alcohols specimens should be kept in the dark, to avoid the formation of precipitates, which occur under the influence of light. Pieces of tissue placed in Müller's fluid should not be more than 1 cm. in thickness.

Two excellent modifications of Müller's fluid have been devised by Zenker and Orth for the purpose of hastening the fixation and of securing a more faithful preservation of structural detail.

**2. Zenker's Fluid.—**

| Potassium bichromate, | 2.5 grams. |
|---|---|
| Sodium sulphate, | 1 gram. |
| Mercuric chloride, | 5 grams. |
| Distilled water, | 100 cc. |

To this stock solution 5 per cent. of glacial acetic acid is to be added just before use of the fluid.

Zenker's fluid fixes tissues in from three to twenty-four hours. The pieces should not be more than 5 mm. thick, and after fixation should be washed for several hours in running water and then hardened in alcohol.

This solution possesses the disadvantage that a precipitation of mercury or some mercurial compound is likely to take place within the tissues. This deposit may be, at least in great measure, removed from the tissues by adding a little tincture of iodine to the hardening-alcohols. The iodine combines with the mercury and produces a soluble compound, which is dissolved out by the alcohol. As the iodine disappears from the alcohol the latter becomes bleached, and fresh tincture must be added until the alcohol remains permanently tinged. If, after sections of the tissue have been prepared, they are found to contain a mercurial deposit, this can be removed by treatment with dilute iodine tincture or with Lugol's solution.

**3. Orth's Fluid.—**

| Potassium bichromate, | 2.5 grams. |
|---|---|
| Sodium sulphate, | 1 gram. |
| Distilled water, | 100 cc. |

This stock solution is Müller's fluid. Before use, 10 cc. of formaldehyde (40 per cent.) is to be added to every 100 cc. of the Müller's fluid.

Orth's fluid fixes in three or four days. The pieces of tissue should not be more than 1 cm. thick. The time for fixation can be shortened if smaller pieces are used and the process is carried on at a slightly elevated temperature; *e. g.*, in an incubator kept at 37° C. (98.6° F.). After fixation the specimens should be washed in running water, as in the previous methods.

4. **Mercuric Chloride Solution.**—A saturated solution of corrosive sublimate in 0.5 per cent. salt solution is prepared by heating an excess of sublimate crystals in the salt solution and allowing the mixture to cool. The clear fluid is decanted from the crystals when desired for use. The penetration and action of the solution are favored by the addition of 5 per cent. of glacial acetic acid at the time of using. The thickness of the pieces of tissue should not exceed 5 mm., and much thinner pieces are better. Fixation takes place within six hours, after which the tissues may be washed in running water, or placed at once in 70 per cent. alcohol. If acetic acid has been used, it is best to wash in water before immersing in alcohol. Tincture of iodine should be added to the alcohol for the reasons given in the description of Zenker's fluid.

5. **Formaldehyde.**—This gas is capable of being absorbed by water to form a 40 per cent. solution, but its volatility renders such a solution liable to deterioration. The strength employed for fixation is usually 4 per cent., and may be prepared by adding 10 cc. of 40 per cent. formaldehyde to 90 cc. of distilled water. A 0.75 per cent. solution of common salt may be substituted for the distilled water with possible advantage.

Formaldehyde penetrates deeply and quickly into the tissues, which may be 1 cm. in thickness, and accomplishes fixation within twenty-four hours, but the preservation of structural detail is not very perfect. The solution is useful where the general characters of the tissues are to be determined and the details of the cells are of comparatively little consequence. After fixation the tissues may be washed in water, or placed directly in 70 per cent. alcohol; or frozen sections may be at once prepared. Satisfactory sections may be obtained from small pieces of tissue if they are put in the formaldehyde solution for an hour or two and then cut with the freezing-microtome. After they have been washed for a short time in water they may be stained by any of the more usual methods.

6. **Flemming's Solution.**—This is a solution containing osmic acid, chromic acid, and acetic acid. It does not keep well, and it is best to prepare it just before it is to be used. For this purpose the following stock solutions may be kept on hand:

A. 2 per cent. solution of osmic acid in 1 per cent. chromic acid.
B. 1 per cent. solution of chromic acid in distilled water.

Osmic acid is sold in sealed tubes containing 1 gram. To prepare the stock solution "A," the tube should be washed on the outside and a deep file-scratch made near its centre. It should then be broken into a bottle containing 50 cc. of a 1 per cent. solution of chromic acid in distilled water. The halves of the tube should be dropped into the bottle and its contents shaken at intervals until solution is effected. This solution had best be kept in the dark to avoid decomposition of the osmic acid. When required for use, prepare the Flemming's solution by mixing:

   Solution "A,"       4 cc.
   Solution "B,"       15 "
   Glacial acetic acid,     1 "

Flemming's solution is especially useful for fixing the finer details of structure within the cell. It was devised for the preservation of the mitotic figures formed during karyokinesis, but its range of usefulness far exceeds that limited application. Its power of penetration is very slight and the pieces of tissue selected for fixation must be small. They should not exceed 2 mm. in their least measurement, and thinner pieces are apt to give better results. Owing to the presence of osmic acid, Flemming's solution stains fat a dark-brown or black color, and may be used as a reagent for the identification of fatty substances.

Tissues should be left in Flemming's solution for about twenty-four hours, though twice that length of time would cause little if any harm. They must then be thoroughly washed in running water for twenty-four hours or longer, and hardened in alcohol. Since Flemming's solution is usually employed for the study of the individual cells, it is desirable to prepare very thin sections of the tissues that have been hardened in it. For this purpose embedding in paraffin is the best method.

The foregoing fixing solutions will meet most of the requirements of the practitioner of medicine, but it frequently happens that he would like to obtain speedy results from a microscopical examination without running the risk of loss of material or of poor results. When this is the case he may use absolute alcohol as a fixing-agent, thus taking advantage also of its ability to harden tissues and fit them for rapid embedding in collodion.

7. **Absolute Alcohol.**—If fresh tissues are placed in strong alcohol, say 95 per cent., they are hardened; but during the process there is an opportunity for the albuminous fluids in the tissues to escape to a certain extent, and for shrinkage to take place in consequence. If absolute alcohol be employed, it causes such rapid coagulation that this leaching of the tissues does not take place. It is necessary, however, that the alcohol should remain of nearly its original strength, otherwise the water in the tissues will dilute it sufficiently to destroy this coagulating action.

An excellent means for maintaining the strength of the alcohol is to immerse in it a few lumps of quick-lime. Take a small jar that can be hermetically closed by a tightly fitting cover (a museum jar, holding six or eight ounces, will answer). Place the lime in the bottom and then nearly fill with absolute alcohol. A few pieces of crumpled filter-paper are placed upon the lime and covered with a smooth piece placed so as to slant a little. The latter should lie near the surface of the alcohol, but be entirely submerged. Small pieces of the tissue to be fixed are placed upon the filter-paper where they will be covered by the alcohol. The alcohol immediately coagulates the albuminous substances on the surface of the pieces and then gradually replaces the water in the specimen, coagulating the deeper-seated albumins as it penetrates the mass. The expelled water sinks to the bottom of the jar, owing to its greater specific gravity, and is at once taken up by the lime. It is essential for the success of this method that the lime should be exceedingly quick. It must show immediate signs of slaking if even a drop of water be placed upon it.[1]

It will be seen that this method not only fixes the tissues, but quickly dehydrates them. The real dehydrating-agent is, however, the lime, the alcohol serving merely as a vehicle for conveying the water from the specimen to the lime. If the pieces of tissue are

---

[1] A jar of absolute alcohol, prepared as above, may be used for purposes of fixing or hardening until the lime has become slaked or the alcohol so impregnated with dissolved fat that the latter interferes with embedding in collodion. When the latter is the case the hardened collodion is opaque or opalescent.

small, not over 5 mm. thick, they will be hardened by remaining in the absolute alcohol over night, and mounted sections may be ready for examination by the next afternoon.

8. **Fixation by Boiling.**—Throw small pieces of the tissue, not larger than 1 cm., into boiling 0.75 per cent. salt solution. Keep them at the temperature of boiling for two minutes. Then throw them into cold water. They may then be cut with the freezing-microtome, or may be placed in 70 per cent. alcohol for hardening. This method is excellent for the detection of albuminous exudates within the tissues, but it causes so much shrinkage that it is not useful for general purposes.

## Methods of Hardening.

Solutions of chromates, as Müller's fluid, will, after a time, confer a pretty firm consistency upon tissues, and even render them brittle. Tissues fixed in corrosive sublimate are also very much hardened. But the usual practice is to harden specimens in alcohol after fixation. To obtain the best results this hardening should be done gradually, since immersion in strong alcohol is apt to produce undesirable shrinkage, affecting the various tissue-elements in different degree.

Seventy per cent. alcohol (736 cc. 95 per cent. alcohol to 264 cc. water) is weak enough to begin with. After the tissues have been in alcohol of that strength for twenty-four to forty-eight hours, according to the size of the pieces, they are placed in 80 per cent. alcohol (842 cc. 95 per cent. alcohol to 158 cc. water) for an equal length of time, and then in 95 per cent. alcohol. From the 95 per cent. alcohol they are placed in absolute alcohol, if it be desired to embed them in either collodion or paraffin. If they are not intended for immediate use, they may be kept indefinitely in 80 per cent. alcohol.

During the hardening it is best not to allow the tissues to rest on the bottom of the vessel containing the alcohol, as they are liable to slight maceration in the alcohol, which there becomes diluted with water from the specimen. They can be kept off the bottom by means of a little crumpled filter-paper. Specimens that have been fixed in a chromate solution should be kept in the dark while being hardened; those that have been fixed in corrosive sublimate should be hardened in alcohols to which a little tincture of iodine (sufficient to give them a sherry color) has been added. When absolute alcohol is used, its strength should be maintained by contact with quick-lime (see directions for fixing tissues in absolute alcohol).

## Methods of Impregnation.

When tissues are so porous or friable that sections are likely to tear or disintegrate it is desirable to impregnate them with some embedding-material. The most useful substances for this purpose are collodion, or celloidin, and paraffin. Whichever of these is used, it is necessary to remove the water from the specimen before the impregnation can be accomplished, for both collodion and paraffin are insoluble in water. Tissues that have been hardened in alcohol are to a certain extent already dehydrated. The residual water may be removed or reduced to a trace by treatment with absolute alcohol, in which collodion is soluble.

The "celloidin" manufactured by Schering is an excellent preparation of gun-cotton, but almost equally good results may be obtained by using the more economical soluble cottons employed by photographers. Two solutions in a mixture of equal volumes of ether and absolute alcohol (both, if possible, of Squibb's preparation) should be kept in stock: one, a weaker solution, having about the consistency of thin mucilage; the other, a stronger solution, resembling a syrup.

Collodion is soluble in absolute alcohol, so that tissues containing only that fluid are ready for impregnation without further preliminary treatment. When thorough impregnation is desired the tissues should be immersed in equal parts of ether and absolute alcohol for a few days, and then in the weaker solution of celloidin or collodion for a number of days or weeks—the longer the better;[1] but such complete impregnation is often unnecessary, and soaking for a day or two will often suffice if the sections to be made need not be very thin. It is not possible, in any event, to make very thin sections from tissues embedded in collodion; but sections of large area may be obtained, which is often of greater importance. For very thin sections it is better to use paraffin for the embedding-material, although the resulting sections will have to be smaller.

Paraffin is insoluble in alcohol of all strengths. It is therefore necessary to remove the absolute alcohol from the tissues before they can be impregnated with paraffin. This may be done by immersing the tissues in some liquid that is a solvent for paraffin

[1] Impregnation may be greatly hastened if done at the body-temperature in a hermetically closed vessel.

and is also miscible with alcohol. For this purpose, xylol, chloroform, or oil of cedar-wood may be used. Xylol yields the most rapid results, but its use is contraindicated when it is desired to retain fatty substances that have been colored with osmic acid, as the xylol extracts them. If their preservation within the tissues is important, chloroform should be used; but the sojourn even in that liquid should be as short as possible. Oil of cedar-wood probably causes less change in tissues than chloroform, but the method is more protracted, and, requiring longer treatment in the paraffin-oven, probably has little ultimate advantage over chloroform for general purposes.

If xylol is used, the tissues are transferred from the absolute alcohol to xylol, on which they at first float. Subsequently they sink, and are gradually rendered transparent as the alcohol is expelled by the xylol. When there are no opacities left the specimen is ready for the paraffin-oven. These changes take from two to twenty-four hours.

The treatment with chloroform is similar to that with xylol, but after the tissues have been cleared in chloroform (six to twenty-four hours) they are immersed in a saturated solution of paraffin in chloroform for about the same length of time. They are then ready for the paraffin-oven.

When oil of cedar-wood is used the pieces should be soaked in two successive portions of the oil, about twelve hours in each, to insure removal of the alcohol.

The foregoing steps are all preliminary to the actual impregnation with paraffin.

It is important that the paraffin used for impregnation and embedding should have a wax-like, and not a crystalline, texture, and that its melting-point should be such that its consistency will be favorable for cutting at the average temperature of the laboratory. Grübler, of Leipzig, furnishes excellent qualities of paraffin. For a room-temperature of 20° C. (68° Fah.) a variety melting at 50° C. (122° Fah.) will give good results. If the laboratory is warmer, a paraffin of higher melting-point should be used.

Impregnation is accomplished by placing the bits of tissue in a bath of melted paraffin maintained at a temperature only slightly above that of fusion, say 52° C. (125.6° Fah.), if the paraffin melts at 50° C. (122° Fah.). This may be accomplished in a water-jacketed oven provided with a thermoregulator, or upon a plate of brass or copper, resting on a tripod and heated at one end by a burner. When the latter method is employed the paraffin is melted in a little glass dish, which is moved along the plate until a point is found at which the paraffin remains melted at the bottom, but is covered at the edges of the surface with a thin layer of congealed paraffin.

The length of time that the specimens should remain in the melted paraffin will vary with the character of the tissues and the method of getting rid of the alcohol which has been employed. It should not be protracted longer than necessary for complete impregnation, as heat is injurious to the tissues. When xylol has been used two hours will usually suffice if the pieces of tissue are small, and especially if they are transferred to a fresh paraffin-bath after about an hour. This renewal of the paraffin is still more important if oil of cedar-wood has been used. Chloroform requires a little more time than xylol, and should be transferred to fresh paraffin once or twice.

When impregnation has taken place and the final bath of paraffin no longer has the slightest odor of the clearing-agent the pieces of tissue are removed from the bath with warmed forceps and placed on bits of writing-paper, to which they adhere. A designation of the specimen may be written on these papers, and the tissues kept in this condition until required for cutting. They must then be embedded.

## Methods of Embedding.

The object of embedding is to surround the piece of tissue from which sections are to be cut with a mass of the embedding-substance, which then not only supports the tissue when it comes in contact with the knife, but also affixes it to a block or other support which can be fitted into the clamp of the microtome.

Microtomes designed for cutting paraffin usually have special supports for the embedded specimen, but blocks of hard wood may be used in their place.

For the support of tissues embedded in collodion blocks of plate-glass are probably both better and cheaper than those made of other materials. They may be easily prepared from waste pieces of plate-glass, about a quarter of an inch thick, and " obscured " or ground on one surface. The glass may be cut into blocks of any desired size by scoring the smooth side with a diamond and then splitting the pieces apart with a sharp blow from a wedge-shaped hammer. The embedded specimen is affixed to the rough surface of these blocks by means of collodion, and the blocks may be numbered with a lead pencil upon the rough surface. The writing will be preserved from obliteration by the specimen subsequently placed upon it, and can be read through the glass.

1. **Embedding in Collodion (or Celloidin).**—Tissues of firm consistency and moderately uniform structure, such as liver, kidney, and the majority of tumors which have been hardened, may be embedded without previous impregnation. Before this can be done, however, they must be either dehydrated with absolute alcohol, or soaked for a few hours in a mixture of equal volumes of ether and alcohol (95 per cent. alcohol will answer, if absolute alcohol is not to be had). For this rapid method the bottom of the piece of tissue must be flat and parallel to the plane of the desired sections. When the necessary trimming of the specimen is completed moisten it with absolute alcohol or the ether-alcohol mixture, then dip it in the thick solution of gun-cotton and place it at once upon the ground surface of the glass block (previously labelled). In a few minutes the collodion will have evaporated sufficiently for the formation of a distinct pellicle upon its surface. When this has become firm enough to withstand gentle pressure immerse the block and specimen in several times their volume of 80 per cent. alcohol. This will harden the collodion, and in the course of a few hours the specimen will be ready for cutting.

Tissues impregnated with collodion had best be embedded by a slower process than the foregoing, although that method will answer where only a slight support of the tissue-elements within the specimen is needed. A gradual concentration of the collodion within the tissues may be brought about in the following manner:

Smear the inside of a small, straight-sided glass dish with a *trace* of glycerin and then fill it with enough moderately thick collodion to cover the pieces of tissue with a layer about one-quarter of an inch deep. Now place the specimens that have been in thin collodion in the dish, with the surfaces from which sections are to be cut resting on the bottom. Place the dish in a larger vessel with higher sides and loosely cover the latter. The ether and alcohol in the collodion will gradually evaporate, and their vapors will first fill the outer vessel and then overflow its sides. The depth of the outer vessel keeps these vapors in contact with the surface of the collodion, preventing the formation of a pellicle, which would retard evaporation and also favor the formation of bubbles in the collodion. After an interval of one or more days the collodion will have a gelatinous consistency. It should be allowed to become so hard that it has considerable firmness, but is still soft enough to receive an impression of the ridges in the skin when pressed with the finger. The outer vessel is then partly filled with 80 per cent. alcohol so that the whole of the inner dish is submerged.

By the next day the collodion will be hard enough for removal from the dish. With a small scalpel, held vertically, divide the hardened mass of collodion into portions, each of which contains one of the pieces of tissue (for several pieces may be embedded in the same dish, provided care be taken to preserve their identity). Remove the pieces and trim down the collodion around the specimens, leaving a margin of about an eighth of an inch. Trim the top surfaces of the collodion parallel with the bottom surfaces, then dip the trimmed surface into a little absolute alcohol contained in a watch-glass, in order to dehydrate it. This will take about two minutes. Label glass blocks with lead-pencil, place a drop of thick collodion on the writing, and transfer the embedded specimens immediately from the absolute alcohol to the drop of collodion, pressing it into contact with the glass. When a good pellicle has formed on the collodion drop the whole block into 80 per cent. alcohol. If the block of hardened collodion containing the specimen be sufficiently dehydrated on the surfaces coming in contact with the drop of collodion, and the latter have not time for the formation of a pellicle before it receives the block, there will be no difficulty in cementing the embedded specimen to the roughened surface of the glass. It is best not to cut sections until the day after the specimen has been affixed to the glass block. These blocks, with attached specimens, may be preserved indefinitely in 80 per cent. alcohol.

The thin coating of glycerin on the inside of the embedding-dish serves the purpose of preventing the collodion from sticking to the glass.

2. **Embedding in Paraffin.**—The specimen should first be trimmed so as to have one surface parallel to the plane of the future sections. If it is surrounded by too much paraffin to permit of ready inspection, it may be placed on a piece of filter-paper and warmed until the superfluous paraffin is absorbed by the paper. The trimmed surface is then laid upon a small glass plate that has been smeared

with a mere trace of glycerin, and metallic right-angles, similarly smeared on the inside, are placed around the specimen in such a way as to form a box with a clear space at least an eighth of an inch broad between its sides and the specimen. Melted paraffin, at a temperature only slightly exceeding that necessary to keep it fluid, is then poured into the box, filling it. The paraffin should now be made to cool as rapidly as possible, in order to prevent its becoming crystalline. For this reason it is well to prepare the box formed by the plate of glass and the metallic right-angles in the bottom of a deep soup-plate or some similar vessel. After the box has been filled with melted paraffin cold water may be poured into the plate until its surface is nearly on a level with the top of the box, and when the top of the paraffin has congealed the plate may be filled with cold water. After a few minutes the box may be taken apart and the block of paraffin left in the water to become cold.

These paraffin-blocks may be labelled with a needle and kept indefinitely in the dry condition, at a temperature below that at which the paraffin softens. When they are to be used the bottom of the block should be trimmed parallel with the top, sufficient paraffin being removed to obliterate the hollow which formed when the paraffin solidified. This trimmed surface is then made to adhere to the paraffin-support of the microtome, or a block of hard wood, by means of a heated scalpel.

It often happens that little air-bubbles are present in the paraffin close to the specimen, or that cracks exist between the specimen and the surrounding paraffin, owing to the retention of a little air at the time of embedding. These defects can be remedied by melting the paraffin with a heated needle. It is important that the paraffin should everywhere be in perfect contact with the specimen. When this repairing, if necessary, has been done and the paraffin has become cold again, the block should be trimmed so that the specimen, or at least its upper part, is contained in a little cubical mass resting on the main block, with a margin of paraffin, about 1 mm. thick at the places where the edges of the cube are nearest to the specimen. Those edges should be straight and at right angles to each other, and the sides of the trimmed cube should be vertical. In trimming the block only thin slices should be removed at a time, in order to avoid cracking the paraffin forming the small cubical mass enclosing the specimen.

These manipulations prepare the specimen for cutting.

## Methods of Cutting.

It is possible to obtain useful sections from fresh or hardened tissues by free-hand cutting with a sharp razor; for this purpose the razor should either be very hollow ground, so as to have a thin blade, or the lower surface should be ground flat. In stropping the razor, or microtome-knife, the stroke should be from point to heel during both the forward and return motions. In cutting, the edge should be used from heel to point, and this same motion should be used in honing. A wire arrangement is usually furnished with microtome-knives, which is intended for use while honing or stropping. It serves to raise the back of the knife when the flat side is sharpened, and should always be employed. Care must be taken not to press the knife against the strop, as this is liable to turn or blunt the edge. A few light strokes on the strop immediately after each day's use will keep the knife sharp and coat it with a little grease, protecting it from rust. A microtome-knife should never be allowed to rest with its edge on any hard surface; the mere weight of the knife is sufficient to spoil its edge.

In cutting free-hand sections of fresh tissues the upper surface of the razor should be kept flooded with normal (0.75 per cent.) salt solution. The sections float in this fluid and are kept from tearing. Each section should be removed by a single stroke of the razor. When hardened specimens are cut, 80 per cent. alcohol should be used instead of salt solution.

Free-hand sections cannot be made either so thin or uniform as sections prepared with a microtome, and these instruments are now so cheap that they are universally used. There are three principal forms: 1, freezing-microtomes; 2, paraffin-microtomes; 3, microtomes for cutting sections of tissues embedded in collodion. The last are often fitted with attachments intended for use in cutting frozen sections, and can also be used for paraffin. But the best results are obtained by using instruments especially designed for each purpose.

1. **Frozen Sections.**—Freezing is usually employed when sections of fresh tissues are to be made, but hardened tissues may be cut with a freezing-microtome if the alcohol be first removed by soaking for a considerable time in water. The tissue may be placed upon the plate of the microtome in a little water or neutral salt solution; but a better method is first to soak the tissue in a syrupy solution of

gum-arabic, and to moisten the plate with the same before freezing. This solution freezes in less coarsely crystalline form than water or salt solution.

When the tissues are frozen, thin sections are removed with a quick forward and slightly oblique stroke of the knife. The motion is intermediate between that of a plane and a single stroke of a saw. The sections are floated from the knife in a dish of water or normal salt solution; or they may be fixed in a 4 per cent. solution of formaldehyde. The frozen tissue must not be too hard. Should that be the case, the upper surface may be moistened by means of a camel's-hair brush, dipped in water or salt solution, or warmed with the breath.

2. **Collodion-sections.**—The block upon which the embedded specimen is fastened is secured in the clamp of the microtome in such a position that the sections will be made in the desired plane. The knife is then adjusted on its carrier in an oblique position, so that the greatest possible length of its edge will be utilized in cutting. The upper surface of the knife is flooded with 80 per cent. alcohol, and slices are removed with the knife until the desired level of the specimen has been reached. Sections are then made as thin as is compatible with obtaining complete sections from the whole surface. The sections float in the 80 per cent. alcohol, with which the knife should be kept flooded, and may be removed with a camel's-hair brush. At no time should either the knife or the specimen be allowed to dry. The sections may be kept indefinitely in 80 per cent. alcohol, or they may be dropped into water if they are to be used within a short time.

After use, the knife should be carefully wiped, stropped, and placed in its case. The microtome should be dried and the tracks moistened with a little oil of sweet almonds or paraffin oil, to prevent rusting.

3. **Paraffin-sections.**—The knife should be fixed perpendicular to the direction of cutting, its edge acting like that of a plane. Its surfaces must be clean and dry; adherent paraffin can be removed with a cloth moistened with xylol.

The paraffin-block containing the specimen to be cut is firmly clamped with one of its narrow edges parallel to the edge of the knife. The block is now raised and moderately thick slices removed until the desired level is reached, when the thin sections desired may be cut. It not infrequently happens that the sections roll before the edge of the knife. This is probably due to the paraffin being too hard. In that case the cutting should be done in a warmer room. This rolling will, however, cause little trouble in the use of the sections unless it be desired to have them adhere to each other at the edges to form ribbons, in which the succession of the sections is preserved.

Before paraffin-sections can be stained it is necessary to remove the paraffin. If the tissues are sufficiently coherent, this can be done by dropping the sections into xylol or chloroform; but if this would cause a disintegration of the sections, they must be affixed to slides or cover-glasses by means of a cement which shall hold the different parts of the tissues in their proper relative positions after the paraffin has been removed. The simplest cement for this purpose is Mayer's albumin mixture, prepared as follows: beat up the white of an egg and allow the froth to liquefy. Then add an equal bulk of glycerin and a few pieces of camphor (for the preservation of the mixture). This cement is applied to the clean surface of a slide, or, better, a cover-glass, in a very thin layer with the side of a camel's-hair brush, care being taken to leave no air-bubbles. The paraffin-sections are removed from the knife with a fine camel's-hair brush or a small, but rather stiff, feather inserted into a handle, and placed upon the coating of cement. They are then flattened out with the brush or feather and pressed against the glass to remove superfluous cement. If the sections have rolled, unrolling will be facilitated by warming the sections with the breath. The cover-glasses are set aside to dry a little, and are then heated to render the albumin insoluble. This requires some practice. The manipulation is intended to accomplish the following results: the paraffin melts at a lower temperature than that at which the albumin is coagulated, and this fact is utilized to remove all excess of the cement, which is washed away from the tissues by the flow of melted paraffin. The residual albumin is sufficient to make the section adhere to the glass when subjected to a high enough temperature to cause its coagulation. The albumin should be dried to a considerable extent before it is converted by the heat into its insoluble form, otherwise it will coagulate in opaque masses. To bring about the desired results the cover-glass, held in a pair of forceps, is waved over a flame until the paraffin is seen to melt. That temperature is maintained for a few moments, and then the cover-glass is heated until vapors are distinctly seen to rise from its surface. Great care must be taken not to scorch the sections. When the sections have been cemented to them the cover-glasses are placed in absolute alcohol to dehydrate them, and are then treated with xylol, chloroform, or some other solvent of paraffin. The solvent is then removed by another bath of absolute alcohol, and the alcohol removed by water, when the sections are ready for staining.

When the sections do not require affixing to cover-glasses they may be dropped into the solvent for the paraffin, and the latter removed with absolute alcohol, for which water is then substituted, preparing the sections for staining. It sometimes happens that when sections are transferred from absolute alcohol to water the diffusion-currents are so strong that the sections are destroyed. When this is the case the transition must be made more gradually, baths of 80 per cent., 50 per cent., and 30 per cent. alcohol being interposed between the absolute alcohol and the water.

## Methods of Staining.

A large number of methods have been devised for bringing out the structure of tissues. Many of the methods are of almost universal application, while others require special methods of fixation or other preliminary treatment of the tissues. Some are calculated to render the general features of structure more evident than they would be if the tissues were not stained; others stain certain elements some characteristic color, and, to that extent, serve the purpose of microchemical reagents. Only a few of the more useful methods can be described here; for others the reader is referred to the larger text-books and the technical journals.

1. **Hæmatoxylin and Eosin.**—Hæmatoxylin, the coloring-principle of logwood, has proved a very useful stain for the nuclei of cells. It is not a pure nuclear stain, but also tints the cytoplasm of cells and the intercellular substances. It is most commonly employed in combination with alum. Such combinations of coloring-matter with a base are called "lakes."

A hæmatoxylin-lake may be used alone, or its use may be preceded or followed by the employment of a counterstain with some diffuse color not affecting the nuclei. For counterstaining, eosin or neutral carmine is usually employed. Both stain the tissues a diffuse red, varying in depth according to the nature of the tissue-elements in the section.

There are several formulæ for the preparation of alum-hæmatoxylin, but that devised by Böhmer will answer all purposes, and is very simple:

1. Hæmatoxylin crystals,      1 gram.
   Absolute alcohol,          10 cc.
2. Alum,                      20 grams.
   Distilled water,           200 cc.

Cover the solutions and allow them to stand over night. The next day mix them and allow the mixture to stand for one week in a wide-mouthed bottle lightly plugged with cotton. Then filter into a bottle provided with a good cork. The solution is then ready for use. Nearly all solutions of alum-hæmatoxylin require an interval of time for "ripening," and their staining-powers improve with age.

Alum-hæmatoxylin is intended for staining sections from tissues that have been fixed and hardened. It is especially useful when the fixing-solution employed contained chromates, but may be used after almost any method of fixation, if the time of staining is of the right length and the sections are previously freed from acidity by thorough washing.

If the following directions are closely adhered to, the student can hardly fail to obtain good results in the use of Böhmer's hæmatoxylin:

Transfer the sections from the 80 per cent. alcohol in which they have been kept to a dish of distilled water. At first they will float on the surface of the water; this is a favorable moment for removing all folds and wrinkles. The sections should be manipulated with platinum needles, prepared by fusing a bit of platinum wire into the end of a glass rod. Such needles can be cleaned by heating the wire red in a flame.

When the sections sink to the bottom of the dish of water, and remain there, it may be assumed that they are free from alcohol.

Filter about 5 cc. of the hæmatoxylin into a watch-glass or butter-dish and transfer the sections from the water to the dye.

Let the sections stain for three minutes by the watch, and then transfer them to a dish of distilled water. At first the sections will have a reddish tint, but as the washing proceeds the color will turn to a pure blue. During the washing the water should be renewed, until finally it acquires no color from the sections and the latter have lost all traces of a red tint. This washing may take several minutes, or even a few hours; but if good, permanent stains are desired, it is of great importance that it be thorough. This washing completes the actual staining with hæmatoxylin, and the sections are then ready for counterstaining with eosin or for dehydration.

The eosin solution used for diffuse staining is prepared by dissolving 1 gram of eosin in 60 cc. of 50 per cent. alcohol. Of this solution, about ten drops are added to 5 cc. of distilled water in a small dish; the sections are stained for about five minutes and then washed in distilled water. They are then ready for dehydration and mounting. The diluted eosin should be thrown away after use, but the hæmatoxylin can be filtered back into the stock-bottle.

Since the hæmatoxylin solution improves with age, no exact directions can be given as to the length of time sections should remain in a particular solution. Three minutes will usually yield good results; but if it is found that the color is too dark, a shorter time should be employed, and *vice versâ*. One soon becomes familiar with the staining-powers of the particular solution used. The dishes that have contained hæmatoxylin should be washed soon after use, or may be subsequently cleaned with a little hydrochloric acid, all traces of which should then be removed by thorough washing in water.

The above method for staining with hæmatoxylin and eosin is highly recommended for general routine work.

2. **Neutral Carmine.**—

| | |
|---|---|
| Carmine, "No. 40," | 1 gram. |
| Distilled water, | 50 cc. |
| Ammonia, | 5 " |

The solution is allowed to remain exposed to the air until the odor of ammonia is no longer perceptible. It is then filtered into a bottle, where it is kept till needed.

Neutral carmine gives a diffuse stain, resembling that of eosin, but rather clearer in character. It is employed in a greatly diluted form, according to the following directions:

One drop of the neutral carmine is mixed with about 20 cc. of distilled water. A trace of acetic acid is then added by dipping a platinum needle into the acid and stirring the diluted dye with the acidulated needle. A piece of filter-paper is then placed upon the bottom of the dish, and the sections to be stained are transferred from distilled water to the dye and distributed upon the paper in such a way that they do not lie over each other. The dye acts very slowly, twenty-four hours being none too long for good results. If the staining be hastened by using a stronger solution, it suffers in sharpness. After staining, the sections are thoroughly washed in distilled water, and may then be subjected to a nuclear dye, such as hæmatoxylin. The proper acidulation of the diluted dye is of importance for the success of this method. If the solution is not sufficiently neutralized, the sections will not be stained; if it is too acid, precipitation of the carmine will take place.

3. **Alum-carmine.—**

| | |
|---|---|
| Alum, | 5 grams. |
| Distilled water, | 100 cc. |
| Carmine, "No. 40," | 2 grams. |

The alum is dissolved in the water with the aid of heat, the carmine then added, and the mixture kept at the boiling-point for about half an hour. It is then allowed to cool and filtered into the stock-bottle. Two or three drops of deliquesced carbolic acid may be added to prevent the development of fungi.

Sections are stained in the undiluted, but filtered, dye for at least five minutes. There is no danger of over-staining. It is a pure nuclear stain, coloring the chromatin red. After staining, the sections are either washed, and are then ready for dehydration, or they may receive a counterstain with picric acid, coloring the tissues a diffuse yellow. This may be most readily accomplished by adding a few small crystals of picric acid to the first dish of dehydrating alcohol (see p. 428).

4. **Borax-carmine.—**

| | |
|---|---|
| Borax, | 4 grams. |
| Distilled water, | 100 cc. |
| Carmine, "No. 40," | 3 grams. |
| Alcohol, 70 per cent., | 100 cc. |

The borax is dissolved in the water by warming, and the solution allowed to cool; the carmine is then stirred in and the alcohol added. After standing twenty-four hours the solution is filtered into the stock-bottle, a process that is exceedingly slow.

Borax-carmine is used for the staining of little masses of tissue before they are embedded. It is a nuclear dye, giving the chromatin a red color. It is useful when paraffin-embedding is to be employed and it is desirable to restrict the manipulation of the sections to a minimum.

Small pieces of hardened tissues, not over 5 mm. thick, are transferred from distilled water to the undiluted dye and allowed to stain for twenty-four hours, or longer. After staining they are immediately placed in an acid alcohol, prepared by adding 5 drops of concentrated hydrochloric acid to 100 cc. of 70 per cent. alcohol. The tissue should not rest on the bottom of the vessel containing the alcohol, but upon crumpled filter-paper, so that the extracted excess of coloring-matter may sink to the bottom. If the acid alcohol around the specimen becomes colored, fresh portions of alcohol should be used. The treatment with acid alcohol is continued until no more color is given off from the specimen. It is then transferred to 90 per cent. alcohol, in which it should remain for twenty-four hours, after which it can be subjected to the dehydration necessary for embedding.

**5. Orth's Lithio-carmine.—**

Carmine, "No. 40,"   3 grams.
Lithium carbonate, saturated aqueous solution, 100 cc.

The solution of lithium carbonate is prepared by occasionally shaking a mixture of distilled water and an excess of lithium carbonate. Twenty-four hours will suffice for the production of a strong enough solution. The supernatant liquid is then filtered. Carmine readily dissolves in this solution. For preservation a crystal of thymol may be added.

Lithio-carmine stains sections in about five minutes, and there is no danger of overstaining. Like borax-carmine, it requires after-treatment with acid alcohol. The sections should be transferred, without intermediate washing, to 70 per cent. alcohol containing 1 per cent. of concentrated hydrochloric acid; they may then be dehydrated, and, if desired, counterstained with picric acid during the dehydration.

**6. Unna's Methylene-blue.—**

Methylene-blue,   1 gram.
Potassium carbonate,   1 "
Distilled water,   100 cc.

When required for use, this solution should be diluted with distilled water to about one-tenth of its strength. It is a good stain for bacteria, and may also be used for staining the nuclei of tissues either by itself, or after using eosin as a diffuse stain. An aqueous solution of eosin, 5 per cent., is used for this purpose, the sections being stained for about five minutes. They are then washed to remove the excess of eosin, and stained in the diluted methylene-blue for about an hour. After this they are again washed and treated with absolute alcohol, which discharges the excess of blue. They are then cleared with xylol and mounted in dammar or Canada balsam, dissolved in xylol. The preliminary staining with eosin may be omitted, when a contrast- or counterstain is not required.

7. **Aqueous Methylene-blue.**—This is usually prepared at the time when needed by mixing one part of a saturated solution of the anilin-color in 95 per cent. alcohol with nine parts of distilled water. It is frequently employed as a general stain for bacteria.

Other anilin-colors, such as fuchsin, gentian-violet, methyl-violet, and Bismarck-brown, may be kept in concentrated alcoholic solution, to be diluted in a similar manner just before use. When these solutions are used for staining sections or cover-glass preparations the adherent dye is washed off with water, after which the intensity of the stain is reduced by the use of alcohol, 95 per cent. or absolute, which bleaches the portions of the specimen which retain the color with the least tenacity. If the action of the alcohol be maintained for too long a time, the color may be discharged from all parts of the specimen. The method of overstaining a specimen, and then discharging the color from those parts where it is not desired, is a common one. The process of discharging the color is called the "differentiation" of the stain, because it serves to distinguish those elements which hold the color strongly from those which part with it easily.

8. **Carbol-fuchsin.**—

Saturated alcoholic solution of fuchsin,     10 cc.
Aqueous solution of carbolic acid crystals, 5 per cent.,     90 cc.

This solution should always be carefully filtered before use.

9. **Anilin-gentian-violet.**—A. Ehrlich's formula:

Saturated alcoholic solution of gentian-violet,     1.5 cc.
Freshly prepared anilin-water,     8.5 cc.

The anilin-water is prepared by shaking a few drops of anilin with distilled water, allowing the mixture to stand for about ten minutes, and then filtering through well-moistened filter-paper. The filtrate should contain no globules of the anilin. In order to avoid this the filtration should be stopped before all the watery part of the mixture has run through the paper, otherwise oily drops of anilin will follow.

Precipitates are likely to occur in this gentian-violet solution when it is first prepared. After twenty-four hours these are less abundant. The solution deteriorates soon after that time, and should not be used more than a week after its preparation.

B. Stirling's formula:

| | |
|---|---|
| Gentian-violet, | 5 grams. |
| Alcohol, | 10 cc. |
| Anilin, | 2 cc. |
| Distilled water, | 88 cc. |

This solution keeps better than the preceding. Both must be filtered carefully through moistened filter-paper immediately before being used.

10. **Gram's Solution.**—This is a differentiating agent used in connection with anilin-gentian-violet:

| | |
|---|---|
| Iodine, | 1 gram. |
| Potassium iodide, | 2 grams. |
| Distilled water, | 300 cc. |

The specimens are first overstained with the gentian-violet solution. They are then washed in water and placed in Gram's solution for from three to five minutes. While in this solution they turn a brown color, and the combination between the coloring-matter and some of the elements of the specimen is loosened. The specimen is then transferred to 95 per cent. alcohol, in which it remains until no more color is given off. If sufficient color has not been removed, the treatment with Gram's solution and alcohol may be repeated. After this differentiation the specimen may be dehydrated, cleared, and mounted; or a contrast-stain may be used before those manipulations. This is a useful method for staining bacteria in sections of tissue when the species of bacteria are such as resist the decolorizing action of the iodine. In this respect different species of bacteria differ greatly, and the method is commonly employed in bacteriological work to distinguish those species which retain the stain, or are "positive to Gram," from those which are decolorized or "negative to Gram."

11. **Van Giesson's Picric Acid and Acid Fuchsin Stain.**—

Aqueous solution of acid fuchsin, 1 per cent.,     5 cc.
Saturated aqueous solution of picric acid,     100 "

This solution serves to stain fibrous intercellular substances. It is used in the following manner:

1. Slightly overstain with alum hæmatoxylin; *e. g.*, Böhmer's hæmatoxylin.
2. Wash thoroughly in distilled water.
3. Stain in Van Giesson's dye for five minutes.
4. Wash in water.
5. Dehydrate in 95 per cent. alcohol.
6. Clear in oil of origanum.
7. Mount in xylol-balsam or xylol-dammar.

The tissues should have been fixed in a corrosive-sublimate solution or one containing chromates; *e. g.*, Müller's fluid, Zenker's fluid, or sublimate solution. The connective-tissue fibres are stained red by the acid fuchsin. The reason for overstaining with hæmatoxylin is that subsequent treatment with picric acid discharges some of that color.

12. **Benda's Iron-hæmatoxylin Stain.**—This is a powerful stain well adapted to the staining of paraffin-sections that have been affixed to cover-glasses. It stains nuclei and intercellular substances, as well as the protoplasm of cells, various shades of gray, and the color is very permanent. The outline of the method is as follows:

1. Mordant the sections (after affixing to cover-glasses, if that method is used) in a mixture of equal parts of liquor ferri sulfurici oxydati of the German Pharmacopœia and distilled water for twenty-four hours.
2. Rinse in distilled water, and then wash in three changes of tap-water.
3. Stain in aqueous solution of hæmatoxylin, prepared by mixing 10 drops of a concentrated alcoholic solution of the crystals with 10 cc. of distilled water. Stain for from one-half to twenty-four hours.

4. Rinse in distilled water.
5. Differentiate in equal parts of glacial acetic acid and distilled water.
6. Wash thoroughly in distilled water.
7. Dehydrate in absolute alcohol.
8. Clear in xylol, carbol-xylol, or some essential oil.
9. Mount in balsam.

**13. Pal's Modification of Weigert's Stain for the Medullary Sheath of Nerves.**—This method is useful for the study of the central nervous system, and may, with advantage, be preceded by staining with neutral carmine. The tissues should have been fixed in a chromate solution; *e.g.*, Müller's fluid.

1. Soak sections several hours in 1 per cent. chromic acid solution in water.
2. Stain twenty-four to forty-eight hours in:

Hematoxylin crystals, 1 gram,
Absolute alcohol, 10 cc.
Lithium carbonate (saturated aqueous solution), 7 "
Distilled water, 90 "

The hymatoxylin crystals may be dissolved in the alcohol and the solution kept in stock, the proper proportions of lithium carbonate solution and water being added at the time of use.

3. Wash in water to which a little lithium carbonate solution has been added (about 2 cc. to each 100 cc. of water). The sections should acquire a deep-blue color.
4. Differentiate in 0.25 per cent. solution of potassium permanganate in distilled water, till the gray matter—*e.g.*, of the spinal cord—becomes brownish-yellow (one-half to five minutes).
5. Decolorize the gray matter in the following solution:

Oxalic acid, 1 gram,
Potassium sulphite, 1 "
Distilled water, 200 cc.

6. Wash thoroughly in distilled water.
7. Dehydrate in 95 per cent. alcohol.
8. Clear in carbol-xylol, oil of bergamot, or oil of origanum.
9. Mount in xylol-balsam or xylol-dammar.

This method stains the myelin-sheaths of the medullated nerve-fibres a dark blue, nearly black, color. If it has been preceded by a stain with neutral carmine, the axis-cylinders of the nerve-fibres will be stained red, and the nuclei of the nerve-cells will also appear red.

14. **Golgi's Methods.**—These methods have yielded most excellent results in the study of the central nervous system, the distribution of the peripheral nerves, and the delicate terminations of the ducts of glands; *e. g.*, the bile-capillaries. The methods must be regarded as special procedures in such studies, and can but be referred to here. They all depend upon hardening in some chromium salt, with or without the addition of osmic acid, and the subsequent impregnation with silver nitrate. A precipitate is thus produced on or within certain of the elements in the specimen, giving them a dark-brown or black color. The methods are capricious, and not all of the tissue-elements of like character in the specimen are rendered prominent. This is an advantage, but necessitates a degree of care in the interpretation of the results. Furthermore, irrelevant precipitates may form in the tissues which have no definite relations to any structure. Considerable practice is, therefore, required for the successful employment of all these methods, not only for a satisfactory execution of the manipulations, but also in the study of the results. The methods have no value for the study of cell-structure, since the whole cell is either covered or filled with the precipitates formed during the impregnation with silver.

Golgi has divided his methods into three groups: the slow, the rapid, and the mixed. For the details of these methods and of the various modifications introduced by different investigators the student is referred to the journals on microscopy. It must suffice to state here that the slow method begins with a hardening of the tissues in a 2 per cent. solution of potassium bichromate, which is gradually raised to 5 per cent. This hardening takes from fifteen days to three months. In the rapid method the tissues are first hardened in a mixture of 4 parts of a 2 per cent. solution of potassium bichromate and 1 part of a 1 per cent. solution of osmic acid. The tissues remain in this mixture for from two to six days, when they are ready for impregnation. For either method the pieces of tissue should not be thicker than 1.5 cm.

## Methods of Dehydration.

The final manipulation in nearly all the methods for staining described above is a washing of the sections in water. This water must be removed before permanent mounts can be made. Dehydration is accomplished by treating the sections with alcohol. If they are impregnated, or have been embedded in collodion or celloidin, they must not be dehydrated in absolute alcohol, as that dissolves the collodion. In such cases 95 per cent. alcohol is employed, the sections being treated with two baths of alcohol. When sections have been stained with carmine a contrast-stain may be obtained by adding a few small crystals of picric acid to the first dish of dehydrating alcohol. The excess of picric acid is then removed by the alcohol in the second dish. Absolute alcohol may be used for dehydration when the sections have not been embedded in collodion or celloidin.

When anilin-dyes have been used to stain sections it must be borne in mind that alcohol not merely dehydrates, but also differentiates the stain. If the sections are left too long in the alcohol, they may lose more color than is desired.

Sections that are to be mounted in glycerin or glycerin-jelly require no dehydration, but can be mounted directly from water.

## Methods of Clearing.

Clearing is necessary when specimens are to be permanently mounted in Canada balsam or dammar. Its object is to impregnate the section with some liquid that will drive out alcohol and also be miscible with the resin used for mounting. Of these clearing-agents there is a large number, from which a choice must be made according to the method of embedding that has been employed and the nature of the dye with which the tissues have been stained. Clearing-agents also differ in their miscibility with water, some requiring dehydration with absolute alcohol, others clearing well when 95 per cent. alcohol has been used for dehydration.

1. **Xylol.**—This is an excellent clearing-agent when the sections have been well dehydrated with absolute alcohol. It does not injure anilin-dyes. It is, perhaps, the best clearing-agent for sections of tissue stained with borax-carmine before cutting, when no counter-stain is employed. Xylol then both removes the paraffin in the section and clears it.

2. **Carbol-xylol.—**

    Carbolic acid crystals (melted),     1 vol.
    Xylol,     3 vols.

This mixture is much more tolerant of water than pure xylol. Sections dehydrated in 95 per cent. alcohol may be cleared with this reagent, which does not dissolve collodion. The carbolic acid used should be pure, but need not be the more expensive synthetic product.

3. **Oil of Bergamot.—**This light-green essential oil clears well and does not dissolve collodion. It may be used when 95 per cent. alcohol has been employed for dehydrating.

4. **Oil of Origanum.—**The oleum origani cretici should be used. It is of light-brown color and clears sections dehydrated in 95 per cent. alcohol or stronger. It slowly discharges anilin-colors.

5. **Oil of Cloves.—**This clearing-agent dissolves collodion and discharges anilin-colors. It may be used when it is desired to get rid of the collodion used for embedding after the sections have been stained. This removal is favored by dehydration in absolute alcohol before clearing.

6. **Oil of Cedar-wood.—**This, when pure, has a very light-yellow color and smells like cedar-wood. It should be free from the more pungent odor of the oil derived from the leaves. This essential oil does not discharge anilin-colors, and is, therefore, useful when those dyes have been employed. It clears slowly, but well, and may be used after dehydration with 95 per cent. alcohol.

## Methods of Mounting.

Sections that have been treated by the foregoing methods of preparation are fitted for mounting in a solution of some resin. The most commonly employed are Canada balsam and dammar. The best solvent for these resins is xylol, though chloroform and benzol are sometimes used for this purpose. All traces of turpentine should be removed from the balsam before its solution, to avoid the discharge of stains with hæmatoxylin or anilin-dyes which turpentine occasions.

When sections are transferred from alcohol to a clearing-agent they float upon the surface of the latter, and can then be flattened and all folds removed. As the alcohol is extracted the sections

sink in the clearing-agent. In order to transfer them from the
clearing-agent to a slide, the first step in mounting, a good method
is to slip a strip of cigarette-paper under the section, withdraw it
along with the section (using a platinum needle as aid, if necessary),
drain off the superfluous fluid, and then lay the cigarette-paper on
the slide, section side down. Light pressure will now squeeze out
considerable of the clearing-agent, when the paper can be stripped
from the section and slide, leaving the section nearly dry and without folds or wrinkles. With a little care, this method of transferring
the section to the slide rarely fails. When such is the case the
manipulations must be repeated.

A drop of the mounting-medium is now placed upon the section
and a cover-glass laid on and gently pressed down until it comes in
contact with the section and the excess of balsam or dammar is
expelled from beneath the cover. If the sections tend to raise the
cover, the latter may be weighted with a bullet placed in its centre.
Freshly mounted specimens are not so favorable for examination
with high powers as those that have been mounted for a few hours
or days. This is because the refractive indices of the clearing-agent
and mounting-medium are not identical. When these have become
thoroughly mixed, or the former has evaporated, the specimen is
impregnated with and surrounded by a homogeneous medium that
does not scatter the light passing through it.

Canada balsam has a somewhat higher refractive index than
dammar. It therefore renders the sections a little more transparent
and more completely obliterates the structure-picture. When it is
desired to retain as much of the structure-picture as possible,
which is usually the case, dammar should be chosen for the mounting-medium. It dries a little more slowly than balsam, but soon is
sufficiently dry at the edges of the cover-glass to preserve the section from injury. If the slides are kept in a horizontal position,
in a warm place (40° to 50° C.; 104° to 122° F.), for a couple of
days, they will be dry enough for storage, but for several weeks
must be handled with care.

Stained sections may be examined in glycerin, having been
mounted by the same manipulations as those used for mounting
in balsam, without previous dehydration or clearing. Such mounts
are, however, difficult of preservation. The various cements that
have been recommended for fastening the edges of the cover-glass
to the slide are usually inefficient, as the changes of temperature

that are inevitable cause the glycerin to make its way between the glass and cement, loosening the latter.

A better medium than glycerin for sections that cannot be subjected to the action of alcohol for the purpose of dehydration is glycerin-jelly. This is prepared by soaking the best French gelatin in cold water until it has imbibed all it will readily take up, then melting the gelatin, after pouring off the excess of water, and adding an equal bulk of glycerin. A little carbolic acid may be added to the mixture to preserve it. The manipulations for mounting are similar to those given above, the sections being transferred from water to the slide. The glycerin-jelly may be melted and a drop placed upon the section, or a little lump of the solid jelly may be placed upon a cover-glass, melted by gentle heat, and the cover-glass then inverted over the section on the slide. After the jelly has dried at the edges of the cover-glass they may be painted with xylol balsam, dammar, or some cement.

### The Rapid Preparation of Sections for Diagnosis.

The most expeditious means of obtaining sections of fresh tissues is to cut them without preliminary treatment with reagents, either free hand with a razor, or with the aid of a freezing microtome (page 415). Such sections may be stained with methylene-blue (aqueous solution, page 423), or they may be examined in neutral salt solution. If they are to be stained, spread them out on a slide, pour a few drops of the methylene-blue solution over them, and, after a few moments, wash off the dye with water and cover the section. If such rapid work is not necessary, the sections can be fixed in formalin (page 416), and, after washing out that reagent, stained. Such sections may be hardened and dehydrated, by placing them in dishes of increasingly strong alcohols, and finally mounted in dammar; but the results are by no means so good as when fixation and hardening are done before sections are cut.

When time is not pressing the following method will give good results:

1. Fix and harden pieces not over $\frac{1}{4}$ inch thick in absolute alcohol on quick-lime over night (page 407).

2. Dip the specimen in thick collodion and embed it on a glass block by the rapid method (page 412). When the block has been in 80 per cent. alcohol for three or four hours it may be cut; but it is better to let the collodion harden for twenty-four hours.

3. Stain with hæmatoxylin and eosin (page 418), cutting short the time of washing after the hæmatoxylin, if in a hurry.
4. Dehydrate in 95 per cent. alcohol; two successive baths.
5. Clear in carbol-xylol.
6. Mount in xylol-dammar.

Very serviceable sections can be prepared in less than twenty-four hours by this method, and the specimens, though not of the best quality, will be permanent, and may be kept for future reference.

## Special Methods.

The foregoing methods are for the preparation of tissues from which sections must be made before they are fit for examination under the microscope. The physician is, however, frequently called upon to examine other objects, when the following directions will be found useful.

1. **Examination of Urinary and other Sediments.**—For the collection of the sediment vessels with vertical walls should be used, not conical glasses. A test-tube answers very well. The sediment should be allowed to settle for several hours in a cool place, to avoid decomposition; or, better, the sediment should be precipitated by means of a centrifuge. It should be borne in mind that urine becomes alkaline during decomposition, and that the ammonia produced causes changes in the characters of the crystalline or other inorganic constituents of the sediment, and also renders the identification of the organic constituents difficult or impossible.

When the sediment has collected at the bottom of the vessel a portion should be removed with a pipette for examination. Place the finger over one end of the pipette before introducing it into the liquid, to retain the air, then place the other end in contact with the sediment and allow the air to escape slowly by raising or moving the finger a little. Close the upper end of the pipette and withdraw it. Now carefully wipe the outside of the pipette and let the fluid escape until a good sample of the sediment is at the end of the tube. Place a drop of this sediment on a slide and cover. Examine the specimen with a low power at first, taking care to use a very small diaphragm. In this way the presence of urinary casts may be more rapidly determined than if a high power is used. When there is doubt as to a given object being a cast examine it with a higher power. After the specimen has been examined for casts and other objects large enough to be identified with a low power, use the high power for the detection of red blood-corpuscles, pus, etc. Objects in urinary sediments may be stained with aqueous methylene-blue, Gram's solution of iodine, or alum-carmine; or their chemical nature determined by means of microchemical reactions.

2. **Preparation of Cover-glass Smears.**—These are used for the examination of blood, pus, sputa, cultures of bacteria, etc., when it is desired to employ stains. They are also employed occasionally for the study of some of the constituents of soft tissues.

A small drop, or fragment, of the specimen is placed between two cover-glasses. If the specimen is sufficiently fluid, it will at once spread out into a thin layer between the covers. When this is not the case pressure may be used. The covers are then *drawn* apart, not lifted, leaving a coating upon both. They are allowed to dry spontaneously, after which the film is fixed by passing the cover-glasses three times through a flame, care being taken not to scorch the film, which should not come in contact with the flame. Heat applied through the glass to the dry film will render it insoluble and affix it to the cover. The constituents of the film may then be stained on the cover-glass, the latter being either floated on the dye or immersed in it as though it were a section. Hæmatoxylin and eosin may be employed; but anilin-dyes, such as methylene-blue, carbol-fuchsin, anilin-gentian-violet, etc., are more commonly used.

3. **Examination of Sputa for Tubercle Bacilli.**—The small cheesy particles in the sputa are most likely to contain tubercle bacilli. Cover-glass smears are stained by the following method:

*a.* Stain fifteen minutes in freshly filtered carbol-fuchsin at the room-temperature, or heat until vapors rise from the surface of the dye, and maintain that temperature for about three minutes.

*b.* Wash off the excess of dye with water.

*c.* Differentiate in dilute sulphuric acid, prepared by adding 5 cc. of pure acid to 95 cc. of distilled water, until the cover-glass has only a faint tinge of pink when the acid is washed off with water.

*d.* Wash in water to remove the acid.

*e.* Counterstain with aqueous methylene-blue for two minutes.

*f.* Wash in water.

*g.* Dry the cover-glass and mount it, film side down, on a drop of xylol-dammar.

The tubercle bacilli will be stained red, while other bacteria and

28

the nuclei of cells will be blue. This method, like all others used for the detection of the tubercle bacillus, depends upon the fact that that bacillus takes up colors with reluctance, but, after staining, holds them tenaciously. The specimen is therefore first stained with a strong dye, is then decolorized with some agent that will discharge the color from all bacteria except the tubercle bacillus (and spores, which, however, have a different shape from that of the tubercle bacillus), and afterward stained with a weaker dye of another color which is imparted to the bacteria that have been decolorized.

4. **Examination of Urethral Pus for the Gonococcus.**—The gonococcus is shaped a little like a coffee-bean, and usually occurs in pairs with the flattened surfaces of the individual cocci facing each other. In pus it is frequently situated within the leucocytes, while the other varieties of pyogenic cocci usually lie outside of the pus-corpuscles. The gonococcus is decolorized by treatment with Gram's iodin solution followed by alcohol; the more common cocci found in suppuration are not decolorized. These differences in shape, situation, and behavior toward dyes serve to distinguish the gonococci from the other cocci that may be present. The smears, fixed by heat, are stained as follows:

*a.* Stain for five minutes in freshly filtered anilin-gentian-violet.

*b.* Wash off excess of dye with water.

*c.* Immerse in Gram's solution for two minutes.

*d.* Decolorize in 95 per cent. alcohol till no more color is given off.

*e.* Stain two minutes in aqueous fuchsin, prepared in a manner similar to that used for aqueous methylene-blue. Bismarck-brown may be used for this counterstain in place of the fuchsin.

*f.* Wash in water, dry, and mount in dammar or balsam. The gonococci will be stained by the *second* dye used; other cocci belonging to the pyogenic group will be a dark purple, they having retained the color first imparted to all the bacteria by the gentian-violet. In this case the gonococci are distinguished from the other cocci by taking advantage of the fact that they are "negative to Gram," while the others are "positive."

5. **Examination of Blood-smears.**—Hæmatoxylin, followed by a strong counterstain with eosin, will furnish useful specimens for most purposes. The differentiation of the various granules in the white corpuscles described by Ehrlich requires special methods, for a description of which the reader is referred to special works on the

blood or clinical microscopy. The malarial plasmodia are best detected in perfectly fresh blood, examined immediately with an immersion-lens, when their changes of form serve to make them more easily recognizable than when they are sought in smears. In the latter they may be stained by the following method:

*a.* Fix the film by means of heat, or, better, by immersion in absolute alcohol for half an hour. (In the latter case wash off the alcohol with water before staining.)

*b.* Stain for several hours in Chenzinsky-Pehn's stain:

Concentrated alcoholic solution of methylene-blue,    10 cc.
0.5 per cent. solution of eosin in 70 per cent. alcohol,    5 cc.
Distilled water,    10 cc.

The solution should be filtered before, and preserved from evaporation during, the staining.

*c.* Wash in water, dry, mount in xylol-dammar.

The malarial plasmodia will be stained blue, the body of the red corpuscles red, the nuclei of the leucocytes blue, and eosinophile granules, within those cells, red.

**6. Examination of Bacteria in Cover-glass Preparations.**—If the bacteria are already in a fluid, a drop is placed upon a cover-glass, spread over its surface, allowed to dry spontaneously, and then fixed by heat, as described above. If cultures on solid media are to be examined, a drop of water is first placed upon the cover-glass, and a little mass of the bacteria disseminated through it, and then the mixture is spread in a thin layer by means of the platinum needle. It is then dried and fixed, as in the preceding case. Such preparations may be stained with methylene-blue, carbol-fuchsin, by Gram's method (anilin-gentian-violet, followed by Gram's iodine solution, and then alcohol), or with some other anilin-dye. For the diphtheria or typhoid bacillus an alkaline methylene-blue (see Unna's formula) serves well.

**7. Examination in Hanging Drop.**—This method is useful for the observation of objects suspended in a fluid. It is extensively used in bacteriology for the study of living bacteria. A drop of the fluid is placed on the centre of a cover-glass, which is then inverted over the concavity in a hollowed slide. The edges of the cover-glass should then be sealed with a drop of water or oil, to prevent evaporation of the hanging drop.

**8. Microchemical Reactions.**—These reactions are resorted to to determine the chemical nature of objects under the microscope. Every stain is the result of a microchemical reaction, but as yet the knowledge obtained by staining tissues cannot always be expressed in chemical language.

The manipulations are usually so conducted that the reaction can be directly observed under the microscope. The object to be studied is placed in the middle of the field. The reagent used is then placed at one edge of the cover-glass, whence some of it will flow beneath the latter. To facilitate the entrance of the reagent a narrow strip of filter-paper may be brought in contact with the opposite edge of the cover, withdrawing some of the fluid from beneath it. It is best to sharpen the end of the strip which comes in contact with the cover-glass, so that the absorption of fluid shall be slow; otherwise the currents induced will be likely to wash the object from the field of vision. The following tests, applied in this way, may be of use:

*a.* Urates. Insoluble in 1 per cent. acetic acid; soluble, on the application of heat, in water (or urine). The slide must be removed from the microscope when heat is applied to it.

*b.* Earthy phosphates. Dissolve on the addition of 1 per cent. acetic acid. Are not dissolved by heat.

*c.* Calcium oxalate. Insoluble in 1 per cent. acetic acid; soluble in 1 per cent. hydrochloric acid.

*d.* Carbonates. Soluble in 1 per cent. acetic acid or hydrochloric acid, with evolution of gas-bubbles.

*e.* Albuminoid granules. Become indistinct, and finally invisible, on the addition of 1 per cent. acetic acid or 1 per cent. potassium hydrate; not blackened by osmic acid.

*f.* Fatty granules. Not affected by 1 per cent. acetic acid or 1 per cent. potassium hydrate. Stained black or dark brown by osmic acid.

*g.* Starch. Stained dark blue to black by iodine solutions. Use Gram's solution.

*h.* Cellulose. Stained yellow by iodine solutions. If the water be then removed and concentrated sulphuric acid introduced, the color becomes blue. The walls of most vegetable cells are composed of cellulose.

*i.* Teichmann's test for hæmoglobin. This test depends upon the conversion of the hæmoglobin or its derivatives into hæmin, which crystallizes in rhombic plates of a reddish-brown color. The hæmin is produced by heating with a little salt and strong acetic acid. Evaporate a drop of neutral salt solution to dryness on a slide. Place the substance to be tested upon it and cover. Fill the space between cover and slide with glacial acetic acid and heat over a flame till bubbles begin to form. Maintain that heat for a few minutes, replacing loss by fresh additions of acetic acid. Let the slide cool slowly, and, when cold, examine. If the results are negative, repeat the heating with acetic acid. The acid should not actually boil, but should be kept at the point of incipient ebullition.

*j.* Tests for amyloid substance. Sections of fresh tissue may be soaked for some time in Gram's solution, then washed and examined in water. Amyloid substance is stained reddish-brown, the tissues yellow. Sections of tissues fixed in alcohol, corrosive sublimate, or formaldehyde, may be stained in a solution of 1 per cent. methyl-violet dissolved in distilled water, without the addition of alcohol. The sections are then washed in 1 per cent. hydrochloric acid for the purpose of differentiating the stain. After thorough washing in several changes of water they may be mounted in glycerin-jelly. The amyloid substance is stained reddish-violet, the other tissues blue.

*k.* Test for iron in pigmentations. The iron from the hæmoglobin of the blood is sometimes present in the pigmentation resulting from old extravasations, in the form of hæmosiderin. The same compound is also sometimes found in the tissues in cases of pernicious anæmia. The presence of iron in this pigmentation may be demonstrated by the following method:

(*a*) The tissues should be fixed in alcohol.

(*b*) Soak the section in a 2 per cent. solution of potassium ferrocyanide for ten minutes.

(*c*) Transfer to Orth's acid alcohol (page 422) for five or ten minutes.

The sections may now be examined in a glycerin-mount with a wide diaphragm, or they may be counterstained, for which purpose treat as follows:

(*d*) Wash with water.

(*e*) Stain with Orth's lithio-carmine.

(*f*) Dehydrate and mount in xylol-dammar.

The iron in the section is converted into Prussian blue; the nuclei of the cells, when the counterstain has been employed, are red.

*l.* Examination of sputa for elastic fibres. In pulmonary disease involving a destruction of pulmonary tissue and the appearance of fragments in the expectoration, elastic fibres from the alveolar walls may frequently be found in the sputa :

Fill a test-tube one-third full of sputa, add five or six drops of 36 per cent. potassium hydrate solution, and boil the mixture for three or four minutes. Add an equal bulk of distilled water. Divide the contents of the tube between the two tubes of the centrifuge and precipitate their contents. If elastic fibres were present, they will be found either in the sediment or in the scum on the top of the fluid.

### 9. Methods of Maceration.—

*a.* One-third alcohol.

| | |
|---|---|
| 95 per cent. alcohol, | 35 cc. |
| Distilled water, | 65 " |

This dilute alcohol is excellent for the separation of epithelium from the surfaces of mucous membranes. The fresh tissues are placed in the alcohol for a day or two, after which the cells can easily be detached and separated by shaking. The cells are well preserved, and may be stained with methylene-blue or alum-carmine.

*b.* Potassium hydrate.

| | |
|---|---|
| Potassium hydrate, pure by alcohol, | 36 grams. |
| Distilled water, | 64 cc. |

The solution should be cold before use. It cannot be filtered through paper; but if not clear, should be decanted from any sediment, or a fresh solution prepared. Maceration takes place very quickly in this solution. The tissues can usually be teased apart within fifteen to thirty minutes. They must be examined in the potash solution *without dilution*, as the addition of water quickly destroys the tissue-elements. For this reason the specimens to be macerated should be placed in several times their bulk of the potash solution; otherwise the water they contain will dilute the potash. Permanent mounts cannot be made.

*c.* Chromic acid. A solution of 1 part of the acid in 10,000 parts of distilled water will facilitate the teasing apart of tissue-elements which have macerated in it for one to several days. After careful washing on the slide alum-carmine alone, or followed by picric acid, may be used for staining.

## SPECIAL METHODS.

10. **Methods of Decalcification.**—Tissues which contain calcified nodules or bone must be freed from lime-salts before they can be cut. It is difficult to do this rapidly without injury to the softer tissue-elements. When good results are desired, and the necessary time can be afforded, the tissues should first be fixed and hardened, small pieces being selected. Zenker's fluid fixes well for this purpose, but Orth's fluid or alcohol may be used. If Zenker's or Orth's fluid is used, the tissues must be washed in water and hardened in alcohol for at least a day before they are decalcified (see Methods of Fixing and Hardening, pp. 403, 408).

Decalcification is accomplished by treatment with acids. Five per cent. nitric acid will decalcify small pieces of bone in from one to five days. The progress of the decalcification may be determined by pricking the tissue with a needle, but after it appears to be soft it is well to continue the action of the acid for a day or two, lest some undissolved particles should remain and injure the edge of the microtome-knife. A saturated aqueous solution of picric acid is sometimes used for decalcifying. Its action is very slow, though not injurious to the tissues, which require no preliminary treatment, the picric acid acting as a fixing and decalcifying agent.

After decalcifying in nitric acid the tissues should be thoroughly washed in running water for twenty-four hours and then rehardened in alcohol, after which they may be embedded. After decalcifying in picric acid the tissues are placed in 70 per cent. alcohol and hardened without previous washing in water.

When rapid decalcification is necessary nitric acid and phloroglucin, which restrains the destructive action of the acid, may be used. The solution is prepared by dissolving 1 gram of phloroglucin in 10 cc. of pure nitric acid. To this 100 cc. of 10 per cent. nitric acid are added. Decalcification takes place within a few hours in this solution, which contains about 20 per cent. of nitric acid. The tissues should then be washed and hardened.

Another rapid method which combines decalcification with hardening is to place the fresh tissues in a large bulk of 5 per cent. nitric acid in 80 per cent. alcohol. After decalcification has taken place the tissues are hardened in alcohols of increasing strength, large quantities being used in order to remove the acid. Before staining, the sections should be washed thoroughly in water to get rid of any residual traces of acid.

# INDEX.

ABSCESS, 312
  cold, 319
Absorption, 295
Achromatin, 34
Acidophilic cells, 119
Active hyperæmia, 298
Acute inflammation, 297
  parenchymatous inflammation, 268
  nephritis, 272
Adeno-carcinoma, 390
Adeno-fibroma, 377
Adenoma, 376
Adipose tissue, 78
Adrenal bodies, 186
Adventitia, 112
Akromegaly, 191
Albumin, Mayer's, 417
Albuminoid degeneration, 266
Alcohol, absolute, 407
Alveoli, pulmonary, 171
Amœba, 28
Amyloid infiltration, 281
  substance, tests for, 437
Anæmic infarcts, 332
Angiomata, 373
Angiomatous tumors, 373
Anilin-water, 424
Areolar tissue, 76
Arteries, 110
  helicine, 223
Association-fibres of cerebrum, 249
  of spinal cord, 239
Atrophy, 284
  functional, 284
  pressure-, 285
  senile, 287
Attraction-spheres, 35
Axis-cylinder, 97

BACTERIA, examination of, 435
  Basement-membrane, 58
Basophilic leucocytes, 126
Bergamot, oil of, 429
Bladder, 164
Blood, 122
  -plates, 126
  -smears, preparation of, 434
Bodies, adrenal, 186
  Malpighian, 154

Bodies, Malpighian, of spleen, 177
  Pacinian, 252
  pearl-, 390
  polar, 35, 217
Body, pituitary, 139
Bone, 68
  canaliculi of, 68
  general character of, 68
  Haversian canals of, 68
  lacunæ of, 68
  -marrow, 71, 119
  red, 119 .
  yellow, 119
  regeneration of, 338
Bowman, glands of, 255
Bowman's capsule, 159
Bronchi, 169
Bronchioles, 170
Broncho-pneumonia, 317
Brownian movement, 29
Brunner, glands of, 141
Bulb, olfactory, 258
  glomeruli of, 257

CACHEXIA strumipriva, 183
  Calcareous infiltration, 282
Calcium oxalate, tests for, 436
Callus, 309
Canada-balsam, 428, 430
Capillaries, 113
Capsule, Bowman's, 159
  Glisson's, 146
Capsules, supra-renal, 186
Carbol-fuchsin, 423
Carbonates, tests for, 436
Carbo-xylol, 429
Carcinoma, 382
  colloid, 388
  medullary, 384
  scirrhous, 384
  simple, 384
Cardiac glands, 136
  muscles, 89
Carmine, alum-, 421
  borax-, 421
  lithio-, 422
  neutral, 420
Carotid glands, 194
Cartilage, 64

441

Cartilage, elastic, 67
  fibro-, 66
  general character of, 64
  hyaline, 66
  matrix of, 65
  ossification of, 64
  regeneration of, 338
Catarrhal inflammations, 316
  pneumonia, 317
Cedar-wood, oil of, 429
Cell, or cells, 27
  acidophilic, 119
  compound granule-, 316
  decidual, 215
  of Deiters, 260
  -division, 40
    amitotic, 40
    centrosome in, 34
  ganglion-, 95
  giant-, 40, 119
  glia-, 101
  goblet-, 52, 139
  hair-, 260
  migratory, 124
  mitral, 258
  of Müller, 262
  nerve-, 95
  organs of, 31
  plasma-, 120
  prickle-, 55
  of Purkinje, 243
  reproduction of, 34
  of Sertoli, 225
  stellate, 245
  sustentacular, of retina, 261
  in testis, 227
  wandering, 124
Cellulose, tests for, 436
Centrosome, 31
Cerebellum, 243
Cerebrum, 246
  association-fibres of, 250
  commissure-fibres of, 249
  projection-fibres of, 249
Cheesy degeneration, 274
Chemotactic substances, 309
Chemotaxis, 309
Chondroma, 350
Chromatin, 34
  reduction of, 226
Chromolysis, 294
Chromoplasm, 34
Chromosomes, 37
Chronic inflammations, 322
  interstitial, 324
  parenchymatous, 269
Chyle, 126
Cicatricial tissue, 308
Ciliated epithelium, 53
Circulatory system, 108
Cirrhosis of liver, 323
Clarke, column of, 241
Clearing, methods of, 428

Clearing, methods of, bergamot, oil of, 429
  carbol-xylol, 429
  cedar-wood, oil of, 429
  cloves, oil of, 429
  origanum, oil of, 429
  xylol, 428
Clefts of Lantermann, 99
Cloves, oil of, 429
Coagulation, explanation of, 127
  -necrosis, 294
Coccygeal gland, 193
Collateral fibres of spinal cord, 239
Colliquative necrosis, 295
Collodion, 409, 412
Colloid, 181
  carcinoma, 388
  degeneration, 278
Colon, 142
Colostrum, 219
  -corpuscles, 219
Column of Clarke, 241
Columnar epithelium, 52
Commissure-fibres of cerebrum, 249
Compensatory hypertrophy, 289
Congestion, passive, 326
Connective tissue, 63
  tumors, 347
Contractile substance, 83
Cord, spinal, 236
Corium, 196
Corpora amylacea, 224
Corpus album, 210
  cavernosum, 222
  hæmorrhagicum, 210
  luteum, 210
  spongiosum, 222
Corpuscles, colostrum-, 219
  genital, 253
  of Krause, 253
  of Meissner, 253
  red, 123
  tactile, 252
  white, 124
Croupous inflammation, 317
  membrane, 319
Crypts of Lieberkühn, 139
Cubical epithelium, 49
Cuticle of epithelium, 50
Cuticularized epithelium, 54
Cutting, methods of, 415
  free-hand, 415
  frozen sections, 415
  celloidin sections, 416
  collodion sections, 416
Cylindroma, 356
Cystoma, 392
Cytoplasm, 29, 32

DECALCIFICATION, methods of, 439
  Decidual cells, 215
Degenerations, 265
  albuminoid, 266
  cheesy, 274

Degenerations, colloid, 278
 fatty, 266
 hyaline, 280
 keratoid, 280
 mucous, 277
 of nerves, 283
 parenchymatous, 266
Dehydration, methods of, 428
Deiters' cells, 260
Dendrite, 234
Dermoid cysts, 392
Developmental hypertrophy, 230
Diapedesis, 301
Diaster-phase of karyokinesis, 38
Digestive organs, 128
Diphtheritic inflammation, 313
 membrane, 294
Discus proligerus, 209
Dispirem-phase of karyokinesis, 38
Ductless glands, 62, 180
Duodenum, 137

ECTODERM, 20
 Ectoplasm, 29
Elastic cartilage, 67
 fibres, 73
Eleidin, 193
Elements, sarcous, 93
Elementary tissues, 41
Embedding, methods of, 411
 celloidin, 412
 collodion, 412
 paraffin, 413
Embolism, 330
Embryonic layers, 22
Encapsulation, 296
Endoderm, 20
Endoneurium, 100
Endoplasm, 29
Endothelioma, 355
Endothelium, 45
 functions of, 48
 general characters of, 45
 regeneration of, 336
Energy, kinetic, 18
 potential, 18
Eosin, 420
Eosinophilic leucocytes, 126
Epicardium, 109
Epidermis, 197
Epididymis, 225
Epiglottis, 168
Epineurium, 100
Epithelial tumors, 376
Epithelioma, 391
Epithelium, 49
 ciliated, 53
 columnar, 52
 cubical, 49
 cuticle of, 50
 cuticularized, 54
 functions of, 41, 57
 activities of, 57

Epithelium, general characters of, 49
 germinal, 207
 glandular, 50
 pavement-, 51
 regeneration of, 336
 stratified, 54
 transitional, 56
Erectile tissue, 222
Erythroblasts, 119
External genitals, 217
Exudate, inflammatory, 301

FALLOPIAN tubes, 210
 Fatty degeneration, 266
 infiltration, 574
Fibres, association-, of cerebrum, 250
 of cord, 239
 collateral, of cord, 239
 commissure-, of cerebrum, 249
 connective-tissue, staining, 425
 elastic, 73
 moss-, 246
 nerve-, 96
 staining, 426
 projection-, of cerebrum, 249
 Sharpey's, 70
 white, 73
 yellow, 73
Fibrin, 126
Fibrinous inflammation, 313
Fibro-cartilage, 66
Fibroma, 347
Fibrous tissues, general character of, 72
 regeneration of, 336
Figures, mitotic, preservation of, 406
Fixation, methods of, 403
 alcohol, absolute, 407
 boiling, 408
 Flemming's solution, 406
 formaldehyde, 405
 mercuric chloride solution, 405
 Müller's fluid, 403
 Orth's fluid, 404
 Zenker's fluid, 404
Flemming's solution, 406
Follicles, Graafian, 207
 lymph-, 143
Formaldehyde, 405
Fractures, healing of, 308
Functional atrophy, 284
 hypertrophy, 288

GALL-BLADDER, 151
 Ganglion-cells, 95, 234
Gangrene, 296
 dry, 296
 moist, 296
Genital corpuscles, 253
Gentian-violet, 423
Germinal epithelium of ovary, 207
Giant-cell sarcoma, 367
Giant-cells, 40, 119
Giannuzzi, crescents of, 131

Gland, mammary, 218
 thyroid, 181
Glands of Bowman, 255
 of Brunner, 141
 cardiac, of stomach, 136
 carotid, 194
 coccygeal, 195
 ductless, 62, 180
 lymphatic, 114
 parotid, 131
 pyloric, 136
 salivary, 131
 sebaceous, 201
 secreting, 58
 sublingual, 131
 submaxillary, 131
 sweat-, 198
Glandular epithelium, 50
Glioma, 394
Glisson's capsule, 146
Glomeruli, olfactory, 258
Glomerulus, 158
Glycerin, 430
 jelly, 431
Glycogenic infiltration, 275
Goblet-cells, 52, 139
Gonococcus, staining of the, 434
Graafian follicles, 207
 development of, 208
Gram's solution, 424
Granulation-tissue, 304
Granules, albuminoid, tests for, 436
 fatty, tests for, 436
Granulomata, 318

HÆMANGIOMA, 374
 Hæmatoidin, 328
Hæmatoxylin, 418
Hæmoglobin, tests for, 436
Hæmorrhage, 328
Hæmosiderin, 328
Hair, 199
 -cells, 260
 cuticle of, 200
 -follicles, 199
 development of, 204
Hanging-drop preparations, 435
Hardening, methods of, 408
Haversian canals, 68
Hearing, 259
Heart, 109
Helicine arteries, 223
Henle, tubes of, 155
Hepatization of lung, gray, 313
 red, 313
Hyaline cartilage, 66
 degeneration, 280
Hyaloplasm, 29
Hyperæmia, active, 298
 inflammatory, 298
 passive, 286, 326
Hyperplasia, 288
 inflammatory, 290

Hypertrophy, 288
 compensatory, 289
 developmental, 290
 functional, 288
 inflammatory, 290
Hypophysis cerebri, 189

IMPREGNATION, methods of, 409
 celloidin, 409
 collodion, 409
 paraffin, 409
Infarcts, 332
 anæmic, 332
 hæmorrhagic, 332
Infiltration, amyloid, 281
 calcareous, 282
 fatty, 274
 glycogenic, 275
 serous, 276
Infiltrations, 265
Inflammation, acute, 297
 parenchymatous, 268
 catarrhal, 316
 chronic, 322
 interstitial, 324
 parenchymatous, 269
 croupous, 317
 diphtheritic, 318
 fibrinous, 313
 serous, 315
Inflammatory exudate, 301
 hyperæmia, 298
 hyperplasia, 290
 hypertrophy, 290
 repair, 303
 stasis, 298
Infundibula of lung, 171
Interstitium, 106
Intestine, small, 141
Intima, 110
Involuntary muscles, 88, 91
Iron-hematoxylin stain, 425
 tests for, 437

KARYOKINESIS, 34
 diaster-phase, 38
 dispirem-phase, 38
 monaster-phase, 37
 significance of, 39
 spirem-phase, 35
Karyolysis, 294
Keloid, 360
Keratin, 198
Keratoid degeneration, 280
Kidney, cortex of, 153
 pelvis of, 163
 Malpighian bodies of, 154
Kidneys, 153
Kinetic energy, 18
Krause, corpuscles of, 253

LACTEALS, 114
 Lautermann, clefts of, 99

INDEX. 445

Larynx, 168
Layers, embryonic, 22
Leiomyoma, 370
Leucocytes, 124
 basophilic, 126
 emigration of, 300
 eosinophilic, 126
 large mononuclear, 125
 polynuclear neutrophilic, 125
Lieberkühn, crypts of, 139
Lipoma, 350
Liver, 146
 cirrhosis of, 323
 functions of, 151
 lobules of, 147
Lobar pneumonia, 313
Lung, functions of, 173
 gray hepatization of, 313
 infundibula of, 171
 red hepatization of, 313
Lymph, 122
 -nodes, 114
Lymphadenoid tissue, 76
Lymphatic glands, 114
Lymphatics, 114
Lympho-angioma, 376
Lymphocytes, 125
Lympho-sarcoma, 363

MACERATION, methods of, 438
 alcohol, 438
 chromic acid, 438
 potassium hydrate, 438
Malpighian bodies of kidney, 154
 bodies of spleen, 177
Mammary gland, 218
Marrow, 71, 119
Matrix of cartilage, 65
Maturation of the ovum, 217
Mayer's albumin, 417
Measurements, microscopical, 398
Medullary carcinoma, 384
 sheath, 97
Meissner, corpuscles of, 253
Melano-sarcoma, 369
Membrane, basement, 58
 croupous, 318
 diphtheritic, 294
 pyogenic, 313
Mercuric chloride solution, 405
Mesoderm, 22
Metakinesis, 37
Metaplasia, 291
Metaplasm, 33
Methylene-blue, aqueous, 423
 Unna's, 422
Microchemical reactions, 436
Microscope, care of, 397
 selection of, 397
Microscopical measurements, 398
 technique, 399
Migratory cells, 124
Mitral cells, 258

Monaster-phase of karyokinesis, 37
Mononuclear leucocytes, large, 125
Moss-fibres, 246
Motor plates, 104
Mounting, methods of, 429
 Canada-balsam, 428, 430
 Dammar, 428
 glycerin, 430
 glycerin-jelly, 431
Movement, Brownian, 29
 amœboid, 29
Mucoid marrow, 119
Mucous degeneration, 277
 tissue, 74
Mucus, 278
Müller, cells of, 262
Müller's fluid, 403
Muscular tissues, 83
 tumors, 370
Muscle, cardiac, 89
 regeneration of, 340
 involuntary, 88, 91
 smooth, 83
 function of, 88
 regeneration of, 339
 striated, 91
 regeneration of, 340
Myelin, 97, 98
Myelocytes, 119
Myxœdema, 183
Myxoma, 353

NAILS, 201
 Necrosis, 293
 coagulation-, 294
 liquefaction, 295
 of nucleus, 294
Nephritis, acute parenchymatous, 272
Nerve-cells, 95
 degeneration of, 283
 -fibres, 96
 -terminations, 103
Nervous system, 234
 tissues, 94
 regeneration of, 340
Neurilemma, 97
Neurite, 234
Neuroglia, 101
Neurons, 234
Nodes of Ranvier, 98
Nucleolus, 29, 33
Nucleus, 29
 necrosis of, 294
 structure of, 33

ŒSOPHAGUS, 134
 Olfactory bulb, 258
 layers of, 258
 glomeruli, 258
Organs, 106
Orth's fluid, 404
Origanum, oil of, 429
Ossification of cartilage, 64

Osteoma, 353
Ovary, 207
Ovula Nabothi, 216
Ovum, 20
  maturation of, 217

PACINIAN bodies, 252
  Pancreas, 142
Papilloma, 394
Paraffin, 409, 413
Parathyroids, 185
Parenchyma, 106
Parenchymatous degeneration, 266
  inflammation, acute, 268
    chronic, 269
  nephritis, acute, 272
Passages, alveolar, 170
Passive congestion, 326
  hyperæmia, 326
Pavement-epithelium, 51
Pelvis, renal, 163
Penis, 222
Perichondrium, 65
Perineurium, 100
Periosteum, 71
Peyer's patches, 143
Phagocytosis, 332
Phosphates, earthy, tests for, 436
Pia mater, 251
Picture, color-, 402
  structure-, 402
Pituitary body, 189
Plasma-cells, 120
Pleurisy, 314
Pneumonia, broncho-, 317
  catarrhal, 317
  lobar, 313
Polar bodies, 35, 217
Polynuclear neutrophilic leucocytes, 125
Potential energy, 18
Pressure-atrophy, 285
Prickle-cells, 55
Projection-fibres of cerebrum, 249
Prostate, 224
Protoplasm, 29
Psammoma, 356
Pseudopodium, 29
Pseudo-stomata, 47
Pulmonary alveoli, 171
Purkinje, cells of, 243
Pus, 312
Pyloric glands, 136
Pyogenic membrane, 313

RANVIER, nodes of, 98
  Razor, stropping, 415
Reaction, microchemical, 436
Rectum, 142
Red corpuscles, 123
Regeneration of bone, 338
  of cartilage, 338
  of endothelium, 336
  of epithelium, 336

Regeneration of fibrous tissue, 336
  of muscles, cardiac, 340
    smooth, 339
    striated, 340
  of nervous tissues, 340
  of tissues, 334
Renal pelvis, 163
Repair, inflammatory, 303
Reproductive organs, 207
Respiratory organs, 168
Rete mucosum, 197
  vasculosum, 233
Reticular tissue, 76
Retina, 260
  sustentacular cells of, 261
Rhabdomyoma, 372
Round-cell sarcoma, large, 364
  small, 362

SALIVARY glands, 131
  Salt solution, normal, 399
Sarcolemma, 93
Sarcoma, 359
  giant-cell, 367
  large round-cell, 364
  lympho-, 363
  melanotic, 369
  small round-cell, 362
  spindle-cell, 365
Sarcoplasm, 93
Sarcostyles, 93
Sarcous elements, 93
Scar, 308
Schwann, sheath of, 98
Scirrhous carcinoma, 384
Sebaceous glands, 201
Secreting glands, 58
Secretion, internal, 62
Sections, rapid preparation of, 431
  staining of, 402
Sediments, examination of, 432
Seminal vesicles, 225
Senile atrophy, 287
Serous infiltration, 276
  inflammations, 315
Sertoli, cells of, 228
Sharpey's fibres, 70
Sheath of Schwann, 98
Sight, 260
Simple carcinoma, 384
Skin, 196
  functions of, 203
Smears, cover-glass, 433, 435
Smell, 255
Smooth muscles, 83
Special senses, organs of, 252
Spermatids, 227
Spermatocytes, 227
Spermatogonia, 227
Spermatozoa, 231
Spinal cord, 236
  association-fibres of, 239
  collateral fibres of, 239

## INDEX. 447

Spindle, achromatic, 37
Spindle-cell sarcoma, 365
Spirem, formation of, 35
Spirem-phase of karyokinesis, 35
Spleen, 176
    Malpighian bodies of, 177
Spongioblasts, 262
Spongioplasm, 29
Sputa, elastic fibres in, 438
    tubercle-bacilli in, 433
Staining, methods of, 418
    carmine, alum-, 421
        borax-, 421
        lithio-, 421
        neutral, 420
    eosin, 420
    fuchsin, carbol-, 423
    gentian-violet, 423
    Golgi's methods, 427
    Gram's solution, 424
    hæmatoxylin, 418
    iron-hæmatoxylin, 425
    methylene-blue, 422, 423
    Pal's method, 426
    Van Giesen's stain, 425
Stasis, inflammatory, 298
Starch, tests for, 436
Stellate cells, 245
    large, 245
    small, 245
Stomach, 134
Stomata, 46
    pseudo-, 47
Stratum granulosum, 198
    lucidium, 198
Stratified epithelium, 54
Striated muscles, 91
Stropping, method of, 415
Submaxillary glands, 131
Substance, contractile, 83
Suppuration, 296, 309
Supra-renal capsules, 186
Sustentacular cells of retina, 261
    of testis, 227
Sweat-glands, 198

TACTILE corpuscles, 252
    Taste, 254
    -buds, 254
Teasing, 400
Technique, microscopical, 399
Teeth, 205
Teledendrites, 234
Teleneurites, 234
Tendon, 80
Testes, 225
Tests for urates, 436
    amyloid substance, 437
    calcium oxalate, 436
    carbonates, 436
    cellulose, 436
    granules, albuminoid, 436
        fatty, 436

Tests for hæmoglobin, 436
    iron, 437
    phosphates, earthy, 436
    starch, 436
Tissue, adipose, 78
    areolar, 76
    cicatricial, 308
    connective, 63
    elementary, 41
        recognition of, 43
    erectile, 222
    fibrous, 72
    fixation of, 401
    fixed elements of, 303
    granulation-, 304
    lymphadenoid, 76
    mucous, 74
    muscular, 83
    necrosed, fate of, 295
    nervous, 94
Tissues, cardiac muscular, 89
    preparation of, 399
        by cutting, 400
        by maceration, 400
    regeneration of, 334
    reticular, 76
    smooth muscular, 83
    striated muscular, 91
Thrombo-phlebitis, 329
Thrombosis, 329
Thrombus, 329
Thymus, 192
Thyroid gland, 181
Thyro-iodine, 184
Tongue, 129
Tonsils, 143
Touch, 252
Trachea, 168
Transitional epithelium, 56
Tubercle, 320
    -bacilli, detection of, 433
Tubercular ulcer, 322
Tuberculosis, 319
Tubes, Fallopian, 210
    of Henle, 155
Tumors, 341
    angiomatous, 373
        hemangioma, 374
        lymphangioma, 371
    benign, 342
    classification of, 345
    connective-tissue, 347
        chondroma, 350
        cylindroma, 356
        endothelioma, 355
        fibroma, 347
        keloid, 360
        lipoma, 350
        myxoma, 353
        osteoma, 353
        psammoma, 356
        sarcoma, 359
            giant-cell, 367

Tumors, connective-tissue, sarcoma, large
    round-cell, 364
    lympho-, 363
    melanotic, 369
    small round-cell, 362
    spindle-cell, 365
  epithelial, 376
    adenoma, 376
    adeno-fibroma, 377
      cystic, 377
      intracanalicular, 378
    carcinoma, 382
      adeno-carcinoma, 390
      medullary, 384
      simple, 384
      scirrhous, 384
      colloid, 388
      cystoma, 392
    epithelioma, 391
    glioma, 394
  etiology of, 342
  malignant, 343
  metastasis of, 344
  mixed, 344
  morbid changes in, 344
  muscular, 370
    leiomyoma, 370
    rhabdomyoma, 372
  nomenclature of, 345
  papillomata, 394
Tunica albuginea, 226
  vaginalis, 226
Tunica, granulosa, 209
  media, 112

ULCER, tubercular, 322
  Urates, tests for, 436
Ureter, 164
Urethra, 165
Urinary organs, 153
Uterus, 211

VACUOLES, 30
  contractile, 30
Vagina, 216
Van Giesen's stain, 425
Vas deferens, 225
Vasa efferentia, 233
  recta, 233
Veins, 113
Vesicles, seminal, 225

WARTS, 395
  White corpuscles, 124
  fibres, 73

XYLOL, 428

YELLOW fibres, 73

ZENKER'S fluid, 404

# Catalogue of Books

PUBLISHED BY

# Lea Brothers & Company,

706, 708 & 710 Sansom St., Philadelphia.
111 Fifth Ave. (Cor. 18th St.), New York.

The books in the annexed list will be sent by mail, post-paid, to any Post-Office in the United States, on receipt of the printed prices. No risks of the mail, however, are assumed either on money or books. Intending purchasers will therefore in most cases find it more convenient to deal with the nearest bookseller.

## STANDARD MEDICAL PERIODICALS.

### The Medical News,

THE LEADING MEDICAL WEEKLY OF AMERICA,

Combines most advantageously for the practitioner the features of the newspaper and the weekly magazine. Its frequent issues keep the reader posted on all matters of current interest and in touch with the incessant progress in all lines of medical knowledge. Close adaptation to the needs of the active practitioner is shown by a list of subscribers large enough to justify the reduction in price to **$4.00 per annum**, so that it is now the cheapest as well as the best large medical weekly of America. It contains thirty-two quarto pages of reading matter in each issue.

### The American Journal of the Medical Sciences.

Containing 128 octavo pages each month, THE AMERICAN JOURNAL accommodates elaborate Original Articles from the leading minds of the profession, careful Reviews and classified Summaries of Medical Progress. According to the highest literary authority in medicine, "from this file alone, were all other publications of the press for the last fifty years destroyed, it would be possible to reproduce the great majority of the real contributions of the world to medical science during that period." Price, **$4.00 per annum**.

### COMBINATIONS AT REDUCED RATES.

THE AMERICAN JOURNAL OF THE MEDICAL SCIENCES, $4.00
THE MEDICAL NEWS, $4.00 } Together $7.50

THE MEDICAL NEWS VISITING LIST for 1898 (see below and on page 16), $1.25. With either or both above periodicals, 75 cents.
THE YEAR-BOOK OF TREATMENT for 1898 (see page 16), $1.50. With either or both above periodicals, 75 cents.

In all **$10.75** for **$8.50**.

### The Medical News Visiting List.

This LIST, which is by far the most handsome and convenient now attainable, has been thoroughly revised for 1898. A full description will be found on page 16. It is issued in four styles. Price, each, $1.25. Thumb-letter Index for quick use 25 cents extra. For Special Combination Rates with periodicals and the Year-Book of Treatment see above.

1,11,9.)

**ABBOTT (A. C.).** *PRINCIPLES OF BACTERIOLOGY:* a Practical Manual for Students and Physicians. New (4th) edition enlarged and thoroughly revised. In one handsome 12mo. volume of 543 pages, with 106 engravings, of which 19 are colored. Cloth, $2.75.

**ALLEN (HARRISON).** *A SYSTEM OF HUMAN ANATOMY; WITH AN INTRODUCTORY SECTION ON HISTOLOGY,* by E. O. SHAKESPEARE, M.D. Comprising 813 double-columned quarto pages, with 380 engravings on stone on 109 full-page plates, and 241 woodcuts. One volume, cloth, $23. *Sold by subscription only.*

**A TREATISE ON SURGERY BY AMERICAN AUTHORS.** *FOR STUDENTS AND PRACTITIONERS OF SURGERY AND MEDICINE.* Edited by ROSWELL PARK, M.D. In two magnificent octavo volumes. Vol. I., *General Surgery,* 799 pages, with 356 engravings and 21 full-page plates in colors and monochrome. Vol. II., *Special Surgery,* 796 pages, with 451 engravings and 17 full-page plates in colors and monochrome. *Complete work now ready.* Price per volume, cloth, $4.50; leather, $5.50. *Net.*

**AMERICAN SYSTEM OF PRACTICAL MEDICINE.** *A SYSTEM OF PRACTICAL MEDICINE.* In Contributions by Eminent American Authors. Edited by ALFRED L. LOOMIS, M.D., LL.D., and W. GILMAN THOMPSON, M.D. In four very handsome octavo volumes of about 900 pages each, fully illustrated. *Complete work just ready.* Per volume, cloth, $5; leather, $6; half Morocco, $7. *For sale by subscription only.* Prospectus free on application.

**AMERICAN SYSTEM OF DENTISTRY.** *IN TREATISES BY VARIOUS AUTHORS.* Edited by WILBUR F. LITCH, M.D., D.D.S. In four very handsome super-royal octavo volumes, containing about 4000 pages, with about 2200 illustrations and many full-page plates. Volume IV., *preparing.* Per volume, cloth, $6; leather, $7; half Morocco, $8. *For sale by subscription only.* Prospectus free on application to the Publishers.

**AMERICAN TEXT-BOOK OF ANATOMY.** See *Gerrish,* page 7.

**AMERICAN TEXT-BOOKS OF DENTISTRY.** *IN CONTRIBUTIONS BY EMINENT AMERICAN AUTHORITIES.* In two octavo volumes of 600–800 pages each, richly illustrated:

—— *PROSTHETIC DENTISTRY.* Edited by CHARLES J. ESSIG, M.D., D.D.S., Professor of Mechanical Dentistry and Metallurgy, Department of Dentistry, University of Pennsylvania, Philadelphia. 760 pages, 983 engravings. Cloth, $6; leather, $7. *Net.*

—— *OPERATIVE DENTISTRY.* Edited by EDWARD C. KIRK, D.D.S., Professor of Clinical Dentistry, Department of Dentistry, University of Pennsylvania. 700 pages, 751 engravings. Cloth, $5.50; leather, $6.50. *Net.*

**AMERICAN SYSTEMS OF GYNECOLOGY AND OBSTETRICS.** In treatises by the most eminent American specialists. Gynecology edited by MATTHEW D. MANN, A.M., M.D., and Obstetrics edited by BARTON C. HIRST, M.D. In four large octavo volumes comprising 3612 pages, with 1092 engravings, and 8 colored plates. Per volume, cloth, $5; leather, $6; half Russia, $7. *For sale by subscription only.* Prospectus free.

**ASHHURST (JOHN, JR.).** *THE PRINCIPLES AND PRACTICE OF SURGERY.* For the use of Students and Practitioners. Sixth and revised edition. In one large and handsome 8vo. volume of 1161 pages, with 656 engravings. Cloth, $6; leather, $7.

**A SYSTEM OF PRACTICAL MEDICINE BY AMERICAN AUTHORS.** Edited by WILLIAM PEPPER, M.D., LL.D. In five large octavo volumes, containing 5573 pages and 198 illustrations. Price per volume, cloth, $5; leather, $6; half Russia, $7. *Sold by subscription only.* Prospectus free on application to the Publishers.

**A TEXT-BOOK OF OBSTETRICS BY AMERICAN AUTHORS.** See *Jewett,* page 9.

**ATTFIELD (JOHN).** *CHEMISTRY; GENERAL, MEDICAL AND PHARMACEUTICAL.* Fourteenth edition, specially revised by the Author for America. In one handsome 12mo. volume of 794 pages, with 88 illustrations. Cloth, $2.75; leather, $3.25.

**BALL (CHARLES B.).** *THE RECTUM AND ANUS, THEIR DISEASES AND TREATMENT.* New (2d) edition. In one 12mo. volume of 453 pages, with 60 engravings and 4 colored plates. Cloth, $2.25. See *Series of Clinical Manuals,* page 13.

**BARNES (ROBERT AND FANCOURT).** *A SYSTEM OF OBSTETRIC MEDICINE AND SURGERY, THEORETICAL AND CLINICAL.* The Section on Embryology by PROF. MILNES MARSHALL. In one large octavo volume of 872 pages with 231 illustrations. Cloth, $5; leather, $6.

**BACON (GORHAM) AND BLAKE (CLARENCE J.).** *ON THE EAR.* One 12mo. volume, 400 pages, with 109 engravings and one colored plate. *Just Ready.* Cloth, $2, *net.*

**BARTHOLOW (ROBERTS).** *CHOLERA; ITS CAUSATION, PREVENTION AND TREATMENT.* In one 12mo. volume of 127 pages, with 9 illustrations. Cloth, $1.25.

**BARTHOLOW (ROBERTS).** *MEDICAL ELECTRICITY. A PRACTICAL TREATISE ON THE APPLICATIONS OF ELECTRICITY TO MEDICINE AND SURGERY.* Third edition. In one octavo volume of 308 pages, with 110 illustrations.

**BELL (F. JEFFREY).** *COMPARATIVE ANATOMY AND PHYSIOLOGY.* In one 12mo. volume of 561 pages, with 229 engravings. Cloth, $2. See *Students' Series of Manuals,* p. 14.

**BERRY (GEORGE A.).** *DISEASES OF THE EYE; A PRACTICAL TREATISE FOR STUDENTS OF OPHTHALMOLOGY.* Second edition. Very handsome octavo volume of 745 pages, with 197 original illustrations in the text, of which 87 are exquisitely colored. Cloth, $8.

**BILLINGS (JOHN S.).** *THE NATIONAL MEDICAL DICTIONARY.* Including in one alphabet English, French, German, Italian and Latin Technical Terms used in Medicine and the Collateral Sciences. In two very handsome imperial octavo volumes, containing 1574 pages and two colored plates. Per volume, cloth, $6; leather, $7; half Morocco, $8.50. *For sale by subscription only.* Specimen pages on application.

**BLACK (D. CAMPBELL).** *THE URINE IN HEALTH AND DISEASE, AND URINARY ANALYSIS, PHYSIOLOGICALLY AND PATHOLOGICALLY CONSIDERED.* In one 12mo. volume of 256 pages, with 73 engravings. Cloth, $2.75.

**BLOXAM (C. L.).** *CHEMISTRY, INORGANIC AND ORGANIC.* With Experiments. New American from the fifth London edition. In one handsome octavo volume of 727 pages, with 292 illustrations. Cloth, $2; leather, $3.

**BRICKNER (SAMUEL M.)** *ON THE SURGICAL PATIENT.* Preparing.

**BROADBENT (W. H.).** *THE PULSE.* In one 12mo. volume of 317 pages, with 59 engravings. Cloth, $1.75. See *Series of Clinical Manuals,* page 13.

**BROWNE (LENNOX).** *THE THROAT AND NOSE AND THEIR DISEASES.* New (4th) and enlarged edition. In one imperial octavo volume of 751 pages, with 235 engravings and 120 illustrations in color. Cloth, $6.50.

—— *KOCH'S REMEDY IN RELATION ESPECIALLY TO THROAT CONSUMPTION.* In one octavo volume of 121 pages, with 45 illustrations, 4 of which are colored, and 17 charts. Cloth, $1.50.

**BRUCE (J. MITCHELL).** *MATERIA MEDICA AND THERAPEUTICS.* Fifth edition. In one 12mo. volume of 591 pages. Cloth, $1.50. See *Students' Series of Manuals,* page 14.

—— *PRINCIPLES OF TREATMENT.* In one octavo volume. Preparing.

**BRUNTON (T. LAUDER).** *A MANUAL OF PHARMACOLOGY, THERAPEUTICS AND MATERIA MEDICA;* including the Pharmacy, the Physiological Action and the Therapeutical Uses of Drugs. In one octavo volume.

**BRYANT (THOMAS).** *THE PRACTICE OF SURGERY.* Fourth American from the fourth English edition. In one imperial octavo volume of 1040 pages, with 727 illustrations. Cloth, $6.50; leather, $7.50.

**BUMSTEAD (F. J.) AND TAYLOR (R. W.).** *THE PATHOLOGY AND TREATMENT OF VENEREAL DISEASES.* See *Taylor on Venereal Diseases,* page 15.

**BURCHARD (HENRY H.).** *DENTAL PATHOLOGY AND THERAPEUTICS, INCLUDING PHARMACOLOGY.* Handsome octavo, 575 pages, with 400 illustrations. *Just ready.* Cloth, $5; leather, $6. (*Net.*)

**BURNETT (CHARLES H.).** *THE EAR: ITS ANATOMY, PHYSIOLOGY AND DISEASES.* A Practical Treatise for the Use of Students and Practitioners. Second edition. In one 8vo. volume of 580 pages, with 107 illustrations. Cloth, $4; leather, $5.

**BUTLIN (HENRY T.).** *DISEASES OF THE TONGUE.* In one pocket-size 12mo. volume of 456 pages, with 8 colored plates and 3 engravings. Limp cloth, $3.50. See *Series of Clinical Manuals*, page 13.

**CARTER (R. BRUDENELL) AND FROST (W. ADAMS).** *OPHTHALMIC SURGERY.* In one pocket-size 12mo. volume of 559 pages, with 91 engravings and one plate. Cloth, $2.25. See *Series of Clinical Manuals*, page 13.

**CASPARI (CHARLES, JR.).** *A TREATISE ON PHARMACY.* For Students and Pharmacists. In one handsome octavo volume of 680 pages, with 288 illustrations. Cloth, $4.50.

**CHAPMAN (HENRY C.).** *A TREATISE ON HUMAN PHYSIOLOGY.* In one octavo volume of 925 pages, with 605 illustrations. Cloth, $5.50; leather, $6.50.

**CHARLES (T. CRANSTOUN).** *THE ELEMENTS OF PHYSIOLOGICAL AND PATHOLOGICAL CHEMISTRY.* In one handsome octavo volume of 451 pages, with 38 engravings and 1 colored plate. Cloth, $3.50.

**CHEYNE (W. WATSON).** *THE TREATMENT OF WOUNDS, ULCERS AND ABSCESSES.* In one 12mo. volume of 207 pages. Cloth, $1.25.

**CHURCHILL (FLEETWOOD).** *ESSAYS ON THE PUERPERAL FEVER.* In one octavo volume of 464 pages. Cloth, $2.50.

**CLARKE (W. B.) AND LOCKWOOD (C. B.).** *THE DISSECTOR'S MANUAL.* In one 12mo. volume of 396 pages, with 49 engravings. Cloth, $1.50. See *Students' Series of Manuals*, page 14.

**CLELAND (JOHN).** *A DIRECTORY FOR THE DISSECTION OF THE HUMAN BODY.* In one 12mo. volume of 178 pages. Cloth, $1.25.

**CLINICAL MANUALS.** See *Series of Clinical Manuals*, page 13.

**CLOUSTON (THOMAS S.).** *CLINICAL LECTURES ON MENTAL DISEASES.* New (5th) edition. Crown 8vo., of 736 pages with 19 colored plates. Cloth, $4.25, net. *Just Ready.*

☞ FOLSOM's *Abstract of Laws of U. S. on Custody of Insane*, octavo, $1.50, is sold in conjunction with *Clouston on Mental Diseases* for $5.00 for the two works.

**CLOWES (FRANK).** *AN ELEMENTARY TREATISE ON PRACTICAL CHEMISTRY AND QUALITATIVE INORGANIC ANALYSIS.* From the fourth English edition. In one handsome 12mo. volume of 387 pages, with 55 engravings. Cloth, $2.50.

**COAKLEY (CORNELIUS G.).** *THE DIAGNOSIS AND TREATMENT OF DISEASES OF THE NOSE, THROAT, NASO-PHARYNX AND TRACHEA.* In one 12mo. volume of about 400 pages, fully illustrated. *Preparing.*

**COATS (JOSEPH).** *A TREATISE ON PATHOLOGY.* In one volume of 829 pages, with 339 engravings. Cloth, $5.50; leather, $6.50.

**COLEMAN (ALFRED).** *A MANUAL OF DENTAL SURGERY AND PATHOLOGY.* With Notes and Additions to adapt it to American Practice. By THOS. C. STELLWAGEN, M.A., M.D., D.D.S. In one handsome octavo volume of 412 pages, with 331 engravings. Cloth, $3.25.

**CONDIE (D. FRANCIS).** *A PRACTICAL TREATISE ON THE DISEASES OF CHILDREN.* Sixth edition, revised and enlarged. In one large 8vo. volume of 719 pages. Cloth, $5.25; leather, $6.25.

**CORNIL (V.).** *SYPHILIS: ITS MORBID ANATOMY, DIAGNOSIS AND TREATMENT.* Translated, with Notes and Additions, by J. HENRY C. SIMES, M.D., and J. WILLIAM WHITE, M.D. In one 8vo. volume of 461 pages, with 84 illustrations. Cloth, $3.75.

**CULBRETH (DAVID M. R.).** *MATERIA MEDICA AND PHARMACOLOGY.* In one handsome octavo volume of 812 pages, with 445 engravings. Cloth, $4.75.

**CULVER (E. M.) AND HAYDEN (J. R.).** *MANUAL OF VENEREAL DIS-
EASES.* In one 12mo. volume of 289 pages, with 33 engravings. Cloth, $1.75.

**DALTON (JOHN C.).** *A TREATISE ON HUMAN PHYSIOLOGY.* Seventh edition, thoroughly revised. Octavo of 722 pages, with 252 engravings. Cloth, $5; leather, $6.

—— *DOCTRINES OF THE CIRCULATION OF THE BLOOD.* In one handsome 12mo. volume of 293 pages. Cloth, $2.

**DAVENPORT (F. H.).** *DISEASES OF WOMEN.* A Manual of Gynecology. For the use of Students and General Practitioners. New (3d) edition. In one handsome 12mo. volume, 387 pages and 150 engravings. Cloth, $1.75, net. *Just Ready.*

**DAVIS (F. H.).** *LECTURES ON CLINICAL MEDICINE.* Second edition. In one 12mo. volume of 287 pages. Cloth, $1.75.

**DAVIS (EDWARD P.).** *A TREATISE ON OBSTETRICS.* For Students and Practitioners. In one very handsome octavo volume of 546 pages, with 217 engravings, and 30 full-page plates in colors and monochrome. Cloth, $5; leather, $6.

**DE LA BECHE'S** *GEOLOGICAL OBSERVER.* In one large octavo volume of 700 pages, with 300 engravings. Cloth, $4.

**DENNIS (FREDERIC S.) AND BILLINGS (JOHN S.).** *A SYSTEM OF SURGERY.* In Contributions by American Authors. In four very handsome octavo volumes, containing 3652 pages, with 1585 engravings, and 45 full-page plates in colors and monochrome. *Complete work just ready.* Per volume, cloth, $6; leather, $7; half Morocco, gilt back and top, $8.50. *For sale by subscription only.* Full prospectus free.

**DERCUM (FRANCIS X.), Editor.** *A TEXT-BOOK ON NERVOUS DIS-
EASES.* By American Authors. In one handsome octavo volume of 1054 pages, with 341 engravings and 7 colored plates. Cloth, $6; leather, $7. *(Net.)*

**DE SCHWEINITZ (GEORGE E.).** *THE TOXIC AMBLYOPIAS; THEIR CLASSIFICATION, HISTORY, SYMPTOMS, PATHOLOGY AND TREATMENT.* Very handsome octavo, 240 pages, 46 engravings, and 9 full-page plates in colors. Limited edition, de luxe binding, $4. *(Net.)*

**DRAPER (JOHN C.).** *MEDICAL PHYSICS.* A Text-book for Students and Practitioners of Medicine. Octavo of 734 pages, with 376 engravings. Cloth, $4.

**DRUITT (ROBERT).** *THE PRINCIPLES AND PRACTICE OF MODERN SURGERY.* A new American, from the twelfth London edition, edited by STANLEY BOYD, F.R.C.S. Large octavo, 965 pages, with 373 engravings. Cloth, $4; leather, $5.

**DUANE (ALEXANDER).** *THE STUDENT'S DICTIONARY OF MEDICINE AND THE ALLIED SCIENCES.* Comprising the Pronunciation, Derivation and Full Explanation of Medical Terms. Together with much Collateral Descriptive Matter, Numerous Tables, etc. New edition. With Appendix. Square octavo volume of 690 pages. Cloth, $3; half leather, $3.25; full sheep, $3.75. Thumb-letter Index, 50 cents extra.

**DUDLEY (E. C.).** *A TREATISE ON THE PRINCIPLES AND PRACTICE OF GYNECOLOGY.* For Students and Practitioners. In one very handsome octavo volume of 652 pages, with 422 engravings, of which 47 are colored, and 2 full page plates in colors and monochrome. Cloth, $5.00, net; leather, $6.00, net.

**DUNCAN (J. MATTHEWS).** *CLINICAL LECTURES ON THE DISEASES OF WOMEN.* Delivered in St. Bartholomew's Hospital. In one octavo volume of 175 pages. Cloth, $1.50.

**DUNGLISON (ROBLEY).** *A DICTIONARY OF MEDICAL SCIENCE.* Containing a full Explanation of the Various Subjects and Terms of Anatomy, Physiology, Medical Chemistry, Pharmacy, Pharmacology, Therapeutics, Medicine, Hygiene, Dietetics, Pathology, Surgery, Ophthalmology, Otology, Laryngology, Dermatology, Gynecology, Obstetrics, Pediatrics, Medical Jurisprudence, Dentistry, etc., etc. By ROBLEY DUNGLISON, M.D., LL.D., late Professor of Institutes of Medicine in the Jefferson Medical College of Philadelphia. Edited by RICHARD J. DUNGLISON, A.M., M.D. Twenty-first edition, thoroughly revised and greatly enlarged and improved, with the Pronunciation, Accentuation and Derivation of the Terms. With Appendix. Imperial octavo of 1225 pages. Cloth, $7; leather, $8. Thumb-letter Index, 75 cents extra.

**DUNHAM (EDWARD K.).** *MORBID AND NORMAL HISTOLOGY.* Octavo, 450 pages, with 360 illustrations. Cloth, $3.25, net. *Just Ready.*

**EDES (ROBERT T.).** *TEXT-BOOK OF THERAPEUTICS AND MATERIA MEDICA.* In one 8vo. volume of 544 pages. Cloth, $3.50; leather, $4.50.

**EDIS (ARTHUR W.).** *DISEASES OF WOMEN.* A Manual for Students and Practitioners. In one handsome 8vo. volume of 576 pages, with 148 engravings. Cloth, $3; leather, $4.

**EGBERT (SENECA).** *HYGIENE AND SANITATION.* In one 12mo. volume of 359 pages, with 63 illustrations. *Just ready.* Cloth, $2.25, *net.*

**ELLIS (GEORGE VINER).** *DEMONSTRATIONS IN ANATOMY.* Being a Guide to the Knowledge of the Human Body by Dissection. From the eighth and revised English edition. Octavo, 716 pages, with 249 engravings. Cloth, $4.25; leather, $5.25.

**EMMET (THOMAS ADDIS).** *THE PRINCIPLES AND PRACTICE OF GYNÆCOLOGY.* For the use of Students and Practitioners. Third edition, enlarged and revised. 8vo. of 880 pages, with 150 original engravings. Cloth, $5; leather, $6.

**ERICHSEN (JOHN E.).** *THE SCIENCE AND ART OF SURGERY.* A new American from the eighth enlarged and revised London edition. In two large octavo volumes containing 2316 pages, with 984 engravings. Cloth, $9; leather, $11.

**ESSIG (CHARLES J.).** *PROSTHETIC DENTISTRY.* See *American Text-books of Dentistry,* page 2.

**FARQUHARSON (ROBERT).** *A GUIDE TO THERAPEUTICS.* Fourth American from fourth English edition, revised by FRANK WOODBURY, M.D. In one 12mo. volume of 581 pages. Cloth, $2.50.

**FIELD (GEORGE P.).** *A MANUAL OF DISEASES OF THE EAR.* Fourth edition. Octavo, 391 pages, with 73 engravings and 21 colored plates. Cloth, $3.75.

**FLINT (AUSTIN).** *A TREATISE ON THE PRINCIPLES AND PRACTICE OF MEDICINE.* New (7th) edition, thoroughly revised by FREDERICK P. HENRY, M.D. In one large 8vo. volume of 1143 pages, with engravings. Cloth, $5; leather, $6.

——— *A MANUAL OF AUSCULTATION AND PERCUSSION;* of the Physical Diagnosis of Diseases of the Lungs and Heart, and of Thoracic Aneurism. Fifth edition, revised by JAMES C. WILSON, M.D. In one handsome 12mo. volume of 274 pages, with 12 engravings.

——— *A PRACTICAL TREATISE ON THE DIAGNOSIS AND TREATMENT OF DISEASES OF THE HEART.* Second edition, enlarged. In one octavo volume of 550 pages. Cloth, $4.

——— *A PRACTICAL TREATISE ON THE PHYSICAL EXPLORATION OF THE CHEST, AND THE DIAGNOSIS OF DISEASES AFFECTING THE RESPIRATORY ORGANS.* Second and revised edition. In one octavo volume of 591 pages. Cloth, $4.50.

——— *MEDICAL ESSAYS.* In one 12mo. volume of 210 pages. Cloth, $1.38.

——— *ON PHTHISIS: ITS MORBID ANATOMY, ETIOLOGY, ETC.* A Series of Clinical Lectures. In one 8vo. volume of 442 pages. Cloth, $3.50.

**FOLSOM (C. F.).** *AN ABSTRACT OF STATUTES OF U. S. ON CUSTODY OF THE INSANE.* In one 8vo. volume of 108 pages. Cloth, $1.50. With *Clouston on Mental Diseases* (see page 4), at $5.00 for the two works.

**FORMULARY, THE NATIONAL.** See *Stillé, Maisch & Caspari's National Dispensatory,* page 14.

**FOSTER (MICHAEL).** *A TEXT-BOOK OF PHYSIOLOGY.* New (6th) and revised American from the sixth English edition. In one large octavo volume of 923 pages, with 257 illustrations. Cloth, $4.50; leather, $5.50.

**FOTHERGILL (J. MILNER).** *THE PRACTITIONER'S HAND-BOOK OF TREATMENT.* Third edition. In one handsome octavo volume of 664 pages. Cloth, $3.75; leather, $4.75.

**FOWNES (GEORGE).** *A MANUAL OF ELEMENTARY CHEMISTRY (INORGANIC AND ORGANIC).* Twelfth edition. Embodying WATTS' *Physical and Inorganic Chemistry.* In one royal 12mo. volume of 1061 pages, with 168 engravings, and 1 colored plate. Cloth, $2.75; leather, $3.25.

**FRANKLAND (E.) AND JAPP (F. R.).** *INORGANIC CHEMISTRY.* In one handsome octavo volume of 677 pages, with 51 engravings and 2 plates. Cloth, $3.75; leather, $4.75.

**FULLER (EUGENE).** *DISORDERS OF THE SEXUAL ORGANS IN THE MALE.* In one very handsome octavo volume of 238 pages, with 25 engravings and 8 full-page plates. Cloth, $2.

**FULLER (HENRY).** *ON DISEASES OF THE LUNGS AND AIR-PASSAGES.* Their Pathology, Physical Diagnosis, Symptoms and Treatment. From second English edition. In one 8vo. volume of 475 pages. Cloth, $3.50.

**GANT (FREDERICK JAMES).** *THE STUDENT'S SURGERY.* A Multum in Parvo. In one square octavo volume of 845 pages, with 159 engravings. Cloth, $3.75.

**GERRISH (FREDERIC H.).** *A TEXT-BOOK OF ANATOMY.* By American Authors. Edited by FREDERIC H. GERRISH, M.D. In one imp. octavo volume, richly illustrated. *Preparing.*

**GIBBES (HENEAGE).** *PRACTICAL PATHOLOGY AND MORBID HISTOLOGY.* Octavo of 314 pages, with 60 illustrations, mostly photographic. Cloth, $2.75.

**GIBNEY (V. P.).** *ORTHOPEDIC SURGERY.* For the use of Practitioners and Students. In one 8vo. volume profusely illustrated. *Preparing.*

**GOULD (A. PEARCE).** *SURGICAL DIAGNOSIS.* In one 12mo. volume of 589 pages. Cloth, $2. See *Students' Series of Manuals*, page 14.

**GRAY (HENRY).** *ANATOMY, DESCRIPTIVE AND SURGICAL.* New American edition of 1897, thoroughly revised. In one imperial octavo volume of 1239 pages, with 772 large and elaborate engravings. Price with illustrations in colors, cloth, $7; leather, $8. Price, with illustrations in black, cloth, $6; leather, $7.

**GRAY (LANDON CARTER).** *A TREATISE ON NERVOUS AND MENTAL DISEASES.* For Students and Practitioners of Medicine. Second edition. In one handsome octavo volume of 728 pages, with 172 engravings and 3 colored plates. Cloth, $4.75; leather, $5.75.

**GREEN (T. HENRY).** *AN INTRODUCTION TO PATHOLOGY AND MORBID ANATOMY.* New (8th) American from eighth and revised English edition. Oct. 595 pages, with 215 engravings and a colored plate. Cloth, $2.50, *net.* *Just Ready.*

**GREENE (WILLIAM H.).** *A MANUAL OF MEDICAL CHEMISTRY.* For the Use of Students. Based upon BOWMAN's *Medical Chemistry*. In one 12mo. volume of 310 pages, with 74 illustrations. Cloth, $1.75.

**GROSS (SAMUEL D.).** *A PRACTICAL TREATISE ON THE DISEASES, INJURIES AND MALFORMATIONS OF THE URINARY BLADDER, THE PROSTATE GLAND AND THE URETHRA.* Third edition, revised by SAMUEL W. GROSS, M.D. Octavo of 574 pages, with 170 illustrations. Cloth, $4.50.

**HABERSHON (S. O.).** *ON THE DISEASES OF THE ABDOMEN*, comprising those of the Stomach, Œsophagus, Cæcum, Intestines and Peritoneum. Second American from the third English edition. In one octavo volume of 554 pages, with 11 engravings. Cloth, $3.50.

**HAMILTON (ALLAN McLANE).** *NERVOUS DISEASES, THEIR DESCRIPTION AND TREATMENT.* Second and revised edition. In one octavo volume of 598 pages, with 72 engravings. Cloth, $4.

**HAMILTON (FRANK H.).** *A PRACTICAL TREATISE ON FRACTURES AND DISLOCATIONS.* Eighth edition, revised and edited by STEPHEN SMITH, A.M., M.D. In one handsome octavo volume of 832 pages, with 507 engravings. Cloth, $5.50; leather, $6.50.

**HARDAWAY (W. A.).** *MANUAL OF SKIN DISEASES.* New (2d) edition. In one 12mo. volume, 560 pages with 40 illustrations and 2 colored plates. Cloth, $2.25, *net.* *Just Ready.*

**HARE (HOBART AMORY).** *A TEXT-BOOK OF PRACTICAL THERAPEUTICS*, with Special Reference to the Application of Remedial Measures to Disease and their Employment upon a Rational Basis. With articles on various subjects by well-known specialists. New (7th) and revised edition. In one octavo volume of 775 pages. Cloth, $3.75, *net;* leather, $4.50, *net.*

—— *PRACTICAL DIAGNOSIS.* The Use of Symptoms in the Diagnosis of Disease. New (3d) edition, revised and enlarged. In one octavo volume of 615 pages, with 204 engravings, and 13 full-page plates. Cloth, $4.75, *net.* *Just Ready.*

**HARE (HOBART AMORY), Editor.** *A SYSTEM OF PRACTICAL THERAPEUTICS.* By American and Foreign Authors. In a series of contributions by eminent practitioners. In four large octavo volumes comprising 4600 pages, with 476 engravings. Vol. IV., *now ready.* Regular price, Vol. IV., cloth, $6; leather, $7; half Russia, $8. Price Vol. IV. to former or new subscribers to complete work, cloth, $5; leather, $6; half Russia, $7. Complete work, cloth, $20; leather, $24; half Russia, $28. For sale by subscription only. Full prospectus free on application to the Publishers.

**HARTSHORNE (HENRY).** *ESSENTIALS OF THE PRINCIPLES AND PRACTICE OF MEDICINE.* Fifth edition. In one 12mo. volume, 669 pages, with 144 engravings. Cloth, $2.75; half bound, $3.

—— *A HANDBOOK OF ANATOMY AND PHYSIOLOGY.* In one 12mo. volume of 310 pages, with 220 engravings. Cloth, $1.75.

—— *A CONSPECTUS OF THE MEDICAL SCIENCES.* Comprising Manuals of Anatomy, Physiology, Chemistry, Materia Medica, Practice of Medicine, Surgery and Obstetrics. Second edition. In one royal 12mo. volume of 1028 pages, with 477 illustrations. Cloth, $4.25; leather, $5.

**HAYDEN (JAMES R.).** *A MANUAL OF VENEREAL DISEASES.* In one 12mo. volume of 263 pages, with 47 engravings. Cloth, $1.50.

**HAYEM (GEORGES) AND HARE (H. A.).** *PHYSICAL AND NATURAL THERAPEUTICS.* The Remedial Use of Heat, Electricity, Modifications of Atmospheric Pressure, Climates and Mineral Waters. Edited by Prof. H. A. HARE, M.D. In one octavo volume of 414 pages, with 113 engravings. Cloth, $3.

**HERMAN (G. ERNEST).** *FIRST LINES IN MIDWIFERY.* 12mo., 198 pages, with 80 engravings. Cloth, $1.25. See *Students' Series of Manuals,* page 14.

**HERMANN (L.).** *EXPERIMENTAL PHARMACOLOGY.* A Handbook of the Methods for Determining the Physiological Actions of Drugs. Translated by ROBERT MEADE SMITH, M.D. In one 12mo. vol. of 199 pages, with 32 engravings. Cloth, $1.50.

**HERRICK (JAMES B.).** *A HANDBOOK OF DIAGNOSIS.* In one handsome 12mo. volume of 429 pages, with 80 engravings and 2 colored plates. Cloth, $2.50.

**HILL (BERKELEY).** *SYPHILIS AND LOCAL CONTAGIOUS DISORDERS.* In one 8vo. volume of 479 pages. Cloth, $3.25.

**HILLIER (THOMAS).** *A HANDBOOK OF SKIN DISEASES.* Second edition. In one royal 12mo. volume of 353 pages, with two plates. Cloth, $2.25.

**HIRST (BARTON C.) AND PIERSOL (GEORGE A.).** *HUMAN MONSTROSITIES.* Magnificent folio, containing 220 pages of text and illustrated with 123 engravings and 39 large photographic plates from nature. In four parts, price each, $5. *Limited edition.* For sale by subscription only.

**HOBLYN (RICHARD D.).** *A DICTIONARY OF THE TERMS USED IN MEDICINE AND THE COLLATERAL SCIENCES.* In one 12mo. volume of 520 double-columned pages. Cloth, $1.50; leather, $2.

**HODGE (HUGH L.).** *ON DISEASES PECULIAR TO WOMEN, INCLUDING DISPLACEMENTS OF THE UTERUS.* Second and revised edition. In one 8vo. volume of 519 pages, with illustrations. Cloth, $4.50.

**HOFFMANN (FREDERICK) AND POWER (FREDERICK B.).** *A MANUAL OF CHEMICAL ANALYSIS,* as Applied to the Examination of Medicinal Chemicals and their Preparations. Third edition, entirely rewritten and much enlarged. In one handsome octavo volume of 621 pages, with 179 engravings. Cloth, $4.25.

**HOLDEN (LUTHER).** *LANDMARKS, MEDICAL AND SURGICAL.* From the third English edition. With additions by W. W. KEEN, M.D. In one royal 12mo. volume of 148 pages. Cloth, $1.

**HOLMES (TIMOTHY).** *A TREATISE ON SURGERY.* Its Principles and Practice. A new American from the fifth English edition. Edited by T. PICKERING PICK, F.R.C.S. In one handsome octavo volume of 1008 pages, with 428 engravings. Cloth, $6; leather, $7.

—— *A SYSTEM OF SURGERY.* With notes and additions by various American authors. Edited by JOHN H. PACKARD, M.D. In three very handsome 8vo. volumes containing 3137 double-columned pages, with 979 engravings and 13 lithographic plates. Per volume, cloth, $6; leather, $7; half Russia, $7.50. *For sale by subscription only.*

**HORNER (WILLIAM E.).** *SPECIAL ANATOMY AND HISTOLOGY.* Eighth edition, revised and modified. In two large 8vo. volumes of 1007 pages, containing 320 engravings. Cloth, $6.

**HUDSON (A.).** *LECTURES ON THE STUDY OF FEVER.* In one octavo volume of 308 pages. Cloth, $2.50.

**HUTCHISON (ROBERT) AND RAINY (HARRY).** *CLINICAL METHODS.* An Introduction to the Practical Study of Medicine. In one 12mo. volume of 562 pages, with 137 engravings and 8 colored plates. Cloth, $3.00. *Just ready.*

**HUTCHINSON (JONATHAN).** *SYPHILIS.* 12mo., 542 pages, with 8 chromo-lithographic plates. Cloth, $2.25. See *Series of Clinical Manuals*, page 13.

**HYDE (JAMES NEVINS).** *A PRACTICAL TREATISE ON DISEASES OF THE SKIN.* New (4th) edition, thoroughly revised. Octavo, 815 pages, with 110 engravings and 12 full-page plates, 4 of which are colored. Cloth, $5.25; leather, $6.25.

**JACKSON (GEORGE THOMAS).** *THE READY-REFERENCE HANDBOOK OF DISEASES OF THE SKIN.* Second edition. In one 12mo. volume of 589 pages, with 69 engravings, and one colored plate. Cloth, $2.75.

**JAMIESON (W. ALLAN).** *DISEASES OF THE SKIN.* Third edition. Octavo, 656 pages, with 1 engraving and 9 double-page chromo-lithographic plates. Cloth, $6.

**JEWETT (CHARLES).** *ESSENTIALS OF OBSTETRICS.* In one 12mo. volume of 356 pages, with 80 engravings and 3 colored plates. Cloth, $2.25. *Just ready.*

—— *A TEXT-BOOK OF OBSTETRICS.* By American Authors. One large octavo volume, profusely illustrated. *In press.*

**JONES (C. HANDFIELD).** *CLINICAL OBSERVATIONS ON FUNCTIONAL NERVOUS DISORDERS.* Second American edition. In one octavo volume of 340 pages. Cloth, $3.25.

**JULER (HENRY).** *A HANDBOOK OF OPHTHALMIC SCIENCE AND PRACTICE.* Second edition. In one octavo volume of 549 pages, with 201 engravings, 17 chromo-lithographic plates, test-types of Jaeger and Snellen, and Holmgren's Color-Blindness Test. Cloth, $5.50; leather, $6.50.

**KIRK (EDWARD C.).** *OPERATIVE DENTISTRY.* See *American Text-books of Dentistry*, page 2.

**KING (A. F. A.).** *A MANUAL OF OBSTETRICS.* Seventh edition. In one 12mo. volume of 573 pages, with 223 illustrations. Cloth, $2.50.

**KLEIN (E.).** *ELEMENTS OF HISTOLOGY.* New (5th) edition. In one pocket-size 12mo. volume of 506 pages, with 296 engravings. Cloth, $2.00, *net*. *Just Ready.* See *Students' Series of Manuals*, page 14.

**LANDIS (HENRY G.).** *THE MANAGEMENT OF LABOR.* In one handsome 12mo. volume of 329 pages, with 28 illustrations. Cloth, $1.75.

**LA ROCHE (R.).** *YELLOW FEVER.* In two 8vo. volumes of 1468 pages. Cloth, $7.

—— *PNEUMONIA.* In one 8vo. volume of 490 pages. Cloth, $3.

**LAURENCE (J. Z.) AND MOON (ROBERT C.).** *A HANDY-BOOK OF OPHTHALMIC SURGERY.* Second edition. In one octavo volume of 227 pages, with 66 engravings. Cloth, $2.75.

**LEA (HENRY C.).** *CHAPTERS FROM THE RELIGIOUS HISTORY OF SPAIN; CENSORSHIP OF THE PRESS; MYSTICS AND ILLUMINATI; THE ENDEMONIADAS; EL SANTO NINO DE LA GUARDIA; BRIANDA DE BARDAXI.* In one 12mo. volume of 522 pages. Cloth, $2.50.

—— *A HISTORY OF AURICULAR CONFESSION AND INDULGENCES IN THE LATIN CHURCH.* In three octavo volumes of about 500 pages each. Per volume, cloth, $3. *Complete work just ready.*

—— *FORMULARY OF THE PAPAL PENITENTIARY.* In one octavo volume of 221 pages, with frontispiece. Cloth, $2.50.

**LEA (HENRY C.).** *STUDIES IN CHURCH HISTORY.* The Rise of the Temporal Power—Benefit of Clergy—Excommunication. New edition. In one handsome 12mo. volume of 605 pages. Cloth, $2.50.

—— *SUPERSTITION AND FORCE; ESSAYS ON THE WAGER OF LAW, THE WAGER OF BATTLE, THE ORDEAL AND TORTURE.* Fourth edition, thoroughly revised. In one handsome royal 12mo. volume of 629 pages. Cloth, $2.75.

—— *AN HISTORICAL SKETCH OF SACERDOTAL CELIBACY IN THE CHRISTIAN CHURCH.* Second edition. In one handsome octavo volume of 685 pages. Cloth, $4.50.

**LEE (HENRY)** *ON SYPHILIS.* In one 8vo. volume of 246 pages. Cloth, $2.25.

**LEHMANN (C. G.).** *A MANUAL OF CHEMICAL PHYSIOLOGY.* In one 8vo. volume of 327 pages, with 41 engravings. Cloth, $2.25.

**LEISHMAN (WILLIAM).** *A SYSTEM OF MIDWIFERY.* Including the Diseases of Pregnancy and the Puerperal State. Fourth edition. In one octavo volume.

**LOOMIS (ALFRED L.) AND THOMPSON (W. GILMAN), Editors.** *A SYSTEM OF PRACTICAL MEDICINE.* In Contributions by Various American Authors. In four very handsome octavo volumes of about 900 pages each, fully illustrated in black and colors. *Complete work just ready.* Per volume, cloth, $5; leather, $6; half Morocco, $7. *For sale by subscription only.* Full prospectus free on application to the Publishers.

**LUFF (ARTHUR P.).** *MANUAL OF CHEMISTRY*, for the use of Students of Medicine. In one 12mo. volume of 522 pages, with 36 engravings. Cloth, $2. See *Students' Series of Manuals*, page 14.

**LYMAN (HENRY M.).** *THE PRACTICE OF MEDICINE.* In one very handsome octavo volume of 925 pages with 170 engravings. Cloth, $4.75; leather, $5.75.

**LYONS (ROBERT D.).** *A TREATISE ON FEVER.* In one octavo volume of 362 pages. Cloth, $2.25.

**MACKENZIE (JOHN NOLAND).** *THE DISEASES OF THE NOSE AND THROAT.* In one handsome octavo volume of about 600 pages, richly illustrated. *Preparing.*

**MAISCH (JOHN M.).** *A MANUAL OF ORGANIC MATERIA MEDICA.* Sixth edition, thoroughly revised by H. C. C. MAISCH, Ph.G., Ph.D. In one very handsome 12mo. volume of 509 pages, with 285 engravings. Cloth, $3.

**MANUALS.** See *Students' Quiz Series*, page 14, *Students' Series of Manuals*, page 14, and *Series of Clinical Manuals*, page 13.

**MARSH (HOWARD).** *DISEASES OF THE JOINTS.* In one 12mo. volume of 468 pages, with 64 engravings and a colored plate. Cloth, $2. See *Series of Clinical Manuals*, page 13.

**MAY (C. H.).** *MANUAL OF THE DISEASES OF WOMEN.* For the use of Students and Practitioners. Second edition, revised by L. S. RAU, M.D. In one 12mo. volume of 360 pages, with 31 engravings. Cloth, $1.75.

**MITCHELL (JOHN K.).** *REMOTE CONSEQUENCES OF INJURIES OF NERVES AND THEIR TREATMENT.* In one handsome 12mo. volume of 239 pages, with 12 illustrations. Cloth $1.75.

**MITCHELL (S. WEIR).** *CLINICAL LESSONS ON NERVOUS DISEASES.* In one very handsome 12mo. volume of 299 pages, with 17 engravings and 2 colored plates. Cloth, $2.50. Of the one hundred numbered copies with the Author's signed title page a few remain; these are offered in green cloth, gilt top, at $3.50, *net.*

**MORRIS (HENRY).** *SURGICAL DISEASES OF THE KIDNEY.* In one 12mo. volume of 554 pages, with 40 engravings and 6 colored plates. Cloth, $2.25. See *Series of Clinical Manuals*, page 13.

**MORRIS (MALCOLM).** *DISEASES OF THE SKIN.* In one square 8vo. volume of 572 pages, with 19 chromo-lithographic figures and 17 engravings. Cloth, $3.50.

**MULLER (J.).** *PRINCIPLES OF PHYSICS AND METEOROLOGY.* In one large 8vo. volume of 623 pages, with 538 engravings. Cloth, $4.50.

**MUSSER (JOHN H.).** *A PRACTICAL TREATISE ON MEDICAL DIAGNOSIS,* for Students and Physicians. New (2d) edition. In one octavo volume of 931 pages, illustrated with 177 engravings and 11 full-page colored plates. Cloth, $5; leather, $6.

**NATIONAL DISPENSATORY.** See *Stillé, Maisch & Caspari,* page 14.

**NATIONAL FORMULARY.** See *Stillé, Maisch & Caspari's National Dispensatory,* page 14.

**NATIONAL MEDICAL DICTIONARY.** See *Billings,* page 3.

**NETTLESHIP (E.).** *DISEASES OF THE EYE.* New (5th) American from sixth English edition. Thoroughly revised. In one 12mo. volume of 521 pages, with 161 engravings, 2 colored plates, test-types, formulæ and color-blindness test. Cloth, $2.25. *Just ready.*

**NORRIS (WM. F.) AND OLIVER (CHAS. A.).** *TEXT-BOOK OF OPHTHALMOLOGY.* In one octavo volume of 641 pages, with 357 engravings and 5 colored plates. Cloth, $5; leather, $6.

**OWEN (EDMUND).** *SURGICAL DISEASES OF CHILDREN.* In one 12mo. volume of 525 pages, with 85 engravings and 4 colored plates. Cloth, $2. See *Series of Clinical Manuals,* page 13.

**PARK (ROSWELL), Editor.** *A TREATISE ON SURGERY,* by American Authors. For Students and Practitioners of Surgery and Medicine. In two magnificent octavo volumes. Vol. I., *General Surgery,* 799 pages, with 356 engravings and 21 full-page plates in colors and monochrome. Vol. II., *Special Surgery,* 796 pages, with 451 engravings and 17 full-page plates in colors and monochrome. *Complete work now ready.* Price per volume, cloth, $4.50; leather, $5.50. *Net.*

**PARRY (JOHN S.).** *EXTRA-UTERINE PREGNANCY, ITS CLINICAL HISTORY, DIAGNOSIS, PROGNOSIS AND TREATMENT.* In one octavo volume of 272 pages. Cloth, $2.50.

**PARVIN (THEOPHILUS).** *THE SCIENCE AND ART OF OBSTETRICS.* Third edition In one handsome octavo volume of 677 pages, with 267 engravings and 2 colored plates. Cloth, $4.25; leather, $5.25.

**PAYNE (JOSEPH FRANK).** *A MANUAL OF GENERAL PATHOLOGY.* Designed as an Introduction to the Practice of Medicine. In one octavo volume of 524 pages, with 153 engravings and 1 colored plate.

**PEPPER'S** *SYSTEM OF MEDICINE.* See page 2.

**PEPPER (A. J.).** *SURGICAL PATHOLOGY.* In one 12mo volume of 511 pages, with 81 engravings. Cloth, $2. See *Students' Series of Manuals,* page 14.

**PICK (T. PICKERING).** *FRACTURES AND DISLOCATIONS.* In one 12mo. volume of 530 pages, with 93 engravings. Cloth, $2. See *Series of Clinical Manuals,* p. 13.

**PLAYFAIR (W. S.).** *A TREATISE ON THE SCIENCE AND PRACTICE OF MIDWIFERY.* New (7th) American from the Ninth English edition. In one octavo volume of 700 pages, with 207 engravings and 7 full page plates. Cloth, $3.75, *net;* leather, $4.75, *net. Just Ready.*

—— *THE SYSTEMATIC TREATMENT OF NERVE PROSTRATION AND HYSTERIA.* In one 12mo. volume of 97 pages. Cloth, $1.

**POLITZER (ADAM).** *A TEXT-BOOK OF THE DISEASES OF THE EAR AND ADJACENT ORGANS.* Second American from the third German edition. Translated by OSCAR DODD, M.D , and edited by SIR WILLIAM DALBY, F.R.C.S. In one octavo volume of 748 pages, with 330 original engravings. Cloth, $5.50.

**POWER (HENRY).** *HUMAN PHYSIOLOGY.* Second edition. In one 12mo. volume of 396 pages, with 47 engravings. Cloth, $1.50. See *Student's Series of Manuals*, page 14.

**PURDY (CHARLES W.).** *BRIGHT'S DISEASE AND ALLIED AFFECTIONS OF THE KIDNEY.* In one octavo volume of 288 pages, with 18 engravings. Cloth, $2.

**PYE-SMITH (PHILIP H.).** *DISEASES OF THE SKIN.* In one 12mo. volume of 407 pages, with 28 illustrations, 18 of which are colored. Cloth, $2.

**QUIZ SERIES.** See *Students' Quiz Series*, page 14.

**RALFE (CHARLES H.).** *CLINICAL CHEMISTRY.* In one 12mo. volume of 314 pages, with 16 engravings. Cloth, $1.50. See *Students' Series of Manuals*, page 14.

**RAMSBOTHAM (FRANCIS H.).** *THE PRINCIPLES AND PRACTICE OF OBSTETRIC MEDICINE AND SURGERY.* In one imperial octavo volume of 640 pages, with 64 plates and numerous engravings in the text. Strongly bound in leather, $7.

**REICHERT (EDWARD T.).** *A TEXT-BOOK ON PHYSIOLOGY.* In one handsome octavo volume of about 800 pages, richly illustrated. *Preparing.*

**REMSEN (IRA).** *THE PRINCIPLES OF THEORETICAL CHEMISTRY.* New (5th) edition, thoroughly revised. In one 12mo. volume of 326 pages. Cloth, $2.

**RICHARDSON (BENJAMIN WARD).** *PREVENTIVE MEDICINE.* In one octavo volume of 729 pages. Cloth, $4; leather, $5.

**ROBERTS (JOHN B.).** *THE PRINCIPLES AND PRACTICE OF MODERN SURGERY.* In one octavo volume of 780 pages, with 501 engravings. Cloth, $4.50; leather, $5.50.

—— *THE COMPEND OF ANATOMY.* For use in the Dissecting Room and in preparing for Examinations. In one 16mo. volume of 196 pages. Limp cloth, 75 cents.

**ROBERTS (SIR WILLIAM).** *A PRACTICAL TREATISE ON URINARY AND RENAL DISEASES, INCLUDING URINARY DEPOSITS.* Fourth American from the fourth London edition. In one very handsome 8vo. volume of 609 pages, with 81 illustrations. Cloth, $3.50.

**ROBERTSON (J. McGREGOR).** *PHYSIOLOGICAL PHYSICS.* In one 12mo. volume of 537 pages, with 219 engravings. Cloth, $2. See *Students' Series of Manuals*, page 14.

**ROSS (JAMES).** *A HANDBOOK OF THE DISEASES OF THE NERVOUS SYSTEM.* In one handsome octavo volume of 726 pages, with 184 engravings. Cloth, $4.50; leather, $5.50.

**SAVAGE (GEORGE H.).** *INSANITY AND ALLIED NEUROSES, PRACTICAL AND CLINICAL.* New (2d) and enlarged edition. In one 12mo. volume of 551 pages, with 18 typical engravings. Cloth, $2. See *Series of Clinical Manuals*, page 13.

**SCHAFER (EDWARD A.).** *THE ESSENTIALS OF HISTOLOGY, DESCRIPTIVE AND PRACTICAL.* For the use of Students. New (5th) edition. In one handsome octavo volume of 350 pages, with 325 illustrations. Cloth, $3, *net. Just ready.*

—— *A COURSE OF PRACTICAL HISTOLOGY.* New (2d) edition. In one 12mo. volume of 307 pages, with 59 engravings. Cloth, $2.25.

**SCHMITZ AND ZUMPT'S CLASSICAL SERIES.**
*ADVANCED LATIN EXERCISES* Cloth, 60 cents; half bound, 70 cents.
*SCHMITZ'S ELEMENTARY LATIN EXERCISES.* Cloth, 50 cents.
*SALLUST.* Cloth, 60 cents; half bound, 70 cents.
*NEPOS.* Cloth, 60 cents; half bound, 70 cents.
*VIRGIL.* Cloth, 85 cents; half bound, $1.
*CURTIUS.* Cloth, 80 cents; half bound, 90 cents.

**SCHOFIELD (ALFRED T.).** *ELEMENTARY PHYSIOLOGY FOR STUDENTS.* In one 12mo. volume of 380 pages, with 227 engravings and 2 colored plates. Cloth, $2.

**SCHREIBER (JOSEPH).** *A MANUAL OF TREATMENT BY MASSAGE AND METHODICAL MUSCLE EXERCISE.* Translated by WALTER MENDELSON, M.D., of New York. In one handsome octavo volume of 274 pages, with 117 fine engravings.

**SENN (NICHOLAS).** *SURGICAL BACTERIOLOGY.* Second edition. In one octavo volume of 268 pages, with 13 plates, 10 of which are colored, and 9 engravings. Cloth, $2.

**SERIES OF CLINICAL MANUALS.** A Series of Authoritative Monographs on Important Clinical Subjects, in 12mo. volumes of about 550 pages, well illustrated. The following volumes are now ready: BROADBENT on the Pulse, $1.75; YEO on Food in Health and Disease, new (2d) edition, $2.50; CARTER and FROST's Ophthalmic Surgery, $2.25; HUTCHINSON on Syphilis, $2.25; MARSH on Diseases of the Joints, $2; MORRIS on Surgical Diseases of the Kidney, $2.25; OWEN on Surgical Diseases of Children, $2; PICK on Fractures and Dislocations, $2; BUTLIN on the Tongue, $3.50; SAVAGE on Insanity and Allied Neuroses, $2; and TREVES on Intestinal Obstruction, $2.
For separate notices, see under various authors' names.

**SERIES OF STUDENTS' MANUALS.** See next page.

**SIMON (CHARLES E.).** *CLINICAL DIAGNOSIS, BY MICROSCOPICAL AND CHEMICAL METHODS.* New (2d) and revised edition. In one handsome octavo volume of 530 pages, with 135 engravings and 14 full-page plates in colors and monochrome. Cloth, $3.50.

**SIMON (W.).** *MANUAL OF CHEMISTRY.* A Guide to Lectures and Laboratory Work for Beginners in Chemistry. A Text-book specially adapted for Students of Pharmacy and Medicine. New (6th) edition. In one 8vo. volume of 536 pages, with 46 engravings and 8 plates showing colors of 64 tests. Cloth, $3.00, *net. Just Ready.*

**SLADE (D. D.).** *DIPHTHERIA; ITS NATURE AND TREATMENT.* Second edition. In one royal 12mo. volume, 158 pages. Cloth, $1.25.

**SMITH (EDWARD).** *CONSUMPTION; ITS EARLY AND REMEDIABLE STAGES.* In one 8vo. volume of 253 pages. Cloth, $2.25.

**SMITH (J. LEWIS).** *A TREATISE ON THE DISEASES OF INFANCY AND CHILDHOOD.* New (8th) edition, thoroughly revised and rewritten and greatly enlarged. In one large 8vo. volume of 983 pages, with 273 illustrations and 4 full-page plates. Cloth, $4.50; leather, $5.50.

**SMITH (STEPHEN).** *OPERATIVE SURGERY.* Second and thoroughly revised edition. In one octavo vol. of 892 pages, with 1005 engravings. Cloth, $4; leather, $5

**SOLLY (S. EDWIN).** *A HANDBOOK OF MEDICAL CLIMATOLOGY.* In one handsome octavo volume of 462 pages, with engravings and 11 full-page plates, 5 of which are in colors. Cloth, $4.00.

**STILLÉ (ALFRED).** *CHOLERA; ITS ORIGIN, HISTORY, CAUSATION, SYMPTOMS, LESIONS, PREVENTION AND TREATMENT.* In one 12mo. volume of 163 pages, with a chart showing routes of previous epidemics. Cloth, $1.25.

—— *THERAPEUTICS AND MATERIA MEDICA.* Fourth and revised edition. In two octavo volumes, containing 1936 pages. Cloth, $10; leather, $12.

**STILLE (ALFRED), MAISCH (JOHN M.) AND CASPARI (CHAS. JR.).** THE NATIONAL DISPENSATORY: Containing the Natural History, Chemistry, Pharmacy, Actions and Uses of Medicines, including those recognized in the latest Pharmacopœias of the United States, Great Britian and Germany, with numerous references to the French Codex. Fifth edition, revised and enlarged in accordance with and embracing the new U. S. Pharmacopœia, Seventh Decennial Revision. With Supplement containing the new edition of the National Formulary. In one magnificent imperial octavo volume of 2025 pages, with 320 engravings Cloth, $7.25; leather, $8. With ready reference Thumb-letter Index. Cloth, $7.75; leather, $8.50.

**STIMSON (LEWIS A.).** A MANUAL OF OPERATIVE SURGERY. New (3d) edition. In one royal 12mo. volume of 614 pages, with 306 engravings. Cloth, $3.75.

—— A TREATISE ON FRACTURES AND DISLOCATIONS. In two handsome octavo volumes. Vol. I., FRACTURES, 582 pages, 360 engravings. Vol II., DISLOCATIONS, 540 pages, 163 engravings. Complete work, cloth, $5.50; leather, $7.50. Either volume separately, cloth, $3; leather, $4.

**STUDENTS' QUIZ SERIES.** A New Series of Manuals in question and answer for Students and Practitioners, covering the essentials of medical science. Thirteen volumes, pocket size, convenient, authoritative, well illustrated, handsomely bound in limp cloth, and issued at a low price. 1. Anatomy (double number); 2. Physiology; 3. Chemistry and Physics; 4. Histology, Pathology and Bacteriology; 5. Materia Medica and Therapeutics; 6. Practice of Medicine; 7. Surgery (double number); 8. Genito-Urinary and Venereal Diseases; 9. Diseases of the Skin; 10. Diseases of the Eye, Ear, Throat and Nose; 11. Obstetrics; 12. Gynecology; 13. Diseases of Children. Price, $1 each, except Nos. 1 and 7, Anatomy and Surgery, which being double numbers are priced at $1.75 each. Full specimen circular on application to publishers.

**STUDENTS' SERIES OF MANUALS.** A Series of Fifteen Manuals by Eminent Teachers or Examiners. The volumes are pocket-size 12mos. of from 300–540 pages, profusely illustrated, and bound in red limp cloth. The following volumes may now be announced: HERMAN's First Lines in Midwifery, $1.25; LUFF's Manual of Chemistry, $2; BRUCE's Materia Medica and Therapeutics (fifth edition), $1.50; BELL's Comparative Anatomy and Physiology, $2; ROBERTSON's Physiological Physics, $2; GOULD's Surgical Diagnosis, $2; KLEIN's Elements of Histology (5th edition), $2.00, net; PEPPER's Surgical Pathology, $2; TREVES' Surgical Applied Anatomy, $2; POWER's Human Physiology (2d edition), $1.50; RALFE's Clinical Chemistry, $1.50; and CLARKE and LOCKWOOD's Dissector's Manual, $1.50
For separate notices, see under various authors' names.

**STURGES (OCTAVIUS).** AN INTRODUCTION TO THE STUDY OF CLINICAL MEDICINE. In one 12mo. volume. Cloth, $1.25.

**SUTTON (JOHN BLAND).** SURGICAL DISEASES OF THE OVARIES AND FALLOPIAN TUBES. Including Abdominal Pregnancy. In one 12mo. volume of 513 pages, with 119 engravings and 5 colored plates. Cloth, $3.

—— TUMORS, INNOCENT AND MALIGNANT. Their Clinical Features and Appropriate Treatment. In one 8vo. volume of 526 pages, with 250 engravings and 9 full-page plates. Cloth, $4.50.

**TAIT (LAWSON).** DISEASES OF WOMEN AND ABDOMINAL SURGERY. In two handsome octavo volumes. Vol. I. contains 554 pages, 62 engravings, and 3 plates. Cloth, $3. Vol. II., preparing.

**TANNER (THOMAS HAWKES).** ON THE SIGNS AND DISEASES OF PREGNANCY. From the second English edition. In one octavo volume of 490 pages, with 4 colored plates and 16 engravings. Cloth, $4.25.

**TAYLOR (ALFRED S.).** MEDICAL JURISPRUDENCE. New American from the twelfth English edition, specially revised by CLARK BELL, ESQ., of the N. Y. Bar. In one octavo volume of 831 pages, with 54 engravings and 8 full-page plates. Cloth, $4.50; leather, $5.50. Just ready.

**TAYLOR (ALFRED S.).** *ON POISONS IN RELATION TO MEDICINE AND MEDICAL JURISPRUDENCE.* Third American from the third London edition. In one 8vo. volume of 788 pages, with 104 illustrations. Cloth, $5.50; leather, $6.50.

**TAYLOR (ROBERT W.).** *THE PATHOLOGY AND TREATMENT OF VENEREAL DISEASES.* In one very handsome octavo volume of 1002 pages, with 230 engravings and 7 colored plates. Cloth, $5; leather, $6. *Net.*

—— *A PRACTICAL TREATISE ON SEXUAL DISORDERS IN THE MALE AND FEMALE.* In one octavo volume of 448 pages, with 73 engravings and 8 plates. Cloth, $3. *Net. Just ready.*

—— *A CLINICAL ATLAS OF VENEREAL AND SKIN DISEASES.* Including Diagnosis, Prognosis and Treatment. In eight large folio parts, measuring 14 x 18 inches, and comprising 213 beautiful figures on 58 full-page chromo-lithographic plates, 85 fine engravings, and 425 pages of text. Complete work now ready. Price per part, sewed in heavy embossed paper, $2.50. Bound in one volume, half Russia, $27; half Turkey Morocco, $28. *For sale by subscription only.* Address the publishers. Specimen plates by mail on receipt of 10 cents.

**TAYLOR (SEYMOUR).** *INDEX OF MEDICINE.* A Manual for the use of Senior Students and others. In one large 12mo. volume of 802 pages. Cloth, $3.75.

**THOMAS (T. GAILLARD) AND MUNDE (PAUL F.).** *A PRACTICAL TREATISE ON THE DISEASES OF WOMEN.* Sixth edition, thoroughly revised by PAUL F. MUNDE, M.D. In one large and handsome octavo volume of 824 pages, with 347 engravings. Cloth, $5; leather, $6.

**THOMPSON (SIR HENRY).** *CLINICAL LECTURES ON DISEASES OF THE URINARY ORGANS.* Second and revised edition. In one octavo volume of 203 pages, with 25 engravings. Cloth, $2.25.

—— *THE PATHOLOGY AND TREATMENT OF STRICTURE OF THE URETHRA AND URINARY FISTULÆ.* From the third English edition. In one octavo volume of 359 pages, with 47 engravings and 3 lithographic plates. Cloth, $3.50.

**THOMSON (JOHN).** *A GUIDE TO THE CLINICAL EXAMINATION AND TREATMENT OF SICK CHILDREN.* In one crown octavo volume of 350 pages with 52 illustrations. Cloth, $1.75, *net. Just ready.*

**TODD (ROBERT BENTLEY).** *CLINICAL LECTURES ON CERTAIN ACUTE DISEASES.* In one 8vo. volume of 320 pages. Cloth, $2.50.

**TREVES (FREDERICK).** *OPERATIVE SURGERY.* In two 8vo. volumes containing 1550 pages, with 422 illustrations. Cloth, $9; leather, $11.

—— *A SYSTEM OF SURGERY.* In Contributions by Twenty-five English Surgeons. In two large octavo volumes, containing 2298 pages, with 950 engravings and 4 full-page plates. Per volume, cloth, $8.

—— *A MANUAL OF SURGERY.* In Treatises by 33 leading surgeons. Three 12mo. volumes, containing 1866 pages, with 213 engravings. Price per set, $6. See *Students' Series of Manuals*, page 14.

—— *THE STUDENTS' HANDBOOK OF SURGICAL OPERATIONS.* In one 12mo. volume of 508 pages, with 94 illustrations. Cloth, $2.50.

—— *SURGICAL APPLIED ANATOMY.* In one 12mo. volume of 583 pages with 61 engravings. Cloth, $2. See *Students' Series of Manuals*, page 14.

—— *INTESTINAL OBSTRUCTION.* In one 12mo. volume of 522 pages, with 60 illustrations. Cloth, $2. See *Series of Clinical Manuals*, page 13.

**TUKE (DANIEL HACK).** *THE INFLUENCE OF THE MIND UPON THE BODY IN HEALTH AND DISEASE.* Second edition. In one 8vo. volume of 467 pages, with 2 colored plates. Cloth, $3.

**VAUGHAN (VICTOR C.) AND NOVY (FREDERICK G.).** *PTOMAINS, LEUCOMAINS, TOXINS AND ANTITOXINS,* or the Chemical Factors in the Causation of Disease. Third edition. In one 12mo. volume of 603 pages. Cloth, $3.

**VISITING LIST.** *THE MEDICAL NEWS VISITING LIST* for 1898. Four styles: Weekly (dated for 30 patients); Monthly (undated for 120 patients per month); Perpetual (undated for 30 patients each week); and Perpetual (undated for 60 patients each week). The 60-patient book consists of 256 pages of assorted blanks. The first three styles contain 32 pages of important data, thoroughly revised, and 160 pages of assorted blanks. Each in one volume, price, $1.25. With thumb-letter index for quick use, 25 cents extra. Special rates to advance-paying subscribers to THE MEDICAL NEWS or THE AMERICAN JOURNAL OF THE MEDICAL SCIENCES, or both. See page 1.

**WATSON (THOMAS).** *LECTURES ON THE PRINCIPLES AND PRACTICE OF PHYSIC.* A new American from the fifth and enlarged English edition, with additions by H. HARTSHORNE, M.D. In two large 8vo. volumes of 1840 pages, with 190 engravings. Cloth, $9; leather, $11.

**WEST (CHARLES).** *LECTURES ON THE DISEASES PECULIAR TO WOMEN.* Third American from the third English edition. In one octavo volume of 543 pages. Cloth, $3.75; leather, $4.75.

—— *ON SOME DISORDERS OF THE NERVOUS SYSTEM IN CHILDHOOD.* In one small 12mo. volume of 127 pages. Cloth, $1.

**WHARTON (HENRY R.).** *MINOR SURGERY AND BANDAGING.* Third edition. In one 12mo. volume of 594 pages, with 475 engravings, many of which are photographic. Cloth, $3.

**WHITLA (WILLIAM).** *DICTIONARY OF TREATMENT, OR THERAPEUTIC INDEX.* Including Medical and Surgical Therapeutics. In one square octavo volume of 917 pages. Cloth, $4.

**WILLIAMS (DAWSON).** *MEDICAL DISEASES OF INFANCY AND CHILDHOOD.* In one 12mo. volume of 629 pages, with 18 illustrations. Cloth, $2.50, *net. Just ready.*

**WILSON (ERASMUS).** *A SYSTEM OF HUMAN ANATOMY.* A new and revised American from the last English edition. Illustrated with 397 engravings. In one octavo volume of 616 pages. Cloth, $4; leather, $5.

—— *THE STUDENT'S BOOK OF CUTANEOUS MEDICINE.* In one 12mo volume. Cloth, $3.50.

**WINCKEL** *ON PATHOLOGY AND TREATMENT OF CHILDBED.* Translated by JAMES R. CHADWICK, A.M., M.D. With additions by the Author. In one octavo volume of 484 pages. Cloth, $4.

**WÖHLER'S** *OUTLINES OF ORGANIC CHEMISTRY.* Translated from the eighth German edition, by IRA REMSEN, M.D. In one 12mo. volume of 550 pages. Cloth $3.

**YEAR-BOOK OF TREATMENT FOR 1898.** A Critical Review for Practitioners of Medicine and Surgery. In contributions by 24 well-known medical writers. 12mo., 488 pages. Cloth, $1.50. *Just Ready.* In combination with THE MEDICAL NEWS and THE AMERICAN JOURNAL OF THE MEDICAL SCIENCES, 75 cents. See page 1.

**YEAR-BOOKS OF TREATMENT** for 1892, 1893, 1896, and 1897, similar to above. Each, cloth, $1.50.

**YEO (I. BURNEY).** *FOOD IN HEALTH AND DISEASE.* New (2d) edition. In one 12mo. volume of 592 pages, with 4 engravings. Cloth, $2.50. See *Series of Clinical Manuals*, page 13.

—— *A MANUAL OF MEDICAL TREATMENT OR CLINICAL THERAPEUTICS.* Two volumes containing 1275 pages. Cloth, $5.50.

**YOUNG (JAMES K.).** *ORTHOPEDIC SURGERY.* In one 8vo. volume of 475 pages, with 286 illustrations. Cloth, $4; leather, $5.

www.ingramcontent.com/pod-product-compliance
Lightning Source LLC
Chambersburg PA
CBHW032006300426
44117CB00008B/915